T0395694

Managing Energy Security

This interdisciplinary book is written for government and industry professionals who need a comprehensive, accessible guide to modern energy security.

Introducing the ten predominant energy types, both renewable and non-renewable, the book illustrates the modern energy landscape from a geopolitical, commercial, economic and technological perspective. Energy is presented as the powerhouse of global economic activities. To ensure the uninterrupted supply of energy, nations, industries and consumers need to have options. Efficient energy security planning ensures that when a primary energy source is depleted, compromised or interrupted, an alternative energy source must be readily available. For this reason, the foundations of energy security are built upon the five pillars of Sustainability, Independence, Efficiency, Affordability and Accessibility. The numerous case studies presented in this book demonstrate that energy security may be compromised in the absence of one out of these five ingredients. The book also entertains the Triple-E notion of Energy Efficiency, Environmental integrity and Economies of scale, used by governments and corporations for energy optimization. One of the key strengths of the book is its ability effectively to cover various scientific disciplines, and several energy types, while remaining comprehensible.

This book will be of much interest to security or logistics professionals, economists and engineers, as well as policymakers.

Maria G. Burns is a National Academies scholar. She serves as the Director for the Logistics and Transportation Program, University of Houston, and as Principal Investigator in the DHS Center of Excellence for Borders, Trade and Immigration Research, led by the University of Houston. She is the author of *Port Management and Operations* (CRC Press, 2014) and *Logistics and Transportation Security* (CRC Press, 2015).

Managing Energy Security

An All Hazards Approach to Critical Infrastructure

Maria G. Burns

NEW YORK AND LONDON

First published 2019
by Routledge
52 Vanderbilt Avenue, New York, NY 10017

and by Routledge
2 Park Square, Milton Park, Abingdon, Oxon, OX14 4RN

Routledge is an imprint of the Taylor & Francis Group, an informa business

Library of Congress Cataloging-in-Publication Data
A catalog record for this title has been requested

ISBN: 978-1-4987-7295-2 (hbk)
ISBN: 978-0-429-32007-1 (ebk)

Typeset in Times New Roman
by Swales & Willis Ltd, Exeter, Devon, UK

Contents

Tables

Figures

Foreword

As the global economy develops and the world's population increases, for the last twenty five years the global middle class and its consumer base have expanded at unprecedented rates, driving demand for resources like food, water and energy. These valuable resources and the infrastructure necessary to manage them are vital to sustain life as we know it. This book is about energy infrastructure. It covers the key challenges and opportunities of energy security in-depth from a geopolitical, environmental, technological, and economic/energy-efficiency level.

Houston is well known as the energy capital of the world. Not only do most major energy companies operate in Houston, but the City of Houston and the State of Texas is the fastest growing region in the United States, due in large part to the world's largest refining and petrochemical complex and among the world's largest deposits of oil and gas located along the Texas Gulf Coast and its Port. The Port of Houston is the largest in North America in terms of foreign waterborne tonnage at 245 million tons of cargo per year, and the Houston Ship Channel is the busiest waterway in the US with nearly 10,000 deep-water vessels and over 200,000 barges calling each year, busier than the ports of Los Angeles and New York combined. Nearly 70% of its operations are from crude and liquid bulk vessels, transporting energy resources to and fro all over the world.

I started my career in international supply chain and logistics in 1987 and have held roles in the US and overseas with ocean carriers, international freight forwarders, customs house brokers and beneficial cargo owners. I earned a bachelor's degree in international business administration from California State University at Los Angeles and an MBA from the University of Massachusetts at Amherst, Isenberg School of Management.

As Port Houston's CCO, I'm responsible for the Port Authority's real estate, trade development, economic development, marketing/external communications and media relations departments and the administration of Harris County's Foreign Trade Zone. Energy infrastructure and its security are very important to me, my organization and the economic livelihood of Texans, Americans and our energy partners around the world.

Several years ago, Professor Burns at the University of Houston approached me to be a member of the Advisory Board of the Center for Logistics and Transportation Policy. She was the center's director and wanted to ensure that the Port of Houston was represented and had a voice in the center's important work. She continues to be the program's director, developing programs with the US Coast Guard and conducting research on behalf of the Department of Homeland Security. Professor Burns' work is important to us in particular because of the valuable energy resource infrastructure along the Houston Ship Channel, its impact on the US economy and the importance to the more than two million American men and women it employs across the country as a result of its daily activities.

This book provides a multi-disciplinary study, as it encompasses each of the 10 prevailing energy types from a multi-science, engineering, operational, etc. perspective. Energy security is covered from a physical threat and cybersecurity outlook. I highly recommend this book as an important resource. This book is your essential energy security guide.

John A. Moseley
Chief Commercial Officer
Port of Houston Authority
Houston, Texas

Preface

My goal for writing this book was the need to offer a broader, multidisciplinary picture to energy security professionals and students, so that they better understand the convergence of energy and security.

In our historically significant era, energy and security are becoming increasingly intertwined: The benefits of energy cannot exist without security, whereas homeland security is jeopardized without an energy mix that is sustainable, independent, accessible and affordable. Hence, energy security determines global and national sovereignty, socioeconomic prosperity, and political stability.

A wide array of energy books have been previously written for individual energy segments, or exclusively for energy or security. However, this book was written as a response to the present historical turning point, where governments and energy conglomerates take huge synergistic leaps in science, technology and global operations.

This is an era where governments and energy conglomerates have long been preparing for a more intricate energy mix where both fossil fuels and renewable energy sources serve a nation's strategic energy goals.

Many academic programs around the world have revised their curriculum in response to the increasing significance of energy security, homeland security and supply chain security. This book meets the great demand and offers a multi-disciplinary convergence that includes the most prevailing energy sources (both non-renewables and renewables) covered from an operational, commercial, technological, geopolitical, historical, and regulatory perspective.

Several goals are served in this book: First, to provide energy and security majors with a concrete, in-depth analysis of both energy and security.

Our second goal is to offer professionals and students of security, economics, political science, transportation and other disciplines a panoramic hindsight from a multidimensional, multidisciplinary perspective.

We therefore offer a book that combines the above disciplines, and offers in-depth, advanced scientific information, while remaining comprehensive for every reader.

Each chapter of the book, and each energy segment, has its own voice. However, the reader will soon discover the interconnection among the chapters, and the diverse energy segments that are briefly explained herewith:

1 This is an era of rapid developments as to the new research and development in technologies, processes and regulations. Both renewable and non-renewable fuels are shifting to become greener, lighter and of higher calorific value. At the same time, energy security concerns have reshaped our strategic thinking from a geopolitical, economic, commercial and social perspective. To attain its goals, energy security

must be characterized by sustainability, independence, efficiency, affordability and accessibility (Chapters 1 and 3).

2 To address the individual challenges and opportunities, energy security is encompassed from three viewpoints, i.e., from a government, industry and supply chain perspective. Furthermore, the space of security is distinguished into physical, cyber, and holistic, which again encompasses the entire supply chain (Chapter 2).

3 Throughout the authorship of this book, there have been impressive day-to-day developments in the energy realm: The demand for energy increases dramatically as global population and the industrial output needs increase. Thus, energy harvesting gears towards more unconventional methods (Chapter 3).

4 Innovation is reshaping the oil and gas industry: Shale production and natural gas seem to be increasing market share, while transforming into cleaner, greener fuels. Oil and gas companies strengthen their bottom line through an increasing demand for chemical, plastic and petrochemical products (e.g., plastic resin and synthetic fiber, etc.). Thus, they have established this significant safety cushion for profit that alleviates dramatic oil price fluctuations and market uncertainties. The US has assumed a leadership role in the petrochemical industry, thus raising a relentless comparison with OPEC, which has now agreed to boost production. The final winner in this duel will certainly be the oil and gas industry, which remains the top energy segment with vast reserves and high, sustainable demand (Chapter 4).

5 The coal industry is transitioning into a "clean coal era" and grows with a myriad of applications and byproducts: Thermal coal feeds the global electric grid, while metallurgical coal feeds the construction industry. It is also used for electric cars, the food industry, cosmetics, textiles, potable systems, etc. Gas even adsorbs pollutants including gaseous products and chemical odors (Chapter 5).

6 The nuclear industry consists of two different types of reactions, i.e., fission, the traditional atom-splitting technique, and fusion energy, which uses plasma to fuse atoms and thus produces energy. Fusion, the so-called energy of the future, promises inexhaustible power to be used on planet Earth, space exploration and space colonization (Chapter 6).

7 Over the past decades, wind energy has combined onshore and offshore installations, with several global seaports hosting wind turbines as part of their concession agreements with energy companies. Wind energy is now literally taking off, as with recent research and development initiatives, wind turbines will be suitable for harvesting energy on planet Mars (Chapter 7).

8 Hydropower consists of ocean and tidal energy production, and represents the primary renewable source used for electric power, with 71% production. Impressive hydropower facilities generate power from South Korea to England and France, to Egypt, Ethiopia and Laos, promising to harness the global waters (Chapter 8). Certain new generation commercial ships, e.g., in NYK Japan, are using hydropower to harness the power of the waves while the ship is in motion. This will be a most impressive auxiliary energy in the new generation of ships.

9 Solar photovoltaic (PV) power has two unique particularities: Its installations can be used over buildings, thus giving a unique architectural signature to smart buildings. Solar installations on roof tops could well become part of the smart homes of the future. Solar energy can also provide electricity to the most remote structures of the globe. Panels can be installed in roof tops and thus give light and electricity to houses off the electric grid (Chapter 9).

10 Geothermal energy is equally fascinating, as it captures inexhaustible, unbounded underground energy from extremely high temperatures. Impressive installations are

found in the Antarctic, but also in the so-called “Ring of Fire” encompassing the US, Philippines, Indonesia, Mexico, New Zealand and other countries. The USA and New Zealand will collaborate in an ambitious plan dedicated to advanced geothermal energy production. Naturally, geothermal energy is considered a strong candidate to facilitate space colonization, and planet Mars is a suitable candidate (Chapter 9).

11 Biofuel is humankind’s attempt to recycle all the humble, unwanted materials on earth. There is a noble, most fascinating quality about the biomass industry, as it generates energy out of garbage and makes usefulness out of waste. The biofuel power generation has come a long way with four generations, or four impactful technological eras of producing fuels. Biofuels are used in the space and transportation industry in the form of fuel additives. The US Navy’s “Great Green Fleet” consists of biofuel-powered warships. Also, commercial airlines and courier entrepreneurs have adopted the use of biofuels as the sole fuel in dedicated flights. The use of aquatic flora, especially algae, is a cornerstone in the history of biofuel. Recent scientific breakthroughs by companies like ExxonMobil yielded genetically-modified algae with very high calorific value, suggesting a very promising future for algae-based biofuel (Chapter 10).

12 And while covering the aspects of physical security, the energy industry’s cyber space is also covered in the book. What are the alarming rates of attack on cyber security in the modern energy industry? How do hackers operate, and what is their operational footprint? Most importantly, what are the benefits of companies adopting a proactive security culture, and what are the risks of more optimistic, complacent cultures where companies are “strategically inactive”? In our era of large-scale data sharing via big data and blockchain schemes, cyber security is a significant segment of energy security (Chapter 11).

13 Strategic planning, forecasting, risk management, and training comprise a set of interrelated tools that can be used to attain optimum performance and energy security (Chapter 12).

Prof. Maria G. Burns
Director, LTPP and Faculty SCLT, University of Houston
Honorary Member, US Coast Guard Auxiliary

Acknowledgments

> There is a way to do it better. Find it.
>
> Thomas Edison (1847–1931)

This book is dedicated to my beloved husband and lifelong companion Leonard T. Burns, with the deepest love, respect and appreciation. Also to our beloved parents Lawrence and Frances B. and George and Nancy K., our siblings Kathy and Jim, Stan and Colleen, Larry and Stephanie, Michael and Maureen, their wonderful children, but also to Barbie and Gabriel, with all my love.

A very special thanks goes to my publishers, Routledge, an imprint of Taylor & Francis Group, a trading division of Informa UK Limited, with their most efficient team: To Mr. Andrew Humphrys, Senior Editor, Military, Strategic and Security Studies, Ms. Rebecca McPhee, Editorial Assistant, and Ms. Hannah Ferguson, Senior Editorial Assistant, Ellie Jarvis, Production Editor and Judith Harvey, Editor, I wish to express my deepest appreciation and gratefulness for the wisdom and immense support throughout this exciting voyage. Thank you to the entire editorial team, the illustrators, graphic designers, and book cover designers. Your contribution has been tremendous.

It is an honor to host in this book the distinguished global organizations, corporations, associations, and their most capable "Corporate Ambassadors" who greatly enhanced this book with primary data, images, and interviews, all of which are duly referenced. I hereby wish to thank each and every professional who generously shared information on their corporate achievements and contributed to this publication.

It is a great honor to host a distinguished leader from the global industry who generously provided a foreword for this book: Mr. John Moseley, Chief Commercial Officer at Port of Houston Authority.

I would like to express my immense gratitude to the distinguished individuals and organizations who have offered their invaluable support and contribution. Thank you to the leadership of the following organizations:

1 US Department of Energy (DOE) and its agencies and laboratories including The National Renewable Energy Laboratory (NREL);
2 US Energy Information Administration;
3 US Department of Homeland Security (DHS), and its agencies and directorates, especially the Science and Technology Directorate (S&T), Customs and Border Protection (CBP), Immigration and Customs Enforcement (ICE) and US Coast Guard (USCG);
4 International Energy Agency (IEA), The RVO Netherlands Enterprise Agency, Technology Collaboration Programmes of IEA (Bioenergy);

5 US National Aeronautics and Space Administration (NASA);
6 Exxon Mobil Corporation;
7 World Trade Organization;
8 World Nuclear Association;
9 World Coal Organization;
10 Azuri Life Changing Technologies (UK) Ltd.;
11 Port of Houston Authorities, Texas, USA;
12 US Department of Health and Human Services, Public Health Service Agency for Toxic Substances and Disease Registry;
13 US Environmental Protection Agency (EPA);
14 US Bureau of Ocean Energy Management, and National Research Council (BOEM);
15 US Bureau of Safety and Environmental Enforcement (BSEE);
16 Canadian Environment and Climate Change;
17 Canadian Centre for Occupational Health and Safety;
18 US Center for disease control and prevention (CDC);
19 European Environmental Agency, European Commission's Climate Action.

Please accept my sincere thanks for bringing life to the theories and practices of Energy Security with your extraordinary innovations and accomplishments that set the new era of energy.

I would especially like to thank my employers, colleagues and research companions at the University of Houston for cultivating an environment of productivity and innovation: Chancellor Renu Khator and Senior Vice Chancellor Paula Myrick Short for creating a global "Powerhouse" that leads and enriches energy and security research. A special thanks goes to the College of Technology: Our magnificent Dean Anthony "Tony" Ambler, a most respected leader and visionary; Heidar Malki, PhD Associate Dean for Academic Affairs, and George Zouridakis, PhD Associate Dean for Research and Graduate Studies. Thank you to the leaders, faculty and staff at the CoT's Construction Management Department: Chair Dr. Lingguang Song and the leaders of the Supply Chain & Logistics Technology Program.

My research affiliation with the US Department of Homeland Security has literally changed my life and triggered my "little grey cells," as Agatha Christie's Poirot would say. My immense thanks goes to the DHS leaders, research champions at the BTI Institute, a Homeland Security Center of Excellence. Thank you to Dr. Tony Ambler, BTI director, and thank you to Mr. Kurt Berens, executive director, for their leadership. Thank you to Dr. Mary Ann Ottinger, Associate Vice Chancellor for Research for the UH System, for her tremendous, extraordinary efforts to make this happen.

Thank you to the brilliant leaders at the UH College of Liberal Arts, Department of Political Science: Department Chair Susan Scarrow PhD. Thank you to the current and past Directors of Graduate Studies, Dr. Jennifer Clarke and Dr. Jeffrey Church, together with their wonderful team of professors. Einstein said that "Politics is more difficult than physics," but you have opened the gates of knowledge for me!

Thank you to my colleagues at the National Academies, Transportation Research Board, for giving me the opportunity to chair the Supply Chain Security Sub-committee since 2014.

Finally, thank you to each and every one of my global colleagues, research partners, mentors, mentees and students! Research is like travelling: The more you see, the more you know you haven't seen.

Part I

Managing Energy Security

An Overview

Part I of this book addresses energy security from three distinct perspectives. The first chapter discusses the significance of energy as critical infrastructure. The key types of energy – both non-renewables and renewables – and elements of risk management are duly explained. A historical overview is duly covered.

The second chapter addresses the key threat types of security and safety, and covers government agencies pertinent to energy and energy security, and the respective regulations.

The third chapter provides a mapping of the global energy from a geopolitical and economic landscape, whereas key production and consumption regions are duly covered.

1 Energy as a Critical Infrastructure Sector

By the fall of 1918, it was clear that a nation's prosperity, even its very survival, depended on securing a safe, abundant supply of cheap oil.

Albert Marrin, *Black Gold: The Story of Oil in Our Lives*

Introduction: Defining Energy, Energy Security and Infrastructure

Energy news is among the most prominent topics featured on media headlines, yet covering this complex subject generates a myriad of questions as to the geopolitical, economic, commercial, regulatory, operational, technical and other aspects of energy.

This book aims to encompass such transdisciplinary topics from an energy security perspective, but also to answer questions like the following:

1. What is the definition of energy security, and how is it associated with the geopolitical, socioeconomic, environmental and development strategies of a nation?
2. What is energy mix, and how does it contribute to national energy security?
3. What is the Triple-E concept, and how does energy security fit into the equation?
4. Why do oil price fluctuations affect the global economy?
5. How come industries and nations complain when oil prices peak, yet governments and economic empires collapse when oil prices are low?
6. What are the energy market cycles, and at which stages in the cycle do energy security systems become more vulnerable?
7. What are the layers of defense for national energy security?
8. Why do governments pursue energy independence as a top priority to their political legacy?
9. Why is the 21st century the era of energy independence?

And while this book addresses these and a plethora of other topics, the deluge of political, scientific and economic information seems to outpace the development of scientific and industrial tools adequate to gather and analyze these advances.

Defining Energy Security

The concept of energy security is intricate as it may be perceived from a geopolitical, socioeconomic and commercial development, environmental protection, as well as from an emergency and survival perspective.

The International Energy Agency (IEA) defines energy security as the continuous accessibility to energy sources at a reasonable, inexpensive price. From a long-term perspective, energy security is attained by pursuing appropriate investments in a timely manner to provide power consistent with financial growth and environmentally friendly specifications. From a short-term perspective, energy security focuses on the nation's or industry's capacity to respond instantly to unanticipated shifts in the supply-demand equilibrium. Per IEA, emergency response is essential to safeguard our systems from vulnerabilities caused by such disruptions. (IEA 2016d)

Chapter 3 of this book will duly analyze the geopolitical, economic, and social perspectives of energy security. However, in order to define the concept of energy security, this section will provide a concise introduction to these disciplines.

The Pyramid of Global Energy Security

The Pyramid of Energy Security demonstrated in Figures 1.1a and 1.1b pertains to the interrelated and equally significant notions of energy sustainability, independence, efficiency and affordability.

These aspects are interrelated, hence the greatest challenge and opportunity for energy leaders and decision makers is to strike a perfect balance between different aspects of energy activities at a national, corporate, societal and domestic level.

1 **Energy sustainability**: Pertains to a nation's or an industry's ability to continuously supply and use energy without any threats, intentional or unintentional, that would potentially harm a system's energy supply chain. Most importantly, it ensures that contemporary supply does not sacrifice, eliminate or negatively impact the energy availability or sustainability of a nation's succeeding generations.

Energy sustainability, or the ability to use diverse sources of energy around the globe, is the outcome of each country's economic and currency robustness, as well as their natural energy deposits. Be it natural gas, oil, coal or timber, a country has a fair chance of utilizing these as fundamental energy sources. As an illustration, at least 2.7 billion people, accounting for 38% of the global population, will be using solid

Figure 1.1a The pyramid of global energy security.

Source: the Author.

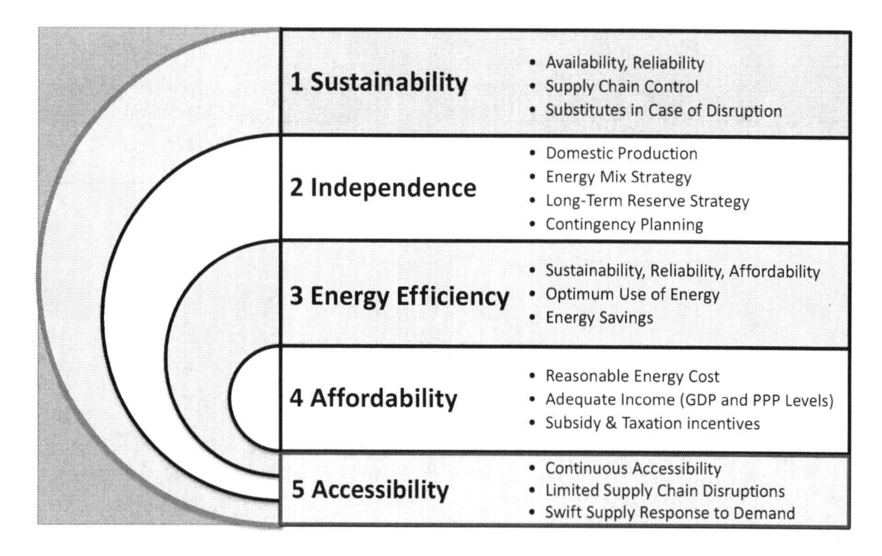

Figure 1.1b The pyramid of global energy security.
Source: the Author.

biomass for cooking, thus causing interior pollution and open fires in inadequately ventilated spaces, with high death rates. The majority of these populations are in developing Asia and sub-Saharan Africa.

If energy availability is present, affordable energy can enhance a nation's development level and purchasing power parity (PPP). Nevertheless, affordability is pointless if the energy supply is questionable, characterized by scarcity and irregular supply due to unforeseen disruptions and delays. Despite the developing technological advancements, several power failures have occurred in different parts of the world, such as the two-day Puerto Rico blackout due to a massive fire at a power plant affecting 3.5 million people (September 2016), or the 140 million people affected by Pakistan's blackout. While most of the modern-day blackouts are triggered by extreme weather conditions, sometimes overloaded electricity interconnectors cause large-scale disruptions. As an example, in September 2016 South Australia suffered a massive blackout due to severe storms which caused the collapse of major transmissions lines. The cost of the disaster was $367 million or $120,000 per minute of outage. It is worth noting that the outage was prolonged as the wind farms were switched off as a safety precaution during the hurricane (Castello, 2016).

To ensure the uninterrupted availability and use of energy, nations, industries and households need to have options. The availability of contingent energy sources ensures that even when the primary energy source is attacked or breaks down an alternative energy source will be readily available. Availability is defined as a nation's or consumers' ability to access energy resources in a timely and sustainable manner, subject to free trade agreements.

Since energy is a global commodity, security should be attained throughout the global supply chains, through infrastructure, superstructure, technologies and regulations that will facilitate the safe, secure, reliable and environmentally viable exploration, transportation and distribution of different energy products. Here, security pertains to reliable, sustainable and readily available energy regardless of the volume and regardless of the price (dollars per unit).

Finally, contingency planning must be in place, containing solutions for substitute supplies and alternative resources, storage and networks, in case of disruptions in the original systems. Based on the above, energy accessibility is achieved in the presence of a reliable, sustainable power supply, whereas affordability may not be relevant or influential enough in the presence of scarcity or disruptions.

2 **Energy independence**: Pertains to a nation's or an industry's autarky in acquiring sufficient energy quantity and type, from reliable sources.

   a Energy dependence from a geopolitical perspective entails a nation's reliance upon another for the provision of energy. Dependence is caused by market monopolies and oligopolies, price leveraging, supply disruptions and political strategies to leverage a nation's supplies in order to obtain financial, diplomatic or other privileges.

   b **Inelastic demand** is the condition where the demand for energy remains steady, regardless of the price fluctuations. When this phenomenon is present, it suggests that the energy suppliers can leverage their position by increasing the energy prices, under the certainty that buyers will still conduct the purchase as they do not have substitute suppliers. A nation's energy dependence results into severe economic, commercial and supply-chain problems.

   c The following factors improve a nation's chances in becoming energy independent:

      i The increasing number of global energy suppliers for non-renewable energy, and the industry's ability to un-tap unconventional energy (deep offshore, severe weather conditions, bitumen and other unconventional energy types, etc.).

      ii The increasing growth of dual or triple fuel burning in technologies, i.e., ocean-going ships burning Intermediate Fuel Oil (IFO), Marine Diesel Oil (MDO), and Liquefied Natural Gas (LNG).

      iii The increasing growth of renewable energy technologies and methods.

Energy independence is a key priority for governments and political agendas; however governments face certain challenges in an effort to ensure energy security and independence.

   a First, there is a matter of leveraging: Most governments typically own 20% of the national energy infrastructure and superstructure, whereas 80% of the nation's grid and supply chains belong to the private sector and international corporations. Hence, governments may be at a leveraging disadvantage when negotiating with entities that expand beyond a nation's borders. The solution for this challenge is the creation of strong public/private sector synergies.

   b Second, governments need to act local, but think global: Strategic decisions related to a nation's energy independence need to take into consideration the global forces in play, and identify the geopolitical, commercial, socioeconomic, and technological developments at a global level.

   c Third, energy independence requires a robust, reliable system that comprises of multi-billion-dollar projects. However, the discovery of new energy reserves in another geographical location may lead to the underutilization of the current infrastructure and necessitate relocation or re-structuring or the creation of new networks.

d The diversion or relocation of such projects has a political and financial cost, as any new infrastructure generates opposition due to land ownership, environmental or budget conflicts.

e Achieving energy security and energy independence is not easy for another reason: To quote Heraclitus (535–475 BC), "The only thing that is constant is change." National and corporate strategies may be hard to sustain in a global flux of changing patterns, as there are too many factors causing ongoing oscillations:

- the oil price fluctuations, OPEC vs NON-OPEC decisions on the future of fossil fuels;
- the shift of energy consumption from oil products to natural gas to renewable energy;
- environmental, security, safety, operational and technology-based policies (UN, IMO, DHS, DOE, EPA, BSEE, and other national agencies);
- technological advancements can shape a new global energy map:
  - safer and cleaner energy can increase popularity of previously stigmatized energy segments,
  - unconventional fossil fuel wells can be unleashed;
- newly discovered energy reserves will reshape the supply chains in terms of distance, trade routes, geopolitical particularities (i.e., wars, diplomatic relations, working patterns), travelling time, fuel quality, availability and overall price among global markets;
- changes in energy trade agreements, the creation of new trading blocs;
- currency fluctuations creating financial instability.

3 **Energy efficiency**: Pertains to the goal of utilizing less input, yet attaining more energy output.

a This is achieved with the use of advanced technologies, constructions and methods that enable more efficient refining, processing and conversion of energy.

b Energy efficiency is attained through advanced engines and power generators, insulated buildings and other structures, fluorescent or LED lighting systems, and so on.

c From an economy perspective, a nation or an industry aims to enhance its economic growth without increasing energy consumption.

d Energy leakage and natural or man-made seepages are common threats to energy efficiency. It has been observed that energy leakage in infrastructures, cables, pipelines, power plants and domestic appliances are responsible for a high share of energy losses. Due to the large-scale infrastructures available at a national and industrial level, it has been observed that identifying the leakage takes more resources, time and efforts than actually rectifying the problem.

4 **Affordability**: One of the United Nations' major sustainable development goals (SDGs) is the "access to affordable, reliable, sustainable and modern energy for all." As of 2017, one out of five people on the planet still has restricted access to present-day electric power, whilst one out of two, i.e., 3 billion out of 7.5. billion persons depend upon charcoal, wood, coal or animal waste for heating and cooking (UN, 2017; *UN Chronicle*, 2014).

Affordability pertains to reasonable energy cost, which derives from adequate revenue at a national or individual consumer's level. Energy efficiency facilitates affordability, as consumers "can do more with less." When affordability is not a result of supply–demand equilibrium, government intervention can induce it by means of tax incentives, subsidies to producers and consumers, budget can be allocated for research and development, and the generation of jobs.

Purchasing power enables the acquisition of sufficient quality and quantity of energy.

a Low-income households and less developed countries may not have adequate funds for the use of modern, greener, more efficient energy sources. As a result, these population segments are prone to energy safety, security, health and environmental damage.
b This has an impact on a nation's economy, health, standards of living, and so on.
c Lack of affordability may be caused by energy regulations pertinent to taxation, energy price quotas, monopolistic or oligopolistic markets, and protectionism. Furthermore, social and environmental levies may increase energy prices, so that consumers pay for energy efficiency infrastructure and technologies.

Affordability directly relates to availability and is subject to the economic and commercial impacts driven by the laws of supply and demand.

Increased energy prices are triggered by:

a increased demand, e.g., due to;

  i population growth
  ii industrial, economic and trade growth
  iii fears of imminent shortages
  iv lack of substitute energy sources, and so on;

b energy shortages caused by;

  i reduced production mutually agreed by the global producing countries, or
  ii supply chain disruptions, such as the closure of the Suez Canal in 1967–1975.

Reduced energy prices are triggered by:

a reduced demand, e.g., due to;

  i shrinking population
  ii diminished industrial, economic and trade growth
  iii availability of national energy reserves
  iv availability of substitute energy sources, and so on;

b energy oversupply caused by;

  i over-production by nations aiming to increase market share, and
  ii the uninterrupted flow of energy.

Market fluctuations may be triggered by several factors, including but not limited to:

a global market share competition, i.e., between OPEC and Non-OPEC country-members;
b price volatility caused by supply–demand disequilibrium and/or overproduction;

c global strategic alliances, trade agreements and production quotas;
d economic factors including:

 i currency wars and rates of exchange fluctuations
 ii Gross Domestic Product (GDP)
 iii purchasing power parity (PPP)
 iv high production costs – high upstream, midstream and downstream costs
 v economic efficiency vs energy efficiency – fuel consumption energy unit per dollar invested in energy.

5 **Accessibility**:

a Pertains to continuous physical availability, through upstream, midstream and downstream, with limited supply chain disruptions or delays.
b From a commercial perspective, an accessible supply system can swiftly respond to market demand, with all the factors of production and input processes ready and available to produce. Safety, security or environmental threats hinder sustainability: Wars, terrorism, sabotage, sociopolitical turmoil, natural disasters, extreme weather conditions. Preventive action and risk management can alleviate any and all of these risks.
c Accessibility may refer to untapped, unconventional resources such as fossil fuels, minerals, etc., that may need unconventional, advanced technologies to be retrieved.

Energy Strategy: Meeting the National Security Goals

Defining the National Energy Strategy

National energy strategy is a coherent action plan or policy framework that has been designed and implemented to achieve the national energy goals throughout the entire supply chain, that is from upstream to midstream to downstream. Public and private partnerships should engage key energy stakeholders, in a manner such that all possible choices, risks and benefits are duly discussed, and the optimum solutions are selected. An energy strategy provides reasonable arguments to justify the country's energy mix, while taking into consideration the retrievable national resources, technologies and disciplines available, and all factors of production required for this purpose. Energy strategy is nation-specific, that is, each nation has distinct geographical, sociopolitical, demographic, economic and regulatory particularities that must be reviewed, substantiated with facts and figures, and finally incorporated into the energy strategic plan.

The Energy Mix

Energy mix is defined as a national, regional or corporate decision to utilize diverse power sources, including primary (non-renewables) and secondary (renewables), in order to generate electricity.

The broader and most significant scope of the energy mix entails a government's strategy to explore, produce, refine, import, store and utilize a number of different energy sources. Hence, alleviating any consequences deriving from supply chain disruptions, policy restrictions, price fluctuations, and future storage purposes.

Each energy sector, in each region and each country, utilizes a distinct energy mix, i.e., consumes energy from diverse energy sources, but also allocates this energy differently to satisfy the national energy demand.

The Use of Energy

Energy is used by:

- **Governments and populations** for homeland security and defense, but also for public utilities including light and power (electricity, natural gas) with functions expanding from traffic lights to public transportation, sewage, and handling of hazardous materials.
- **Industries**, from space exploration to agricultural production; from heavy manufacturing to healthcare.
- **Sciences**, from Chemistry to Biology; from earth science to forensics; from oceanography and geology to astronomy.
- **Business activities**, from stock exchange to banking; from wholesale to retail; from raw materials to public utilities, to global trade and transport.
- **Arts, sports, and crafts**, from architecture to antique restoration; from the visual arts to music and digital art; from social media and entertainment to Olympic games and recreational activities.

According to several global and national energy organizations, such as IEA (International Energy Agency), EIA (Energy Information Administration), OPEC (Organization of the Petroleum Exporting Countries) and so on, energy use is principally classified into five key sectors:

1 Industrial
2 Residential
3 Commercial
4 Electric power
5 Transportation.

As stipulated by the U.S. Energy Information Administration (EIA, 2016a), global residential and commercial energy consumption amounts to 20%, the industrial sector exceeds 50% of worldwide consumption, whereas intermodal transportation (sea, land, air, pipeline) accounts for 30% of consumption.

Figure 1.2 demonstrates the world energy consumption comprising both non-renewables and renewables. As a rule of thumb, industrial and commercial use of energy contributes to a nation's GDP, whereas domestic and public utility power contribute to the standards of living.

The state of a nation's economy, its demographic, industrial, commercial activities, as well as its transportation networks, will determine the allocation of energy resources. Figure 1.3 demonstrates the energy consumption by sector in the U.S.

Energy Mix as Part of a National Energy Strategy

Nations strategically decide on the optimum, i.e., most suitable, country-specific energy mix, which is determined by several factors, including but not limited to:

a national reserves of non-renewable energy;
b optimum selection of renewable energy segments, considering natural conditions such as geography, geologic morphology, climate system, etc.;

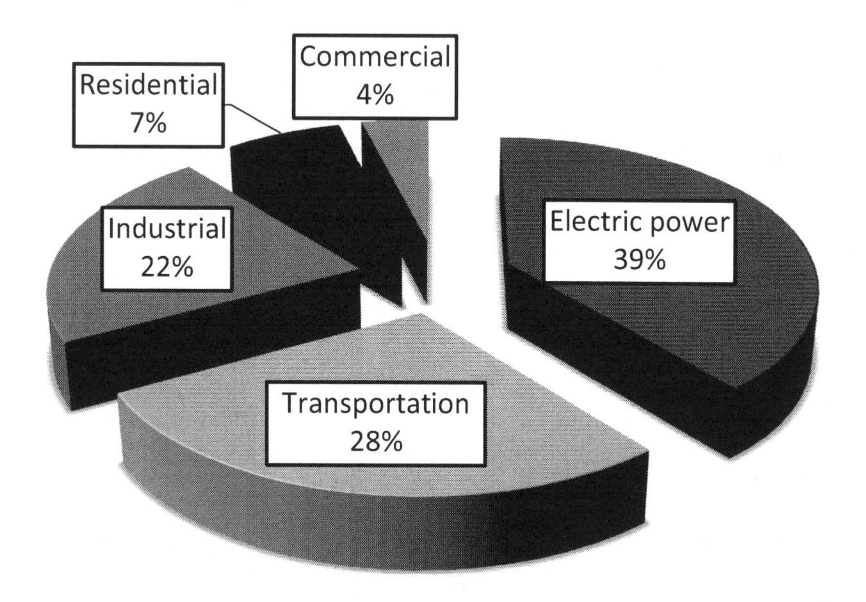

Figure 1.2 U.S. energy consumption per sector (2015).

Source: EIA 2017a. U.S. Energy Information Administration. Available at: www.eia.gov/energyexplained/?page=us_energy_home, last accessed: January 1, 2017.

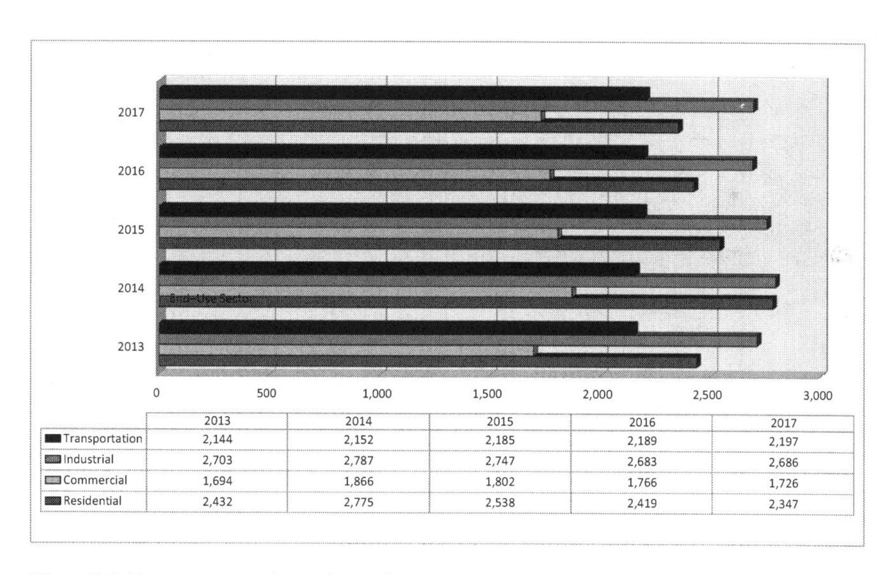

	2013	2014	2015	2016	2017
Transportation	2,144	2,152	2,185	2,189	2,197
Industrial	2,703	2,787	2,747	2,683	2,686
Commercial	1,694	1,866	1,802	1,766	1,726
Residential	2,432	2,775	2,538	2,419	2,347

Figure 1.3 Energy consumption estimates by sector, 2013–2017 (trillion Btu).

Source: U.S. Energy Information Administration, *Monthly Energy Review – Table 2.1, Energy consumption by sector, 4/25/2017*. Available at: www.eia.gov/consumption, last accessed: May 2, 2017.

c existing infrastructure and superstructure for the specific energy type;
d trade agreements with strategic allies for the purchase or sale of energy;
e environmental regulation and "green" technologies available, to monitor and alleviate CO2 emissions, greenhouse gases and other pollutants;
f specific energy types boosting domestic economy, e.g., trade and economic growth, jobs growth, tax incentives, subsidies, innovative technologies and processes, and so on.

A nation's energy mix strategy greatly impacts on energy security from a homeland defense, geopolitical and economic perspective.

Energy security pertains to both protecting homeland security by eliminating vulnerabilities, but also safeguards energy independence due to the accessibility and availability of renewable energy production. Governments and markets can enhance security through stability, and be protected by fuel price fluctuations and unpredictable shortages caused by energy dependency.

During the past two centuries, the predominant energy segments were non-renewable energy types such as coal and petroleum, and, recently, natural and shale gas. Growing energy demand will eventually lead to the decrease and final depletion of conventional energy sources; consequently governments, energy conglomerates and scientists are proactively researching the optimum utilization of innovative technologies, methods and networks needed to progressively replace the non-renewable energy forms.

Innovative technologies will always be needed for the production, storage and conversion of energy. Therefore energy dependency will depend more on the availability and advancement of technologies, and less on the actual availability of primary, natural energy sources.

Non-Renewables, Renewables, and the Energy Mix

While renewable energy seems to be a priority in certain political and social agendas that support green energy through subsidies and tax incentives, techno-engineering advancement has not reached the stage of offering sustainable, cost-effective energy. On the other hand, traditional energy sources such as oil, gas, coal and nuclear energy have a well-established infrastructure, and state-of-the-art technologies can provide reliable and cost-effective energy with limited environmental damage.

A nation's energy mix and the transition of energy from non-renewables to renewables depends on scientific and technological advancements on either side of the fence. For the above reasons, oil, natural gas and coal still comprise over 80% of global market share, showing that the non-renewables have been the primary energy source for several decades now.

Alternative energy key players will require time and technological breakthroughs to evolve accordingly. As each energy segment evolves, nations' energy mix will diversify accordingly.

It appears that most energy sources will retain their market share, whereas the renewables are projected to expand their share. This win–win scenario for all energy segments is realistic, bearing in mind the global trade growth and the population growth. However, it is worth keeping in mind that all of these energy segments invest in innovative technologies in an effort to eliminate their individual weaknesses, be it energy efficiency, sustainable production, safety during production and processing stages, environmental pollution, and so on. While Chapters 4 to 10 of this book explicitly go through the strengths and risks of each energy segment, it is worth noting that a) energy segments that possess a well grid and distribution network retain a competitive advantage, and b) it only takes one technological innovation for an energy segment to eliminate their risk factors and thus to increase their market share.

Reasons for Nations' Decisions to Export Locally Available Energy Types, and Import Others

Based on each nation's perceptions, it is common for nations to traditionally possess specific energy resources and/or technologies, yet they may decide to export them to other nations, or trade them for other energy types, for several reasons, such as:

a To satisfy domestic energy consumption needs, by exporting energy forms less used, and importing energy sources of higher demand. As an example, the U.S. exports metallurgical coal, which is used for steel production, yet utilizes more steam coal, which is used to generate electricity.
b Domestic regulatory restrictions, especially environmental.
c High production costs, not meeting the "energy efficiency" goals.
d To eliminate any hazards to the public, e.g., the high safety, security and environmental measures involved in the operations of nuclear energy plants.
e State decisions to favor one energy type over another due to financial, commercial or social reasons.

Based on the above decisions, countries may decide to become net exporters of one energy type, i.e., by exporting a higher volume of energy compared with what they consume, and net importers in another energy type, i.e., import a higher volume of energy compared with what they export. As an example, the U.S. traditionally generates a high volume of coal each year yet, due to high production costs domestically, several power plants import coal from overseas. In 2015, the U.S. produced 895 million short tons, and utilized approximately 802 million short tons, among which 11 million were imported, primarily from South America. From 2010 to date, the U.S. has been exporting 5% to 12% of its domestic coal production overseas, with a peak export of 126 million short tons in 2012 (12%), and a rapid export decline to 74 million short tons in 2015 (8%) (EIA, 2017b).

Table 1.1 demonstrates the characteristics of energy security from an import/export perspective.

Energy Security, GDP and Overpopulation

Overpopulation has a negative impact upon a nation's GDP and the economy as a whole, as the domestic resources, including energy reserves, are now shared within a greater population segment. As demonstrated in Figure 1.4, there is a negative correlation between the growth rates of energy consumption, world population, and GDP. As world population grows, the demand for energy also rises. On the other hand, modern technologies offer increased energy efficiency levels, which reduce energy consumption per capita.

Energy Scarcity, Population Growth and Supply/Demand Equilibrium

Energy scarcity limits economic growth, and yet increased energy production does not significantly boost economic expansion. A continuous increase in commercial activities and human population have resulted in a spike in the international energy demand. The increased demand of the 2000s coincided with sociopolitical turmoil in over half of the oil-producing

Table 1.1 Energy security from an import/export perspective.

1	Political stability is pivotal in ensuring national security and sovereignty, but also in ascertaining that all economic, regulatory and commercial transactions will run smoothly.
2	Social stability will ensure there are no riots, strikes, acts of sabotage or terrorism, on the importers' and exporters' side.
3	Financial affordability and reliable supply chains will ensure the sustainable flow of goods.
4	A supportive, pro-trade government mechanism is required, enabling stakeholders to explore diverse commercial possibilities.
5	A national energy contingency plan is in place: They support a diversification of energy sources, and sufficient infrastructure and storage capacity to ensure that no disruptions will occur.
6	Innovative technologies are supported in a manner such that energy efficiency meets environmental integrity and economies of scale (the "triple E" concept).
7	Having accurate forecasts and, if possible, inside information about the global markets and trade agreements.
8	Strong domestic market: High standards of living, sufficient salaries and low unemployment rates.
9	Robust infrastructure, storage and transport systems.
10	Contingency planning to eliminate energy disruptions. Sufficient infrastructure and superstructure to securely receive the energy commodity.
11	Energy and electricity available to all population segments.
12	Robust banking system, adequate insurance and underwriters' coverage.
13	Efficient systems to secure Just-In-Time delivery of the appropriate energy mix.
14	Strong currency, favorable exchange rates and taxation.
15	Non-bureaucratic import/export regulations.
16	Adequate resources, i.e., factors of production, and the financial component will enable them to create the desired input.
17	Highly trained personnel and a supportive, pro-export infrastructure system.

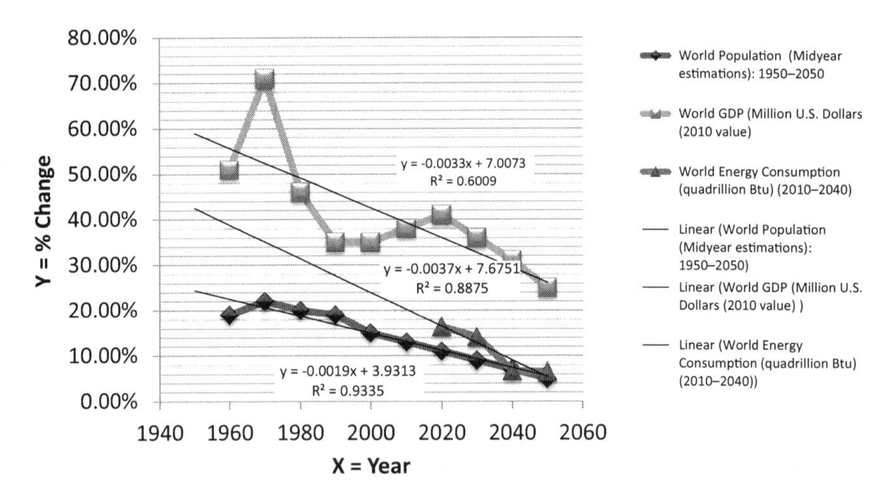

Figure 1.4 Annual growth rates of world population, energy consumption, and GDP (1960–2040 est.).

Sources: the Author, based on data obtained from: U.S. Census Bureau; International Programs; International Data Base; revised: September 27, 2016. Version: Data:16.0804 Code:12.0321. Available at: www.census.gov/population/international/data/idb/worldpoptotal.php. EIA, 2017a. *International Energy Outlook 2016*; Report Number: DOE/EIA-0484(2016). Available at: www.eia.gov/outlooks/ieo.

regions, the weakening of major currencies such as the U.S. dollar, the Euro, the British pound, and the currency manipulation, i.e., the artificial weakening, of other currencies.

Furthermore, population growth boosts energy demand, thus triggering economic growth. The issues of energy scarcity and energy demand will need to be addressed as a matter of urgency, as the United Nations predict the global population will exceed 9 billion by 2050, and a minimum of 10 billion by 2100 (UNESA, 2016). Furthermore, the 2016 energy crisis has indicated that low oil prices may increase the energy volume, but not the revenue (Burns, 2016).

Overpopulation should always be examined in correlation with a nation's natural resources (in this case energy reserves), industrial activities and service-related jobs. Two scenarios are presented herewith: the best case where overpopulation combined with ample energy sources boosts energy security. Furthermore, a high level of industrial or commercial activities leads to comparative advantage and GDP growth.

a **Overpopulation in a best case scenario: Comparative advantage**
When abundant natural resources and industrial activity or services are present within a country, overpopulation stabilizes the national economy as it increases workforce, and enforces the domestic producers' and consumers' markets. Even when high population leads to lower salaries, such as in the cases of China and India, this contributes to the nation's comparative advantage, i.e., the production of specific commodities or services at a cheaper opportunity cost. In this scenario, when there is ample demand for a nation's natural resources, commodities and/or services, overpopulation boosts internal production, exports, and the GDP.

b **Overpopulation in a worst case scenario: Market oversupply**
In the worst case scenario, overpopulation combined with scarcity of energy sources threatens energy security. From a socioeconomic perspective, overpopulation leads to oversupply of labor, which combined with diminished demand may lead to lower salaries, or increased unemployment.

As seen in Figure 1.5, the impact of overpopulation upon national productivity depends on the output/input ratio. For energy independence in particular, when the output/input ratio is positive, and the demand for energy independence is adequately high, nations experience the benefits of overpopulation. This certainly requires a balance among the other factors of production, namely land, capital and energy investment, as well as regional energy exploration, production, processing and other activities. Innovation is the catalyst in the factors of production as it makes the productivity output stand out from that of the competition.

The End of Labor-Based Comparative Advantage

Comparative advantage is an economic model pertaining to a nation's or corporation's capacity to generate low opportunity cost, and hence generate cost-effective output (i.e., products or services). For several decades nations have benefited from cheap labor, and economic empires were built upon populous, low-cost input processes. However, this era may soon be over, as cheaper labor seems to be inevitable for the next generations: The rapid and sustainable rate of population growth, combined with the mass migration crisis, will instigate intense competition among populous nations. The surges of refugees and immigrants from the most populous to the least populous regions will alter the global labor markets. As history has proven, exploitation of immigrants is the first stage, yet it does not last forever. As soon as the new population segments are established and become familiar

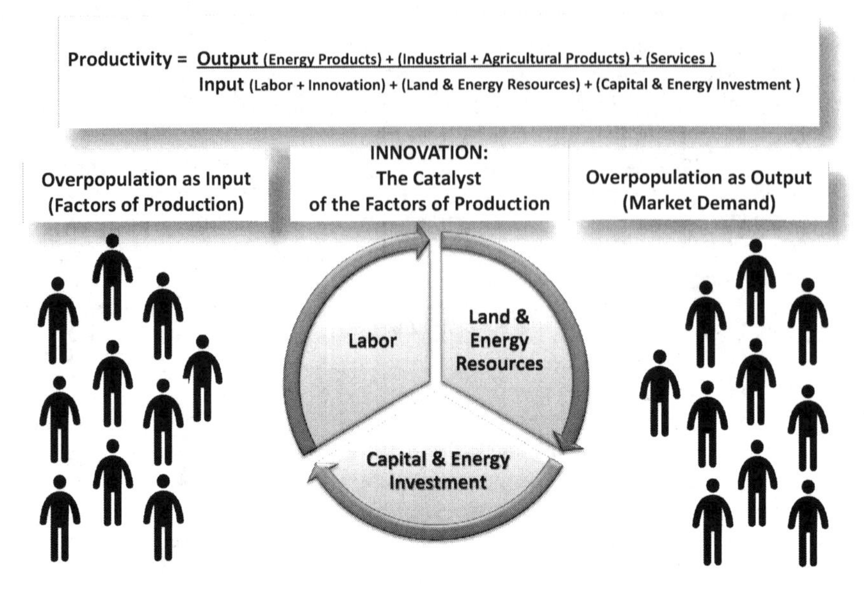

Figure 1.5 Energy independence and overpopulation.
Source: the Author.

with the new socioeconomic realities, they actively participate in the national workforce. Oversupply of labor drives costs down.

Bearing in mind the widespread movement of peoples and the abundance of low-cost labor, labor-based comparative advantage may lose its status. On the other hand, land and natural resources including minerals, agricultural products, water and other resources will become scarce.

The Impact of Energy on Global Trade Growth

At the dawn of the 21st century, energy is considered to be a critical security segment, vital for the wellbeing of nations. Energy drives global trade, industrial activities, national healthcare and social wellbeing, and transportation. It elevates the living standards of a nation and safeguards its sovereignty.

A new energy map is reshaped as humankind moves beyond conventional energy exploration. Old standards, methods, and technologies become obsolete, and modern technologies facilitate the discovery and extraction of increasingly unconventional methods, greener and more efficient energy sources. Hence, we will be able to do more with reduced energy consumption and reduced fuel emissions.

The continuous evolution of the energy business triggered by global trade and world population growth will relentlessly modify the global map with emerging key players, markets, energy types and commodities. Radical changes appear to shake the global socioeconomic and geopolitical status quo with new trade agreements, new global commercial routes, and new political alliances.

According to the World Trade Organization, it is estimated that by 2030 global trade will reach 30 billion metric tons (almost twice as much as in 2017), one-third of which will entail energy products, principally oil and gas. From 2000 to 2017 the global energy demand increased by approximately 33.33%, with China initiating over 50% of this growth.

The global trade volume, as seen in Figure 1.5, Figure 1.6 and Table 1.2, enjoys sustainable growth due to the rapid growth of developing nations, and also due to rapid population growth. As of 2017, global population has exceeded 7.4 billion people. In particular, Figure 1.5 demonstrates the increasing value of world trade from 1948 to 2016, despite the fluctuations, especially the two most recent Kitchin (business) cycles (see Chapter 3 of this book), resulting in market contractions in 2008 and 2014–2016 respectively. Mirroring cycles are observed between the developing and developed economies, possibly due to trade agreements and pertinent commercial interactions.

Value-added or high-value commodities are usually goods that require high energy volumes for processing, manufacturing, transportation, and storage. Manufactured goods, including the automobile industry, electronics, but also medicines and processed food are among the commodities that require high volumes of energy through their value-added processing.

Figure 1.7 and Table 1.2 depict the annual percentile change of world economic growth from 2013 to 2016 for selected economies. It appears that developed and developing Asian economies such as China and India have enjoyed a growth of approximately 8–9%, followed by other emerging economies. What all these economies have in common is their prominent role in the energy markets, as major producers/exporters or consumers/importers.

The broad variation among regions should be interpreted according to the nations' current level of development since the figures reflect percentile change, so it should be noted that advanced economies have already attained high growth levels and their growth and standards of living have long reached optimum levels. On the other hand, less developed and developing economies have not achieved their full potential hence diminished growth rates are observed.

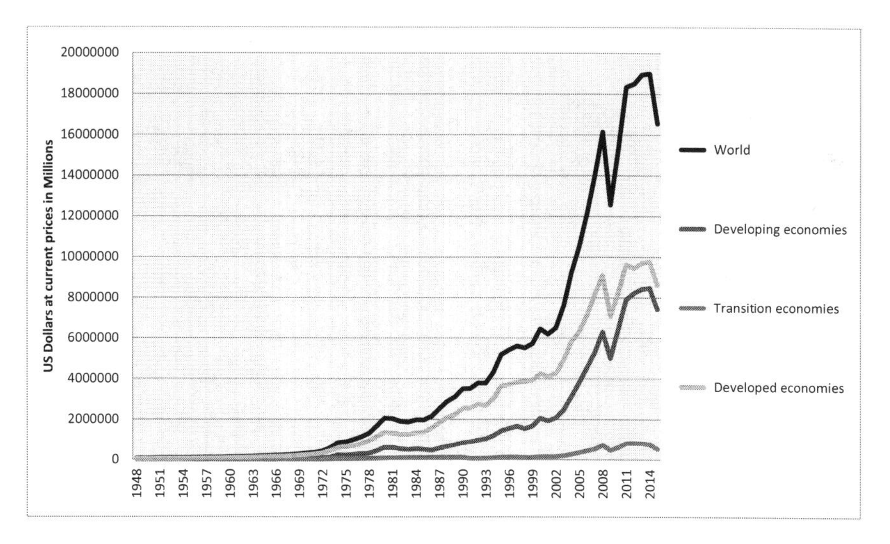

Figure 1.6 Value of global trade 1948–2016 (US dollars at current prices in millions).

Source: UNCTAD, 2016. *Review of Maritime Transport 2016*. Available at: http://unctad.org/en/PublicationsLibrary/rmt2016_en.pdf, last accessed: January 2, 2017.

Note: Calculations for country aggregates based on GDP in constant 2005 dollars.

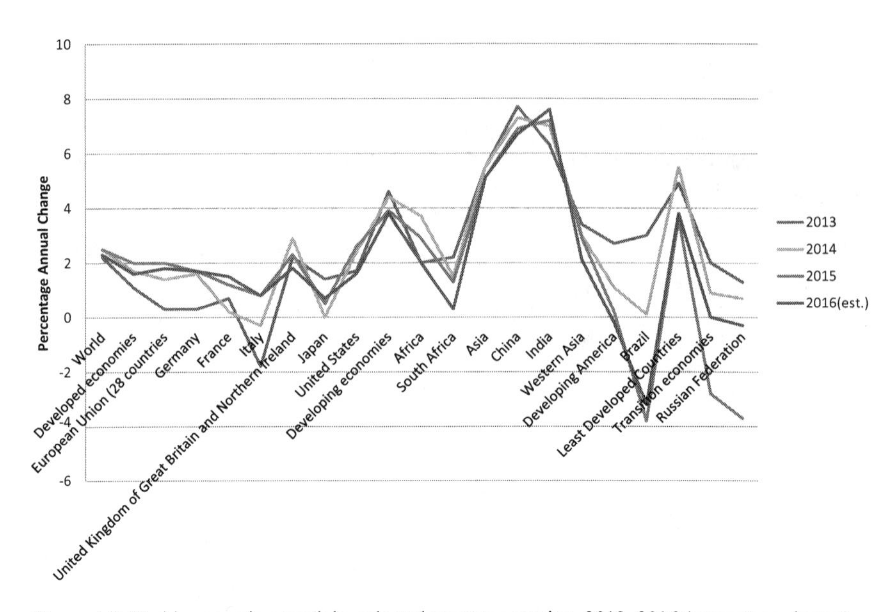

Figure 1.7 World economic growth by selected country grouping, 2013–2016 (percentage change).

Source: UNCTAD, 2016. *Review of Maritime Transport 2016*. Available at: http://unctad.org/en/PublicationsLibrary/rmt2016_en.pdf, last accessed: January 2, 2017.

Note: Calculations for country aggregates based on GDP in constant 2005 dollars.

Table 1.2 World economic growth by selected country grouping, 2013–2016 (percentage change).

Region	*2013*	*2014*	*2015*	*2016 (est.)*
World	2.2	2.5	2.5	2.3
Developed economies	1.1	1.7	2.0	1.6
European Union (28 countries)	0.3	1.4	2.0	1.8
Germany	0.3	1.6	1.7	1.7
France	0.7	0.2	1.2	1.5
Italy	−1.8	−0.3	0.8	0.8
United Kingdom	2.2	2.9	2.3	1.8
Japan	1.4	0.0	0.5	0.7
United States	1.7	2.4	2.6	1.6
Developing economies	4.6	4.4	3.9	3.8
Africa	2	3.7	2.9	2.0
South Africa	2.2	1.5	1.3	0.3
Asia	5.5	5.5	5.1	5.1
China	7.7	7.3	6.9	6.7
India	6.3	7.0	7.2	7.6
Western Asia	3.4	3.0	2.9	2.1
Developing America	2.7	1.1	0.2	−0.2
Brazil	3.0	0.1	−3.8	−3.2
Least Developed Countries	4.9	5.5	3.6	3.8
Transition economies	2.0	0.9	−2.8	0.0
Russian Federation	1.3	0.7	−3.7	−0.3

Source: UNCTAD, 2016. *Review of Maritime Transport 2016*. Available at: http://unctad.org/en/PublicationsLibrary/rmt2016_en.pdf, last accessed: January 2, 2017.

Note: Calculations for country aggregates based on GDP in constant 2005 dollars.

Maritime transportation is responsible for the vast majority of energy transport. In fact, seaborne trade accounts for approximately 90% of global trade, with particular focus on large scale, intercontinental trade routes. Table 1.3 demonstrates these global trade patterns for selected years, from 1970 to 2015. Global trade is hereby measured in millions of tons loaded.

While oil and gas commodities represent almost one third of total cargo volume, the remaining commodities still require energy input for value-added processes such as refining, manufacturing, transportation, storage, construction and other capital-intensive, energy-intensive activities.

Such is the impact of the energy industry, as any market imbalance and any extremity in the supply–demand equilibrium will have severe repercussions on the global economy. This can be verified by the global economic meltdown of 2008, and the oversupply of petroleum in 2015–2016. Both these eras triggered lower oil prices. The recovery year seems to be 2017, where the global economic challenges of 2008, 2014, 2015 and 2016 due to low oil prices appear to recover in 2017 – for most of the world, but not for all. As of 2017, one out of six people live without electricity access. As stipulated by the CIA, this is 1.2 billion among the 7.4 billion people comprising the global population (CIA, 2017).

The Impact of Energy on National Growth

The energy industry is a capital-intensive, technology-intensive sector that promotes growth, innovation, and development to the entire supply chain, and therefore has the power to revitalize nations and transform economies. As the following case studies reveal, a country's destiny is instantly changed once new energy reserves are discovered.

Table 1.3 Developments in international seaborne trade, selected years 1970–2015 (millions of tons loaded).

Year	*Oil and Gas*	*Main bulk commodities (iron ore, coal, grain, bauxite, alumina, phosphate rock)*	*Dry cargo other than main bulk commodities*	*Total (all cargo)*
1970	1,440	448	717	2,605
1980	1,871	608	1,225	3,704
1990	1,755	988	1,265	4,008
2000	2,163	1,295	2,526	5,984
2005	2,422	1,709	2,978	7,109
2006	2,698	1,814	3,188	7,700
2007	2,747	1,953	3,334	8,034
2008	2,742	2,065	3,422	8,229
2009	2,642	2,085	3,131	7,858
2010	2,772	2,335	3,302	8,409
2011	2,794	2,486	3,505	8,785
2012	2,841	2,742	3,614	9,197
2013	2,829	2,923	3,762	9,514
2014	2,825	2,985	4,033	9,843
2015	2,947	2,951	4,150	10,047

Source: UNCTAD, 2016. *Review of Maritime Transport 2016*. Available at: http://unctad.org/en/PublicationsLibrary/rmt2016_en.pdf, last accessed: January 2, 2017.

Note: Calculations for country aggregates based on GDP in constant 2005 dollars.

Case Study 1: Texas, U.S.

The Texans are well acquainted with the energy "gold rush" as they have been into it for over a century now. While for the state of California the Gold Rush came from 1848 to 1855, for the State of Texas the Black Gold Rush started a few decades later, and things are getting better and better!

It was January 10, 1901 when the Lucas No. 1 well blew in at the notorious Spindletop region. This event signaled a brand new era for Texas, which was transformed, almost overnight, from an agricultural, cotton, sugar and livestock-producing region, into a global energy mega power.

To ship the petroleum products to their global markets, a robust supply chain and transportation system was needed: Railroad transport, and then in 1914 the Port of Houston (Figure 1.8).

Life in Texas was all about energy and transport, and continuous migrations "in search of the oil" became the norm in this part of the world. People moved to Texas from the four corners of the earth, making it one of the most international, diverse, multicultural states globally. As new oil towns were created, farmers would sell their land and cattle to follow the trails of oil.

It is impressive to observe the demographic map of Texas change year after year: Oil-towns were literally set up overnight to host the oil workers and their families. It is even more impressive to see these passionate, enthusiastic, hard-working families willing to constantly relocate, depending on the new locations of the oil and gas wells. As soon as the oil wells were depleted, energy workers and their families would leave their homes and the ghost-towns behind in search of new energy wells and new petroleum excitement, with improved technologies and working processes (Figure 1.8).

Shortly before the dawn of 2017, the U.S. Geological Survey discovered a gigantic oilfield in the Permian Basin, West Texas, which they named the greatest "continuous oil" discovery in America. The 20 billion barrels of oil and 1.6 billion barrels of natural gas expand over the Lubbock and Midland cities, hence covering 118 miles (USGS, 2016).

The oil and gas reserves will last for decades, and the local industry can pick and choose the conventional and unconventional wells they need to explore first. Texas will remain a global energy leader for decades to come, yet its people will be too busy to celebrate their successes – they'll be drilling oil and gas, always retaining their down-to-earth, no-nonsense attitude.

Figure 1.8 The impact of energy on national growth: Texas, U.S.

Sources: Compiled by the author. Image 1: Dallas/Fort Worth, Texas, 1891. Source: The Texas General Land Office. Available at: www.glo.texas.gov/history/archives/map-store/index.cfm#search. Image 2: Dallas/Fort Worth, Texas, early 21st century. Source: U.S. District Court, northern district of Texas. Available at: www.txnd.uscourts.gov.

The Texas success story has taught us that the energy industry is for the brave and the humble: It takes relentless efforts and a down-to-earth attitude to keep up with the hard work, even when these concessions are all about continuous growth and success.

Case Study 2: Russia

Russia is the leading global energy producer, with vast oil reserves in Siberia, the Arctic Circle and the Caspian Sea. As seen in Figure 1.7, the country's economic growth has been closely related to the energy reserves, the vast land, and the country's strategic location in Eurasia and the Arctic Circle.

Each country has distinct geographical particularities, and Russia's earlier challenge was the limited sea outlet on its European side. The nation's energy exports greatly depend on Russian's strategic alliances with countries that host the Russian pipelines, and countries with strategically positioned coastlines that have the capacity to transport Russian oil and gas to the global markets.

Chapter 4 of this book presents a case study on the Caspian Sea energy, and duly demonstrates the political dimensions of energy trade and transport, but also the myriad pipeline systems and efficient networks that Russia had to construct in order to export the oil out of the landlocked regions.

Case Study 3: United Arab Emirates

The United Arab Emirates, also called "the Switzerland of the Middle East", is another interesting case study for several reasons (Figure 1.9).

First, due to its very impressive transformation over the decades into a cosmopolitan, highly aesthetic, multicultural energy center. Second, and most important in my opinion, is the country's strategic decision to diversify its economic and industrial activities.

Energy is still king in the UAE; however the maritime and hospitality/tourism sectors have been well developed. In a few decades the UAE with World Dubai Ports managed to promote the country's reputation as a leading maritime stakeholder. Furthermore, the creative city design and luxury hotel industry have made Dubai a popular international destination that embraces the western style of living, without compromising its national and cultural identity.

Energy Infrastructure

Chapters 4 through 10 of this book extensively analyze the strategic, operational and technological elements of infrastructure, classified in accordance with the primary energy categories: Oil, Gas, Coal, Nuclear, Wind, Hydroelectric, Ocean Wave, Tidal, Thermal, Solar, Photovoltaic, Geothermal, and Biofuel.

Nevertheless, this section will cover the common thread among the different energy types.

Energy Security Throughout the Global Supply Chain

Energy infrastructure encompasses the primary amenities and systems that support a nation, region or society's operations. It is defined as the extensive facilitating technological solutions used to:

a move energy sources throughout the supply chain, and
b administer, monitor and control energy streams. It includes any tools, systems or structures such as power plants, refineries, transportation or transmission cables (Burns, 2015; *Your Dictionary*, 2016). It includes the fixed systems necessary for industrial applications, e.g., the energy sector, but also the input used such as crew, technologies or constructions needed for the input, output and distribution (*Merriam Webster*, 2016).

Energy infrastructure systems typically require capital-intensive, high-risk investments; nevertheless they are essential for a nation's financial advancement and security. Initiatives associated with infrastructure upgrades could be financed through private funds, public funds, or via public–private coalitions.

Energy infrastructure is essentially the facilitator between energy input and output. Consequently, it comprises the conventional resources and assets tied in to energy haulage (seaports, rail, pipelines, electric transmission towers and cables, and so on).

Energy Security Threats

Energy security threats are considered to affect both a nation's physical security and its commercial/socioeconomic stability. For this reason, this book offers a two-fold security analysis for the energy sector:

1 Energy as a critical infrastructure sector protected by the U.S. Department of Homeland Security (DHS), and other national departments of security.

According to the U.S. Department of Homeland Security (DHS, 2016), the energy sector is "uniquely critical," as it fuels the economy of the 21st century, and provides an enabling function across all critical infrastructure sectors. A public/private sector coalition is necessary as over 80% of the U.S. energy infrastructure is owned by the private sector. Similarly, a large portion of the global energy infrastructure is owned and controlled by conglomerates.

Table 1.4 demonstrates that energy is considered as a critical infrastructure sector by all major global powers, including the U.S., Canada, the European Union and the UK, Australia, Japan and South Korea. The number in brackets indicates how many critical infrastructure sectors are designated in each country.

2 Energy infrastructure as a facilitator of economic growth through trade and transport.

According to the International Energy Agency, energy security entails the "uninterrupted availability of energy sources at an affordable price" (IEA, 2016c).

Table 1.4 Energy as a critical infrastructure sector in global security plans.

	*USA (*16 sectors)*	*Canada (*10 sectors)*	*European Union (*11 sectors)*	*Australia (*7 sectors)*	*Japan (*10 sectors)*	*South Korea (*7 sectors)*
#	Energy	Energy & Utilities	Energy	Energy	2 Energy Sectors: • Electricity • Gas	Energy

Source: Burns, 2015. *Logistics and Transportation Security: A Strategic, Tactical And Operational Guide To Resilience*. Boca Raton: CRC Press, Taylor & Francis.

> Energy disruptions may affect nations and global supply chains both in the short and long term: Short-term disruptions affect the energy markets' ability to respond to security threats and achieve a supply–demand or production–consumption equilibrium. Long-term disruptions will threaten homeland security through physical unavailability of energy, leading to severe social turmoil and economic crises. Such repercussions eventually lead to socioeconomic instability, which directly and indirectly affect homeland and global security.

Energy infrastructure is critical in supporting a nation's security, safety, economy, and therefore social stability. Its significance exceeds the levels of socioeconomic growth, as a robust energy network ensures social stability (DHS, 2016). Chapter 3 of this book and the subsequent chapters explore the composition of these national and global supply chains, identify the key players, and recommend ways to mitigate security threats.

Energy Critical Infrastructure Sector

The energy critical infrastructure sector is familiar with its weaknesses hence is initiating an important voluntary attempt to enhance its contingency planning, as well as emergency preparedness strategies. Public–private synergies have created efficient big data banks, and the exchange of best practices, vulnerabilities and threats, encompassing cyber security and physical security.

In most developed nations the private sector owns approximately 80% of energy infrastructure and superstructure, thus providing to homes and industries three primary energy categories: Oil, natural gas, and electricity, which is generated to a large extent by coal, nuclear energy and renewables.

Economic Development and Energy Infrastructure

A nation's economic development level may impact its energy infrastructure in the following ways:

- developed nations with saturated economies may suffer from deteriorating infrastructure;
- developing economies may enjoy new state-of-the-art infrastructure;
- less developed countries may have underdeveloped infrastructure.

A nation's infrastructure depends several key factors, including:

a economic growth and investment capabilities;
b technological advances and innovation capabilities;
c geographic particularities such as vast land, rural areas, jungles, desert areas, geomorphology, many islands;
d bureaucracy, bribery or corruption;
e intentional attacks such as sociopolitical turmoil, violence, sabotage, terrorism;
f low density may impose infrastructure challenges with investment restrictions due to low taxation and revenue;
g high density may cause capacity restrictions.

Regular maintenance is required to ensure safe, reliable and smooth energy transport through highways, bridges, railways, seaports, refineries, pipelines and transmission lines. This is an era of instant gratification, where energy logistics and transportation networks are expected to have an optimum performance with no disruptions or delays. And although an efficient flow of goods can be taken for granted, natural disasters, human error or intentional attacks on our energy networks may lead to damaged infrastructure such as pipeline ruptures, explosions, fires, critical equipment breakdown, and so on.

Energy Security: Disruptions From a Supply Chain Perspective

Energy security requires the uninterrupted supply of energy from reliable, sustainable sources, and implies that energy should be readily available regardless of the high demand or supply shortages and production disruptions. Energy security entails overcoming supply chain disruptions that could be triggered by reasons including but not limited to:

a **Political or geopolitical**: Sanctions, warfare, guerrilla action, an act of sabotage or punishment.
b **Security threats**: i.e., intentional damage including cargo theft, hijacking, sabotage, terrorism, and so on.
c **Force majeure or acts of God** pertains to disruptions beyond human control, such as extreme geophysical phenomena. Extreme weather conditions could be halting production, destroying refineries, or sweeping off storage units.
d **Infrastructure or superstructure deficiencies**: These could be related to structural integrity, underdevelopment, machinery breakdown, obsolete technologies, poor maintenance.
e **Transportation-related delays or disruptions** could be related to bureaucracy, poor infrastructure, strikes, weather conditions, warfare and political turmoil at any stage of the supply chain.
f **Product scarcity or scarcity of factors of production**: i.e., producers unable to mitigate high demand during an economic meltdown or during a market recovery and after mass layoffs.
g **Human factors**: Strikes, human error, fatigue, lack of trained personnel.
h The discovery of **new non-renewable energy reserves**: Oil price speculative investment strategies and concerns over unexpected fluctuations have caused the global market meltdown twice in ten years:

 i in July 2008, when the oil price reached an all-time-high with $145.31 a barrel (*Trading Economics*, 2016).
 ii in February 2016, when the oil price hit a low level (depression) with $26.21 a barrel (*New York Times*, 2016).

Reasons may include:

- political decisions such as tax increases, privatization or nationalization of energy organizations, legal framework, energy supply demand equilibrium or lack of
- monopolistic or oligopolistic markets
- protectionism
- currency manipulation or extreme fluctuations
- inflation, deflation
- union-led strikes
- government sanctions or boycott
- production or consumption fluctuations
- poor infrastructure
- supply chain disruptions

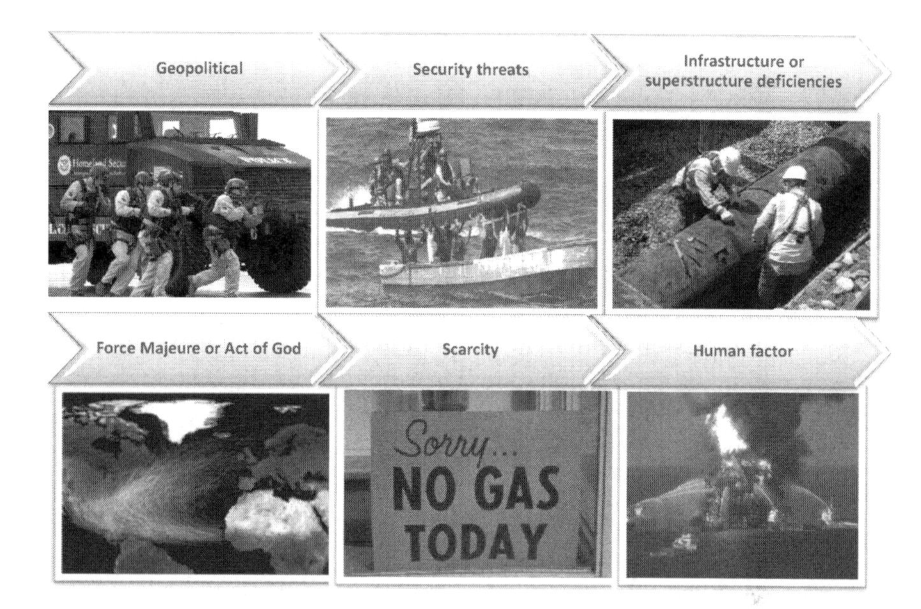

Figure 1.9 Energy security: Disruptions from a supply chain perspective.

Sources: compiled by the Author. Images available at:

1 DHS, 2011. *ICE/HSI, DHS training-using-armored-vehicles*. USCG, Department of Homeland Security. Available at: www.dhs.gov/photo/hsi-using-armored-vehicles-training-ice, last accessed December 2, 2016.
2 USCG, 2010. History, the piracy mission. USCG. DHS. Available at: http://coastguard.dodlive.mil/files/2010/03/Pirates-1024x545.jpg.
3 Oil spill pipeline removal. Available at: https://upload.wikimedia.org/wikipedia/commons/0/03/Worker_tight_-_Enbridge_Oil_Spill_pipeline_removal_2_(4869018519).jpg.
4 Atlantic hurricane tracks. Available at: https://upload.wikimedia.org/wikipedia/commons/3/31/Atlantic_hurricane_tracks.jpg.
5 No Gas sign, Oregon (USA), 1973 oil crisis. Available at: https://upload.wikimedia.org/wikipedia/commons/4/4f/%22NO_GAS%22_SIGNS_WERE_A_COMMON_SIGHT_IN_OREGON_DURING_THE_FALL_OF_1973,_SUCH_AS_AT_THIS_STATION_IN_LINCOLN_CITY_ALONG..._-_NARA_-_555416.jpg.
6 Human factor, Deepwater Horizon, April 21, 2010. Available at: www.uscg.mil/npfc/img/Slideshow/ltimg/DWH.jpg. Fire boat response crews battle the blazing remnants of the off shore oil rig Deepwater Horizon, April 21, 2010. A Coastguard MH-65C dolphin rescue helicopter and crew document the fire aboard the mobile offshore drilling unit Deepwater Horizon, while searching for survivors April 21, 2010. Multiple Coastguard helicopters, planes and cutters responded to rescue the Deepwater Horizon's 126 person crew. (Courtesy: U.S. Coastguard photo.)

- extreme weather conditions increasing consumption and causing scarcity and inelastic demand
- geophysical phenomena
- sociopolitical or warfare hindering accessibility to or from production or consumption locations.

The Impact of Energy in Global Trade, Transport and the Economy

For the past 65 years oil has emerged as the globe's leading energy source, supporting the global community, households and industries (UKOG, 2016).

Emerging energy types such as natural gas, coal and renewables seem to expand their market share. Hundreds of energy companies are directly or indirectly affiliated with the *Forbes* Global 2000, i.e., the world's rich and famous. These are international public corporations with the prime composite ratings according to their listings for market value, annual gross and net revenue, sales, properties and assets (*Forbes*, 2017).

Energy and transportation conglomerates, banking systems, manufacturers, software and communication moguls are among the entities that comprise the global energy supply chains.

A nation's energy strategy facilitates global trade, transport and the economy in the following ways:

a **Adds national and regional value** through infrastructure and superstructure **investments**.

b Energy serves as a **high-value commodity** greatly contributing to development. As seen in Table 1.3, one third of global trade entails energy products.

c **Jobs creation**, government taxation.

d Contributes to **scientific research and technological innovations** that can be transferable to governments and other industries such as transportation, manufacturing, and so on.

e **Promotes domestic and international partnerships**, through trade and transport.

f Energy corporations substantially **invest in human resource training and development**. In fact, their training academies and global conferences are used as prototypes for several other industries. By elevating the workforce standards, both governments and other industries benefit.

g **Inspires and generates new academic and educational programs**, both at a global and national level. It has been observed that energy-producing nations and states develop the most competent academic programs, blending STEM disciplines (i.e., Science, Technology, Engineering and Mathematics) with petroleum engineering and energy, financial, supply chain and transportation programs.

h **Strengthens national security**, as energy corporations are in charge of a significant amount of land, properties and supply chain processes. Hence, private funds, resources and intelligence are used to enhance national security and deter threats.

i **Inspires new financial alliances** and public/private partnerships.

j **Promotes robust banking systems**, by investing in billion-dollar projects.

k While being a **highly regulated** industry, it continually **instigates new regulatory frameworks**. In order for energy corporations to comply with environmental and other regulations, they are **prompted to innovate by means of new technologies, methods and contingency planning**.

Figure 1.10 demonstrates the impact of energy upon global trade, transport and the economy, while demonstrating key factors that strengthen national security, government popularity, standards of living and economic systems.

As demonstrated in Table 1.5, the above elements are interrelated.

The Economic Impact of Energy Market Cycles

Upon examining the history of energy economics, it has been observed that wide fluctuations in the price of oil can lead to significant financial instability. The energy industry is capital-intensive, and its supply chain expands to banking systems, transportation conglomerates, and a wide array of global players. Consequently, oil shocks or oil market meltdowns possess the power to seriously affect the global and national economy. An energy crisis can impact the sociopolitical and economic stability of nations, but also trigger hostility among oil-producing countries. For example, sociopolitical turmoil in energy-rich nations like Libya, Nigeria and Venezuela also contribute to the climate of uncertainty, as supply chain disruptions do not contribute towards reliable, secure energy supply.

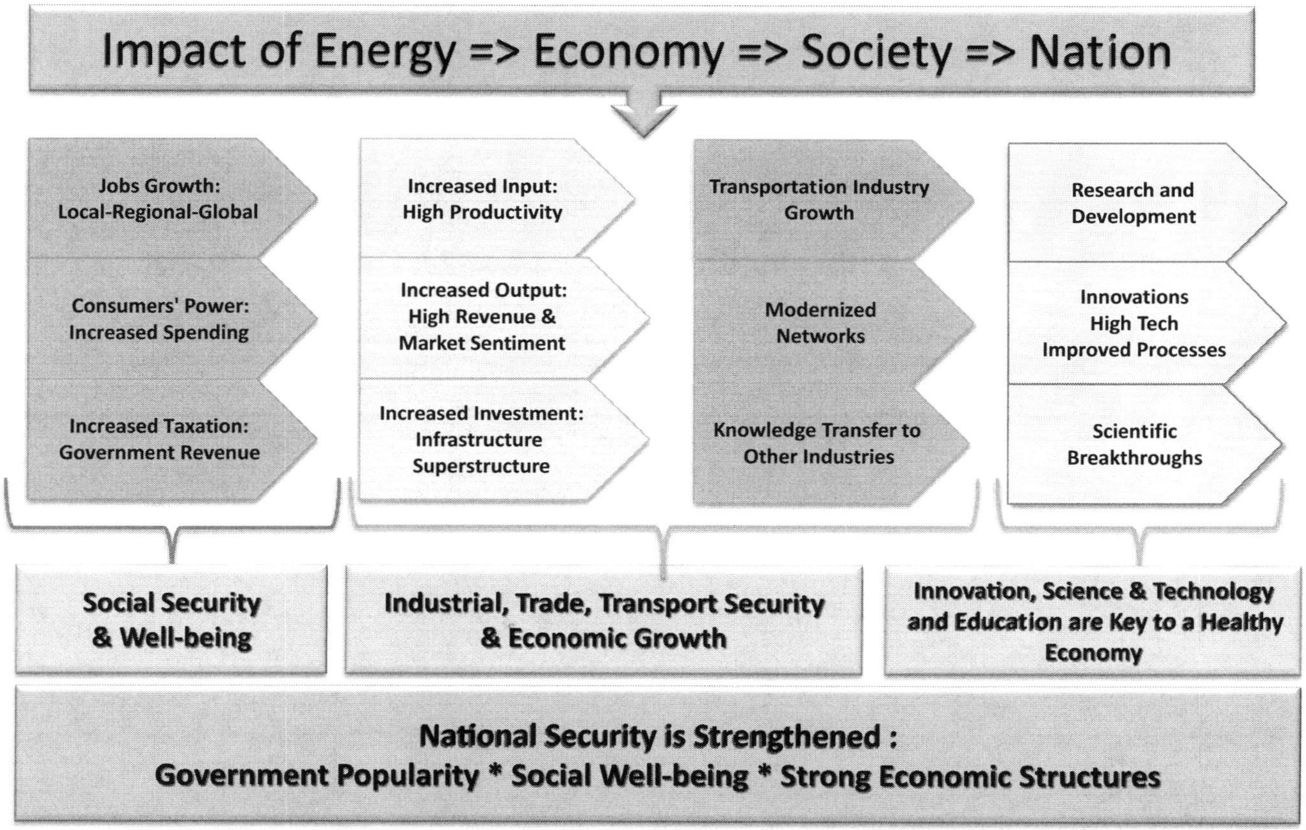

Figure 1.10 The impact of energy upon global trade, transport and the economy.

Source: the Author.

Table 1.5 The impact of energy upon global trade, transport and the economy.

Strengthens national security	*Economic growth*	*Public/private partnerships*	*Supports supply chains*
Regulatory compliance	Innovative technologies	Knowledge transfer to other industries	Manufacturing
Emergency response	Investment banking systems	Strategic alliances	Services
Contingency planning			
Generates jobs	Taxation	Research & development	Retail & distribution

Source: the Author.

Oil Price Spikes and Major Global Recessions

As duly analyzed in Chapter 3, oil price spikes have been strongly associated with major global recessions since World War II. This can be confirmed in events like the Yom Kippur War of 1973 and the closure of the Suez Canal; the Gulf War in 1990; the 1998 Asian economic crisis, the 2008 global economic meltdown, the 2009 Arab Spring, and the most recent 2016 oil price meltdown.

Shortly after the global markets recovered from the 2008 economic meltdown, the 2015 and 2016 oil price collapse generated market uncertainty. Diminished global growth only reached 50% of the ten-year average (only 1.6%, compared with 3.8%), thus creating a market meltdown that can only be compared with the freefall economic crises of 1997–1999, and 2009. Despite the recent market fluctuations, petroleum remains as the primary energy source, comprising 32.9% of world energy market share.

Under the recent political and economic reality, a supply–demand equilibrium seems difficult to achieve. On one hand, rapidly growing economies like China have diminished their industrial activities and reduced oil imports, whereas, on the other hand, emerging oil exporters are increasing their energy output, driving other key players to limit production and market share. These developments are causing market uncertainty, making it difficult to accurately forecast both market trends and factors that could inspire a positive market sentiment.

A great economic and energy-related reform has been observed that threatens to disrupt past cyclical patterns and disprove long-proven theories. Until recently, the golden rule was that a 10% increase in oil prices could trigger a 0.5% reduction in a nation's GDP, whereas similarly a 10% reduction in oil prices could cause a 0.5% increase in GDP. The past year's oil market cycles have shown that this is no longer the case. The all-time highs ($110 per barrel) did not help the economy grow exponentially, and the all-time low ($27 per barrel) did cause job losses and weakened the oil producers' economies, but it did not strengthen the buyers' markets.

In other words, key players' benefits were mainly long-term, i.e., the opportunity to establish and retain their market share position while competing with their global partners.

A Unique Market Cycle

Table 1.6 demonstrates the reasons why this market cycle is unique from a geopolitical and economic perspective. There are three main reasons why the current market cycle is unique.

Table 1.6 A unique oil market cycle.

1	Diminished oil prices did not trigger higher demand for oil.
	In past market cycles, low oil prices triggered high energy demand, which resulted in growth in energy-intensive industries (manufacturing, transportation, etc.).
	The 2016 market showed diminished demand for oil/energy and a modest economic growth of the major global economies. This delayed the energy market recovery.
2	Major importers have now become major exporters.
	Unconventional oil and shale gas has radically changed the map of top global producing countries. As a result, the energy world experiences a radical change of roles in terms of suppliers and consumers.
	Nations like the U.S., the UK, Australia, West Africa's Gulf of Guinea, and so on, enforce their changing roles from buyers to sellers, so the energy price impacts them differently.
	Traditional energy-importing countries like the U.S. (Alaska Oil, Texas and Dakota Shale gas), Canada (Bitumen/Oil Sands), Russia (Siberia), the UK (North Sea), have now untapped large energy fields and not only satisfy their own domestic energy needs, but also export energy to the world, thus altering the market shares among past and current key players.
	Back in the years where these countries were mainly energy importers and consumers, low oil prices were a good sign for their domestic economies.
3	Shift of power from OPEC to non-OPEC nations (Table 1.8).
	Non-OPEC countries now have a greater leverage.
	In the past, oil supply from Non-OPEC nations was lower scale hence competition or friction between OPEC/Non OPEC nations could be easily resolved. Recent technological advancements enabled Non-OPEC nations to tap on unconventional oil (oil sands or bitumen, extra heavy oil, deep offshore) and shale gas have empowered several emerging energy producers to drill deeper.

The Historic 171st OPEC Meeting: "The Vienna Accord"

The year 2017 signified the recovery phase of a global energy cycle: It is the stage when the energy market collapse ended, and a new era of optimism and growth commenced.

The light at the end of the tunnel came in November 30, 2016, during the 171st OPEC Meeting in Vienna, Austria, where OPEC nations agreed to the first oil production cuts in eight years. The "Vienna Accord," named after the city where the assembly took place, has a historical significance as it signified the consensus between OPEC and Non-OPEC countries. The agreement will enable oil producers to induce the recovery of the oil price. Energy producers no longer have to sell their commodities at low prices with diminished return on investment. Energy majors can retain their market share, and an amicable resolution can be attained, leading the fossil fuel industry into a win/win era.

The Impact of Oversupply Upon Prices

Oil prices have a tremendous impact upon the economy for the following reasons:

Diminished revenue for oil majors means the loss of thousands of jobs.

A low oil price does not benefit the transportation industry or the local consumers, as the energy crisis has an adverse bearing on both the transportation industry and household income.

High unemployment rates lead to diminished market sentiment and low purchasing power, which affects the domestic economy and a myriad of businesses.

Traditional OPEC countries like Saudi Arabia can find cheap, good quality oil literally "in their back yard," thus offering them higher profit margins.

Contrary to this, new energy producers must keep on extracting expensive shale gas, and deep offshore oil and gas, for these nations to retain their energy independence, and for the energy conglomerates to retain their hard-fought-for market share. Nevertheless, under the current market they will enjoy minimal returns on investment.

The multi-billion investments required in the national energy sector require a robust support system to provide the financial capital (bank loans), physical capital (assets, such as drilling equipment, infrastructure and superstructure), and the risk-bearing components (insurance and underwriters). When the energy markets encounter diminished profits, the global and national banking, investment and insurance institutions also weaken. This not only increases the risk, as explained on pages 56–60, but also increases the security threats by exposing the system's vulnerabilities.

The Impact of Energy Supply–Demand Disruptions

A supply-shock-triggered energy market meltdown has tremendous impact both at a macro and micro level, as energy is a commodity employed to utilize all other commodities. Any time energy trading markets crash, an energy scarcity evolves. Electricity customers can suffer deliberately caused power shutdowns, thus causing operational disruptions.

Developed countries rely on oil and attempts to reduce its supply has a damaging impact on the financial systems of oil suppliers. From a buyer's perspective, the energy price rises may affect budget for transportation, industries and households.

An energy meltdown is defined as any considerable bottleneck, supply chain or extreme price fluctuation that affects the energy availability and consequently the financial systems.

Energy Shortages and Supply Chain Disruptions

Supply chain disruptions are causing shortages due to:

a pipeline deficiencies;
b incidents;
c terrorist attacks;
d sabotage or militia;
e political changes, government collapse, military occupation, and coup;
f policy changes, such as:

i **Canada's National Energy Program** was enforced in 1980 by the Liberal federal government to some extent in reaction to explosive oil prices. Even though the plan achieved certain targets, decreasing foreign control and increasing oil independence, the program was considered to polarize the country, by increasing the gap between the government and the energy intensive west Canada (Suzuki, 2010).
ii In the U.S.A. with the setup of a cache of secure fuel reserves like the United States Strategic Petroleum Reserve, in case of national emergency (DOE, 2016c).

The Strategic Petroleum Reserve (SPR) is the globe's most significant storage of emergency petroleum. The government-owned oil is kept in over 500 vast subterranean salt domes, located across the shoreline of the Gulf of Mexico. The site was a plausible decision as the Gulf Coast hosts several U.S. refineries and distribution centers for tanker ships, barges and pipelines. As stipulated by the Energy Policy and Conservation Act (EPCA), the American President has the authority to withdraw oil in case of a national energy emergency. The SPR's vast capability of 713.5 million barrels renders it a considerable preventive measure to oil import disruptions and a strategic instrument of America's foreign policy.

iii Chinese energy policy includes specific targets within their five-year plans. China's industrial revolution has been coal-powered (Dupuy and Xuan, 2016). In late March 2016, the Chinese federal government released a significant policy statement referred to as "Document 625" geared towards enhancing and protecting the production of alternative energy, i.e., from wind, solar, hydro, and other environmentally friendly resources. The nucleus of Document 625 is a dictated "assurance" that grid corporations acquire output from alternative generators, with a minimum designated number of hours.

Oil Shortage Theories

Peak Oil or The "M. King Hubbert's Theory"

Peak Oil, the oil depletion theory or the Hubbert curve, named after M. King Hubbert's principle, is the moment in time where the highest petroleum extraction rate is attained, and thereafter a meltdown is anticipated.

The Peak Oil theory was conceived by Dr. M. King Hubbert, a Chief Geologist at Shell Oil, who in 1956 expressed his concerns about the future availability of fossil fuels.

Per Hubbert's theory, energy production rates are intensified when new resources and improved technologies become available, whereas production rates decline when existing energy resources are depleted, and limited new reserves are discovered. In his estimations, Hubbert also took into consideration the high cost of unconventional exploration, diminishing an oil company's financial and commercial motives to drill. He created a series of bell-curve diagrams to depict the production flows of non-renewable energy in the world and the U.S., while making sporadic comparisons with other global energy leaders.

As depicted in Table 1.7, Hubbert's estimations predicted global oil, gas and coal production would reach a peak in the 1970s, 2000s and 2150 respectively.

The "Peak Oil" theory has offered the fundamental principles used to establish the "Oil Depletion" theory, according to which:

- There is a peak and a depletion stage, just like any economic and market cycle.
- The peak and depletion stages occur over several decades, or even centuries, hence giving the industry time to proactively prepare for substitute energy. This gradual replacement and shift into alternative sources occurs naturally, hence eliminating the element of surprise or market panic.
- The Peak Oil theory was conceived by recognizing the oil field production cycles, characterized by stages of expansion, peak, decline, and collapse.
- It pertains to a lasting fall in the obtainable oil production. This, along with rising demand, considerably boosts the global oil prices. Essentially, it is the accessibility and cost of transportation energy (EIA, 2016b).

Table 1.7 The Energy Peak Theory: Comparison of modern global reserves, with estimated peak and depletion years per Dr. M. King Hubbert (1956).

Energy Source	*Global Reserves*		*Estimated Peak Year*	*Estimated Depletion*
	Per Hubbert, 1956	*Per 2017 Estimates*	*Per Hubbert*	*Per Hubbert*
Crude Oil	1,250 billion barrels	1.662 trillion barrels	2000	2200
Natural & Shale Gas	2.5 trillion barrels	6,964.05 cubic feet (197.2 trillion cubic meters)	1975	2075
Coal inc. Lignite	2,500 billion metric tons	892 billion metric tons	2150	2700

Source: the Author, based on data from:

- Hubbert, M. K., 1956, "Nuclear energy and the fossil fuels," in *American Petroleum Institute Spring Meeting: San Antonio, Texas, p. 40 By M. King Hubbert Chief Consultant (General Geology), Presented Before The Spring Meeting Of The Southern District Division Of Production. American Petroleum Institute Plaza Hotel, San Antonio, Texas March 7-8-9, 1956. Publication No. 95* Shell Development Company Exploration and Production Research Division, Houston, Texas, June 1956. To be published in *Drilling and Production Practice* (1956) American Petroleum Institute.
- CIA, 2017. *World Factbook*. Available at: www.cia.gov/library/publications/the-world-factbook, last accessed: January 2, 2017.

Hubbert's theory could be considered controversial for the following reasons:

- The known energy reserves in 1956 were much lower than the known energy reserves 60 years later.
- While Hubbert's estimations took into consideration new technologies and unconventional drilling, he did not predict the high level of technological innovations, and the energy industry's ability to explore extremely unconventional sources.
- Hubbert was inclined towards the development of nuclear energy as a substitute for non-renewables. However, he did not predict the renewable energy options as strong alternatives to conventional energy.
- His estimations on energy production and consumption did not take into consideration oil price fluctuations.
- His estimations were mainly focused on national factors of production and internal parameters. However, he fails to cross-examine external, i.e., consumption factors, such as overpopulation and the radical development of China, India and other economies.
- Over the past 60 years there has been an international change in energy production and consumption. Rapidly changing economies like China and India, and several developing countries like Brazil, Mexico, Indonesia, South Africa, Nigeria and so on, are anticipated to change the global map of energy supply and demand.
- According to the International Energy Agency's *World Energy Outlook* (www.iea.org, 2010–2016), in the years from now to 2030, the developing economies will exceed 87% of the international energy demand and development.

If these factors were included in Hubbert's estimations, the estimated peak and depletion years would be shortened by several decades.

Nevertheless, Hubbert's theory is invaluable for the following reasons:

- It urged governments and industries to re-examine their reservoirs, and to prove or disprove the theory.

- It inspired energy economists and other scientists to conduct systematic, periodic comparisons among energy production and consumption, with the goal of ensuring energy availability in the foreseeable future.
- It promoted a proactive culture, and urged world energy producers to pursue alternative energy sources.
- It indirectly inspired technological innovations and investments, as a coping mechanism in order to prolong the peak years and postpone the depletion years.

Addressing the Peak Oil Phenomenon

To eliminate the severe financial and commercial ramifications deriving from a global energy shortage, it is necessary for governments, industries and communities to proactively arrange for contingencies several years prior to the oil shortage, so that energy conservation, alternative trade agreements and alternative energy sources are readily available. Interestingly enough, contingencies could postpone or alleviate the anticipated peak oil event.

a **Load shedding or rolling blackout**

Load shedding or rolling blackout is a method used to eliminate a total blackout or energy disruptions.

In developed nations, load shedding may be used during an extreme weather phenomenon or other event that requires the control of high energy consumption, either by shutting down power or curtailing the energy availability in specific areas, times of the day, or for specific devices if applicable.

In less developed or developing economies, a rolling blackout may be used when energy scarcity for a number of reasons associated with financial restrictions, i.e., a weak currency, high external debt, low salaries and purchase power parity (ppp).

Regardless of a nation's economic status, rolling blackouts may be the result of political decisions.

Energy scarcity is experienced as:

i need for energy dependence leading to reduced imports and reduced energy output;
ii government's financial inability to invest in robust energy transmission infrastructure;
iii government's lack of contingency planning, heavy reliance on one major energy source;
iv political sabotage or corruption leading to costly, delayed and ineffective infrastructure.

In any of the above cases, rolling blackouts are a regular, widespread phenomenon.

b **Power rationing**

The repercussions of an electrical power shortage may not seriously impact industries with generators in place, but will affect mostly smaller companies and households who rely on electric power for community water supply, heating and cooling systems, food storage and preparation, and so on. Under these conditions, a prolonged energy crunch may become a humanitarian emergency. Governments can mitigate a relatively brief shortage by protectionism, i.e., commodity and price techniques to influence the energy trading markets. Under a prolonged power shortage authorities declare a state of emergency and decide on energy rationing. Power rationing may be defined as the government measures undertaken to restrict free consumption to ensure that the government and the people's basic needs will be met for a sufficient period of time, while hindering entities from purchasing high volumes of these commodities.

Case Study on Power Rationing

Venezuela suffered a power rationing from 25 April to July 4 2016, with power cuts throughout most of Venezuela for four hours a day. The power rationing decision was made when the Guri hydroelectric dam, that generates nearly 66% of the nation's electrical power, underperformed after the El Nino weather phenomenon, leading to record low water levels. The power restrictions emerged in the middle of a financial turmoil and extreme food shortages (BBC, 2016).

Although weather phenomena are unpredictable, some critics think that the government should not depend mainly on hydroelectric power and the specific dam, but instead should have alternative energy-producing sources as part of the nation's contingency planning strategy.

Energy Sources: Renewable vs Non-Renewable

Non-Renewable Energy

A non-renewable energy source pertains to a commodity or asset that cannot be naturally substituted on a quality equivalent to the use of its resource. The majority of fossil fuels, including petroleum, natural gas and coal, are classified as non-renewable energy sources, since their formation requires several billions of years and hence their utilization is not sustainable.

Non-renewable energy sources such as crude oil, natural gas, coal and uranium (used in nuclear energy) are not abundant in nature, cannot be replenished in a brief time period, and may need billions of years to be replenished. They are extracted in a solid, gaseous or liquid form. Uranium is a solid form example. It is a radioactive metal that is mined in solid form and used as a fuel at nuclear power plants. Natural gas and shale gas are examples of gaseous products, whereas petroleum has a liquid form (EIA, 2015).

As the creation of non-renewable resources needs thousands of years to be generated and may eventually lead to depletion, it is important for humankind to make contingency plans for alternative energy sources over the next decades. These contingencies may include several combinations, from extracting unconventional sources to the option of renewable energy resources. The latter option is considered to lead towards large scale energy independence. In 2015, non-renewable energy markets accounted for 86% of the global energy supply, with petroleum representing 33%, coal 29% and natural gas 24%. On the other hand, renewable energy accounted for 10%, with hydroelectric energy covering 7% of the world supply. Finally, nuclear energy has a 4% of global share.

Renewable Energy

Renewable or alternative energy is energy produced from natural sources and processes. They are constantly replenished as a part of a natural cosmic process, hence they cannot be depleted. Solar, wave, tidal, wind, and geothermal energy is available in abundance and hence these are typical renewable energy sources (IEA, 2016a).

The Texas Renewable Energy Industries Alliance (TREIA) defines renewable energy as adopted by the Texas legislature, as:

> Any energy resource that is naturally regenerated over a short time scale and derived directly from the sun (such as thermal, photochemical, and photoelectric), indirectly from the sun (such as wind, hydropower, and photosynthetic energy stored in biomass),

> or from other natural movements and mechanisms of the environment (such as geothermal and tidal energy). Renewable energy does not include energy resources derived from fossil fuels, waste products from fossil sources, or waste products from inorganic sources.
>
> (TREIA, 2016)

Renewable or alternative energy also signifies energy production processes and technologies that are unconventional and hence different from conventional methods and sources that are readily available. Variable renewables represent the natural types of energy comprised of solar, wave, tidal, and wind energy. Their availability is perceived as changeable as it depends on weather conditions, which in turn depends on geographic location, season and month of the year, and so on. Different types of renewable energy will therefore be more or least favorable in different regions.

For example, solar panels require sunny weather conditions and would not be suitable for regions with foggy, cloudy weather. Ideally, future energy solutions should be able to offer a package of energy alternatives, because the performance of certain types of renewables depends on weather conditions, i.e., regional or seasonal climate.

Solar panels convert luminosity (brightness) into energy, and respond to ultraviolet (UV) light and infrared (IR) light which they convert into energy. Light is irrelevant to temperature, hence the ideal temperature for solar power harvesting (photovoltaic panels) is approximately 43°F, or 6.11111°C.

The particularities of variable renewables require meticulous planning and consideration from a national, industrial and consumers' perspective, so that a contingency plan incorporates a region-specific, season-specific package of energy alternatives.

The implementation and cost of renewable energy depends on the regulatory framework currently enforced. Existing subsidies for fossil fuel coupled with the absence of a world rate for coal energy may significantly limit the renewable energy's price competitiveness.

Renewable energy is slowly but steadily growing, and will eventually replace nonrenewable energy forms when these are depleted. It took several decades for the oil and gas industry to replace coal and to develop state-of-the-art technologies. The renewable energy sector is now in a similar position, where technologies are advanced in segments such as solar panels, biomass, geothermal, tidal and hydropower plants or wind farms.

The benefits to reap from this growing energy path entail an infinite supply of greener energy, and thus capability to develop high-scale projects to feed mankind's insatiable need for energy.

Case Studies: Global Energy Organizations

IEA (International Energy Agency)

The International Energy Agency (IEA) is an independent institution comprising 29 member states and it is headquartered in Paris, France. It was founded in 1974 as part of the composition of the Organization for Economic Co-operation and Development (OECD), in the aftermath of the 1973 oil market crunch. It was created by the Agreement on an International Energy Program (IEP Agreement). Its purpose is to make sure that energy is reliable, affordable and environmentally friendly. The IEA was originally created to facilitate nations around the world in putting together an unanimous strategy to eliminate serious energy disruptions throughout its supply chain and to recover after energy crises such as the Iranian revolution or the 2008 global economic meltdown. Even though this continues to be a crucial element of its mission, the IEA has developed over time and broadened its scope.

The Agency's mission has expanded to encompass the "3Es" of successful energy strategy: Energy security, economic growth, and environmental security. It serves as the advocate of the energy industry, offering synergistic solutions while administering well-respected industry reports and statistical analyses. While it works as a policy agent for its member states, it concurrently collaborates with non-member states playing a leading role in the energy sector, such as China, Russia, and India.

The IEA examines the full spectrum of energy issues and advocates policies that will enhance the reliability, affordability and sustainability of energy in its 29 member countries and beyond.

The IEA has four main areas of focus: Energy security, economic development, environmental awareness and engagement worldwide:

- **Energy security**: Supporting diversity, effectiveness and adaptability within all energy sectors;
- **Economic growth**: Aiding free markets to promote economic development and eradicate energy poverty;
- **Environmental awareness**: Investigating policy alternatives to counterbalance the effect of energy output and use on the environment, in particular for dealing with climate change; and
- **Engagement worldwide**: Working directly with member states, in particular with leading economies, to identify solutions to common energy and environmental considerations.

To be a member state of the IEA, a nation also needs to be a member state of the OECD, while there is a procedure to determine whether or not the future member state can satisfy the entry prerequisites (IEA, 2016a; IEA, 2016b).

U.S. Energy Information Administration (EIA)

The U.S. Energy Information Administration (EIA), under the aegis of the U.S. Department of Energy, is an agency in charge of data collection and statistical evaluations for the energy sector. The Department of Energy Organization Act of 1977 recognized the EIA as the principal governing administration expert on energy statistics and data analysis, developing upon programs and institutions first created in 1974 after the oil market collapse of 1973. Situated in Washington, DC, EIA is an agency of approximately 370 government employees, with a yearly budget of $122 million (as of fiscal year 2016).

The EIA accumulates, evaluates, and shares unbiased and independent energy data to showcase robust policymaking practices, competent trading markets, and public familiarization with energy and its interplay with the financial systems and the ecosystem. Its programs include information relating to both conventional energy, i.e., coal, petroleum, natural/shale gas, electric power, but also renewables and nuclear power. The EIA serves as America's leading provider of energy data analysis, hence, as stipulated by U.S. law, its statistics, evaluations, and future market projections are independent, i.e., they do not require authorization or permissions by an administrator or personnel of the U.S. governing administration.

The EIA carries out an all-encompassing data gathering and statistical analysis process that addresses the complete spectrum of energy segments, stakeholders, key players, energy networks and global supply chains. Additionally, the EIA creates explanatory energy analytics and evaluations, monthly predictions of energy market trends, and long-term U.S. and global energy outlooks. The EIA shares its energy information and facts, market analyses, and other tools (IEA, 2016c).

Energy Types: Historical Overview and Timeline

> If you want to find the Secrets of the Universe
>
> think in terms of Energy, Frequency and Vibration.
>
> Nikola Tesla, inventor and futurist

The word "energy" is an exact synonym and derivative of the ancient Greek word "energeia" (ἐνέργεια), a composite word consisting of the preposition "in" and the noun "ergon" (ἔργον), which means "project," "creation," "operation." It is no coincidence that Athena, the Wise Warrior Goddess, had the epithet "Ergane" (energetic), and was worshipped by craftsmen and artisans alike. The word was encountered in the works of Heraclitus (535–475 BC) and Aristotle (384–322 BC).

All the types and classifications of energy refer to an action or resource required for any project, motion, operation or other production, both tangible and intangible.

Through the millennia, the word energy has retained its original meaning: It describes the power produced from utilizing natural, technological or chemical sources as input. Energy enters a system as a force, process or resource, with the goal of generating a specific and well-calculated output. As explained in the section on the history of energy (see pages 38–41 of this chapter), throughout the past centuries notable scientists discovered new types of energy and created a myriad of new technologies in the service of humanity.

As discussed in a previous section, a nation's need for energy security must comprise the elements of sustainability, reliability and accessibility. From medieval times to date, several distinguished scholars enriched the evolution of a robust, sustainable, reliable and accessible energy system with impactful scientific achievements, technologies and enlightened inventions as gifts to humankind.

This section will focus on the overall history of energy, and will cover the names of several notable scientists who hold a prominent place in the history of science and technology.

This section serves as a tribute to the pioneers of energy and science, whose lives were dedicated to a noble cause: The development of scientific and technological instruments as a service to the world.

Units of Measuring Energy Named after Renowned Scientists

To demonstrate the significance of different scientific applications in the global industry, including the energy industry, it is worth observing how units of measuring energy were named after prominent scientists and inventors.

To honor their work, standard units of measure used in physics, electricity and industrial engineering have been named after scientists such as Volt, Ampere, Ohm, Morse, Kelvin, Hertz, and other renowned scientists.

- **James Watt: Watt, measuring joules of energy per second.**

  James Watt (1736–1819): A watt, a unit of power according to the International System of Units (SI), corresponds to one joule of energy conducted per second, or to 1/746 horsepower.

- **Luigi Galvani: Galvanism, the use of metal cables to induce electricity currents.**
  Luigi Aloisio Galvani (1737–1798): Galvani was a professor of Anatomy who discovered the presence of natural electricity currents in animal tissue. He conducted several experiments on frog legs, and used the metal cable to connect nerves and muscles. The term "Galvanism" was generated by Alessandro Volta to express how the direct electricity current is induced through chemical activity (Fox, 1909).
- **Alessandro Volta: Volt, measuring electric potential.**
  Alessandro Volta (full name: Alessandro Giuseppe Antonio Anastasio Volta) (1745–1827) was an inventor, chemist, and physicist who generated theories on energy and electricity while expanding on Galvani's experiments. In his honor, the SI unit of electric potential is named "Volt."
- **Andre Marie Ampere: Ampere, measuring electric current.**
  Andre Marie Ampere (1775–1836) was one of the pioneers of electrodynamics and electromagnetism, and the inventor of the electrical telegraph. To honor the French mathematician and physicist, his name is inscribed on the Eiffel Tower among those of another 71 prominent scientists, mathematicians and scientists. In his honor, the ampere is a SI unit of measurement of electric current.
- **Georg Simon Ohm: Ohm, measuring electrical resistance.**
  Georg Simon Ohm (1789–1854) was a German professor of mathematics. According to Ohm's theory, the electric current via a conductor amongst two points is directly proportionate to the voltage throughout these two points. The numerical equation to explain this relationship is : $I=V/R$, where I represents the electric current via the conductor in amperes, V represents the voltage calculated throughout the conductor in volts, and R reflects the conductor's resistance in ohms.

  From his work the constants of resistance and proportionality are presented: Resistance remains a constant regardless of the current, i.e., Amper and/or Voltage units (Millikan and Bishop, 1917).

  In his 1827 publication *The Galvanic Circuit Investigated Mathematically* (or *The Die galvanische Kette, mathematisch bearbeitet*) he refined the Ohm theory, together with extensive empirical implementation and experiments on electric currents and resistance (Heaviside, 1894).

  At the time of publication his work was rejected by his peers, forcing his resignation from Cologne Jesuits' College. Fourteen years later, his work was globally renowned, and the robustness of his scientific methods was finally appreciated (*Encyclopedia Britannica*, 2016c).

  The ohm, the physical unit measuring electrical resistance was named in his honor.
- **Heinrich Hertz: Hertz, measuring the frequency of electric and magnetic fields per second: See kilohertz (kHz), megahertz (MHz), or gigahertz (GHz).**

Energy Types: A Historic Timeline

Arguably, fire and the invention of the wheel are the two great landmarks of progress that determined the pace of civilization on earth. The control of fire by prehistoric humans signified one of the greatest landmarks in the history of mankind. The Paleolithic era or Old Stone Age commenced 2.5 million years ago, when humans used naturally available items like stones, tree trunks, branches and leaves to create shelters, hunt, fish and eat. At some point during the Paleolithic era, humans used fire by rubbing stones together. The widespread use of fire for cooking, hunting, protecting their caves and shelters and keeping away wild animals, signified a significant breakthrough in human prehistory.

All ancient civilizations created myths that depicted the fire as something precious that only Gods possessed. Traditionally, civilizations have been using myths in order to spread by word of mouth a cultural timeline and people's history that must be remembered. According to an ancient Greek myth, the Titan Prometheus (whose name literally means "provision" or "foresight") stole the fire from the Gods: He ignited a torch from the sun and offered it as a gift to mankind, who until then lived in the dark. The Prometheus myth is the perfect example of how the history of humankind has evolved, as it reflects on the impact of fire throughout the prehistoric and historic times.

The Six Types of Energy: A Brief Historical Overview

Beyond the myth, scientific knowledge and technological innovations in energy have evolved from ancient civilizations through modern inventions.

Understanding how energy works was a huge leap for the civilized world. The first step was to recognize the different types of energy, their distinct behavioral patterns, characteristics, impact, units of performance measurement, and so on.

Here is a historic timeline for the six types of energy, i.e., 1) electric, 2) gravitational, 3) kinetic, 4) magnetic, 5) radiant, and 6) thermal.

1 Electric Energy

The oil and gas refining industry is the most significant commercial and manufacturing consumer of power, with the chemical industry being ranked second. Ninety-six percent of power feedstocks are consumed by the refining, metal, mineral, chemical and paper industries combined. Sixty percent of power is used for heat, electricity generation and propulsion, obviously to feed power plants, production units and technological operations, and 78% of overall energy utilization (EIA, 2017a).

BRIEF HISTORIC OVERVIEW OF ELECTRIC ENERGY SYSTEMS

a 1749: Benjamin Franklin (1706–1790), U.S. Founding Father and 6th President of Pennsylvania, considered lightning to be a naturally made electrical current. He invented the lightning rod or lightning conductor, which is mounted on top of elevated structures (offshore platforms, ships, buildings) and grounds them via an electrode, to protect them in case of a lightning strike. Franklin's original invention in 1753 was used for households, whereas an enhanced version was launched in 1760 (*U.S. History*, 2016).

In July 2016, Governor of Maryland Larry Hogan announced that the Maryland State House, the state's oldest building, was saved by lightning due to "a 208 year-old original Ben Franklin lighting rod" (*Fox News*, 2016).

b The British experimentalist Michael Faraday (1791–1867) sustained Benjamin Franklin's work on electricity, with a focus on electromagnetism and electrochemistry. He invented the first dynamo electric generator, motor and transformer, based on the principles of electromagnetic induction, electrolysis and diamagnetism (*Encyclopedia Britannica*, 2016b; *Famous Scientists*, 2016b).

c 1832: Hippolyte Pixii (1808–1835), a French mechanic and inventor constructed a pioneering alternative current (AC) electrical generator, and several years later enhanced his invention with a commutator to produce sustainable direct current.

d 1878: Joseph Wilson Swan (1829–1914), a British inventor, obtained the world's first patent for the innovative incandescent electric lamp.

e 1879: Thomas Alva Edison (1847–1931) obtained the American patent for an incandescent electric lamp. Over the next years he became America's greatest inventor and entrepreneur, known as the notorious "Wizard of Menlo Park."

f 1880: Thomas Alva Edison patented the direct current (DC) electricity distribution system, pertaining to one direction electric charge (current flow) (*World History Project*, 2016).

Edison established his pioneering engineering research laboratory, with resources and funds to hire some of the most promising engineers and scientists of his time. The outcome of this large-scale scientific research was 1,093 U.S. patents filed in the name of Edison, patents such as the telegraph, phonograph, electric light, motors and power distribution systems, motion pictures cameras (vitascope, kinetoscope), and much more. Edison has received numerous honors and awards from the U.S. Congress, several states and foreign governments. Many of these awards were obtained almost a century after his death.

g The 1880s were an exhilarating era for the improvement of electric power technologies, characterized by ambitious collaborative projects and entrepreneurial consolidations.

h 1882: Nikola Tesla (1856–1943), a Serbian futuristic engineer and inventor, studied in Austria and experimented on a multi-phase voltage structure. Two years later he immigrated to the U.S., where he established his own laboratory and company. While a genius of his times, Tesla's inventions came to light only when investors committed to funding his projects.

i 1882: Sebastian Ziani de Ferranti (1864–1930) was a British-Italian electrical engineer. In 1877, in his early teenage years, he invented a system for arc street lighting, and by the age of 16 he invented the electrical generator/transformer with the support of Lord Kelvin. His collaboration with Lord Kelvin continued at Siemens Brothers, a London city company, where he created the AC power technology, which is based on the use of two conductors, the phase and neutral (*Energy Quest*, 2016).

Figure 1.11 Nikola Tesla in front of the spiral coil of his high-voltage Tesla coil transformer at his East Houston Street laboratory in New York, 1896.

Source: Originally published in "Tesla's important advances" in *Electrical Review*, May 20, 1896, p. 263. Credited in caption to Tonnelé and Co. It appears in Rudjer Boscovich's book *Theoria Philosophiae Naturalis*, 1896. This work is in the public domain in the U.S., Europe and the rest of the world where the copyright term is the creator's life plus 100 years or less.

j 1882: Thomas Alva Edison invented an electric energy and electric incandescent lighting system.
k 1885: Galileo Ferraris (1847–1897), an Italian professor, invented the "three-phase current." In 1889 he invents the two-phase induction motor.

In the late 1880s the three-phase system was individually invented by Galileo Ferraris, Mikhail Dolivo-Dobrovolsky, Jonas Wenström and Nikola Tesla.

l 1887: Nikola Tesla (1856–1943), a Serbian futuristic engineer and inventor, obtained his first patents for a two-phase AC system with four electric power lines, which consists of a generator, a transmission system and a multi-phase motor.
m 1888: Nikola Tesla sold over 40 of his patents to George Westinghouse for $1 million dollars, or a modern-day $24 million dollars. Westinghouse hired Tesla as a consultant, yet their collaboration evolved into a mutually unsatisfying one (KIT, 2016).
n 1887: Sebastian Ziani de Ferranti (1864–1930) developed the Deptford Power Station, which was the greatest station of its era and based in the London area. The station had an electric capacity of 10,000 volts, i.e., four-fold compared with the preceding and contemporary power stations. He was a pioneer for large-scale electricity generating stations (*Encyclopedia Britannica*, 2016e).
o 1888: Mikhail Dolivo-Dobrovolsky (1861–1919), a Polish-Russian engineer working for AEG, invented the three-phase electrical generator and electrical motor.
In 1891 he showcased his inventions in the International Electro-Technical Exhibition in Frankfurt, where his system transmitted electricity over 176 kilometers, by saving 75% of power. During that year he also created the world's unique three-phase hydroelectric power plant (MPower UK, 2016; Neidhöfer, 2008).
p 1889: Jonas Wenström (1855–1893), a Swedish inventor and partner of the Allmänna Svenska Elektriska Aktiebolaget (General Swedish Electric Company), invented the three-phase electric power system. Although the news of his invention travelled the world over, Wenström's limited financial resources and laboratory infrastructure would leash his potential. Other inventors with sufficient funds attain scientific recognition. Wenström died at the age of 38, three years after his three-phase electric power system invention. His legacy as a fine scientist and inventor spreads the world over.
q 1890: More than 12 electric companies merged into three: Edison General Electric, Westinghouse and Thomson-Houston. The War of the Currents initiated intensive scientific and commercial antagonism as to the energy security and efficiency of the direct current (DC), supported by Edison Electric Light Company and the alternating current (AC) supported by the Westinghouse Electric Company.

2 Gravitational Energy

Gravitational potential energy (GPE) or gravitational energy is related to an object's gravitational force. Hence, we can estimate the amount of energy a falling object possesses when we estimate the object's weight and the altitude where it is located.

Its most typical utilization is for an object near the surface of the Earth where the gravitational acceleration may be presumed to be consistent at approximately 9.8 m/s2. Considering that the zero of gravitational potential energy (or the zero of a coordinate system) can be selected at any point, one can calculate the potential energy at a height h over the specific by estimating the effort needed to elevate the object to the specific height devoid of net change in kinetic energy. In other words, the gravitational potential energy

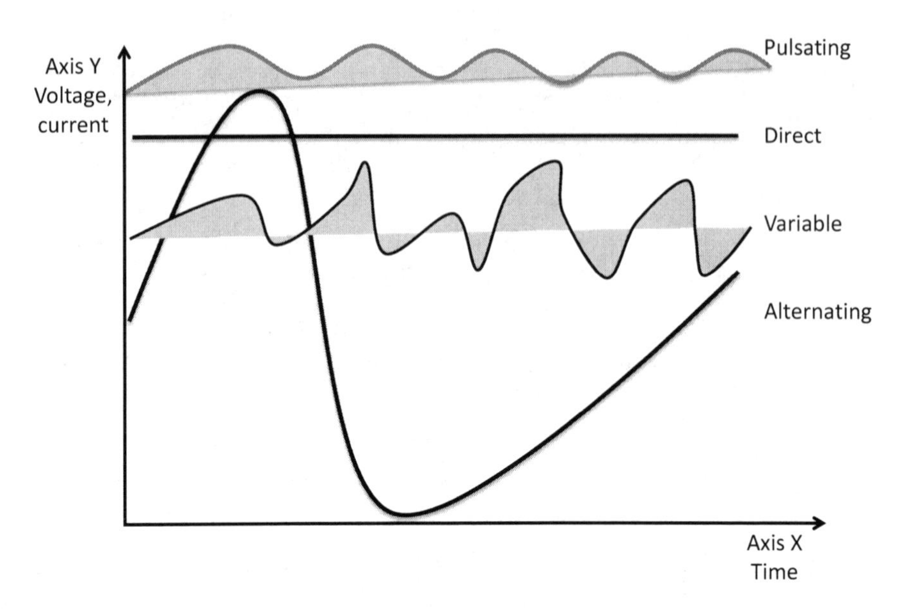

Figure 1.12 Types of current or voltage.
Source: the Author.

(GPE) is equivalent to its weight multiplied by the height to which it is elevated, because the effort needed to elevate it is equivalent to its weight.

The theoretic framework of gravitational energy is widely used in the energy industry as part of engineering and supply chain processes, with applications in geology, drilling, seismology, GIS systems, electromagnetic radiation (scanning and radiowave technologies), drilling, and much more.

Gravitational energy is also part of the industry's HSQE (Health, Safety + Security, Quality and Environment) incident prevention policies, emergency response, and so on.

From an industrial perspective, the applications of gravity and magnetic energy are used in petroleum exploration, e.g., during the upstream/seismic survey process. Modern technologies have facilitated exploration in deeper offshore and other extremely challenging locations and environs such as deep sea, sub-salt structures and many more (Geosoft, 2016).

BRIEF HISTORIC OVERVIEW OF GRAVITATIONAL ENERGY SYSTEMS

a 1589: Galileo Galilei (1564–1642), an Italian inventor and scientific polymath, proves his theory of falling bodies by dropping two balls of dissimilar masses from the Pisa Tower.
b 1687: Sir Isaac Newton (1643–1627), a British natural philosopher, proved the laws of universal gravitation and motion via his work *Philosophiæ Naturalis Principia Mathematica*, known as "Principia." His work will influence scientists throughout the subsequent centuries (Newton 1680/2016, reproduced by NOAA).
c Henry Cavendish (1731–1810), a British aristocrat, experimenter and natural philosopher, contrasted the electrical conductivities of comparable electrolyte solutions.

He formulated the concept of voltage and illustrated the difference between electric quantity and electric potential, as framed by the laws of electrical attraction and repulsion. He discovered hydrogen, subsequently named by Lavoisier, but also performed "the Cavendish experiment," measuring the Earth's density and weight (1797–1798).

Most of Cavendish's scientific contributions were presented to the British scientific circles of his times, yet his inherent introversion prevented him from publishing a great portion of his scientific findings and experiments. And, while he dedicated his life and great fortune to scientific studies and experiments, only a portion of his works were published, although they were spread by word of mouth (*Chemistry Explained*, 2016).

d 1905: Albert Einstein (1879–1955), the famous Nobel Prize winner, in 1921 develops his theory of special relativity and relativity and the mass–energy equivalence formula $E = mc^2$, which stipulates that energy (E) equals the mass (m) multiplied by the speed of light squared. His work addresses a theory of gravitation (NobelPrize, 2016).

3 Kinetic Energy

Kinetic energy is the energy that an object retains in order to reflect motion, and is always related to an object's weight, shape, dimensions and speed. It represents the force or work required to accelerate an object of a certain mass from a motionless state to a certain level of velocity.

Flow energy (FE), flow work or pressure energy, is defined as the quantity or volume of energy required in order to generate a liquid's motion through a specific area against pressure.

Applications of the kinetic and flow energy are found in the energy industry. Especially during the upstream (exploration and production) stages, such theories are used to estimate the energy potential and induce high-pressure high-flow rates in order to attain the daily production quotas.

BRIEF HISTORIC OVERVIEW OF KINETIC ENERGY SYSTEMS

a Aristotle (384–322 BC) was an ancient Greek philosopher and scientist who was Plato's student and the teacher of Alexander the Great. He was the first known scientist to refer to kinetic energy (kinesis in ancient Greek means motion, movement). He used interchangeably the words energy (energeia) and "entelecheia" to describe action, actuality and potentiality. (Aristotle, 1933).

b René Descartes (1596–1650), a French polymath and a founding father of the "Scientific Revolution." He transformed theories into a causal-factor cosmos consisting of surging other material and compounds of moving particular matter in a universe where a matter's motion (kinetic energy) is the only real type of energy (*New Science Theory*, 2016).

c Gottfried Wilhelm Von Leibniz (or Gottfried Leibniz) (1646–1716) was a German polymath, considered as the founding father of kinetic energy. Per Leibniz, "vis viva" the energy or a central dynamic force which is expressed as mv^2.

Leibniz differentiated from Newton's principles of force by distinguishing kinetic energy from momentum. He also differentiated from the Cartesian declaration of mechanics as the principal laws of motion, and in 1676 established his theories on dynamics, based on which kinetic energy is replaced for the preservation of movement.

In 1686 he released his scientific applications on Descartes mechanics, hence commenced the notorious argument regarding the drivers of a moving object, which he named "vis viva" (*Encyclopedia Britannica*, 2016d).

d Christiaan Huygens (1629–1695) was a Dutch polymath who developed the conservation laws in elastic impact, the wave theory of light. He studied the various expressions of kinetic energy and believed that light consisted of waves pulsating in a tempestuous manner towards the direction of the travelling light. His wave theory of light proved that light waves move in three dimensions, whereas light wave heights resemble the coatings of an onion (*Science World*, 2016; *Math Pages*, 2016).

e Johann Bernoulli (1667–1748) was a Swiss mathematician who made significant contributions to mechanical engineering through his research on kinetic energy, and his work *Hydrodynamica* (1738) (History MCS, 2016).

As the Bernoulli Equation demonstrates, the summation of kinetic energy and gravitational potential energy remains constant, where m represents the mass, v represents the velocity, and h represents the height exceeding the datum (EFM Leeds, 2016).

f Joseph Louis Lagrange (1736–1813) was a mathematician of Italian and French origin. He was a founding member of the Royal Academy of Sciences of Turin, where one of his main responsibilities was to publish a scientific periodical named *The Mélanges de Turin*. As a chief contributor in 1759, 1762 and 1766, he published his work on estimating variations and estimating probabilities. In his research on the fundamentals of dynamics he grounds his theoretical proof on the theory of least action and on kinetic energy (Ball, 1908).

In the theory of generalized coordinates he proved that if we define its outline by an adequate number of variables whose sum equals the degrees of freedom influenced by a scheme, the system's kinetic and potential energies may be expressed in relation to these variables, and the differential motion equations are inferred through simple differentiation.

In his theory he explains how the kinetic energy communicated by the specified trends to a material system with certain constraints is a maximum.

4 Magnetic Energy

In modern days we use magnets in practically all industry and household technologies and gadgets, including power generators, computer systems, manufacturing sensors, scanners, monitors, separators, satellite systems, telecommunication, audiovisual systems, closed circuit cameras, and much more. Apparently, the oil and gas industry relies on such technologies (Stanley, 1998).

For several centuries, scientists recognized that subjecting a liquid to a magnetic field results in the alternation in several of the solids contained in the fluid. In the 1700s Magnetic Fluid Conditioning (MFC) was employed by the Russians to handle water scale buildup issues.

John White implemented the magnetic principles in the petroleum industry during his oil drilling operations as a unique method to treat the scale residues issues (*EagleFord Texas*, 2015).

Several fluid instability variables result in the development of insipient solids, residues, and debris that affect technological performance and dependability in several uses of upstream, midstream and downstream, including but not limited to:

- diesel fuel systems;
- crude oil production;

- storage tanks and pipelines;
- lube oil and hudraulic fluid treatment and filtration systems;
- cooling towers and other water systems (Felix Marine, 2016).

Modern petroleum operations employ Magnetic Fluid Conditioning (MFC) to minimize and prevents paraffin and scale accumulation in petroleum production lines and process machinery (Flo-Rite Fluids, 2016).

BRIEF HISTORIC OVERVIEW OF MAGNETIC AND ELECTROMAGNETIC SYSTEMS

In 2750 BC an Ancient Egyptian script referring to electric catfish as "thunders of the Nile" (*Malapterurus electrims*) is the oldest known script referring to a natural electric phenomenon.

Thales of Miletus (624–546 BC) was a Greek philosopher and one of the seven sages of antiquity. He held "hylozoist" views, i.e., proclaiming that matter is vivid (also see Aristotle's *De Anima* (On the Soul). In his works he observed that magnetic particles can move iron and other magnetic particles, but also that when amber is rubbed with wool it becomes magnetic, i.e., attracts small particles (UPS Battery Center, 2016).

Archimedes not only described the principle of buoyancy but is considered as the father of statics, hydrostatics, and the use of leverage mechanisms to lift/move objects. He used lodestones to twitch nails from opponent ships in war to make them collapse and sink.

The Chinese Book of the Devil Valley Master refers to "south-pointer," a lodestone device that works based on the lodestone's orientation in line with earth's magnetic field (Russell, 2003).

Theofrastus (372–287) was an ancient Greek philosopher and a student of Plato and Aristotle. His essay "On Stones" (Περὶ λίθων) was used until the Renaissance. There, he groups minerals and gems in accordance to their properties and refers to the power of attraction found in amber and magnetite.

In 1088 AD *Mengxibitan* (Dream Pool Essays) is an encyclopedia written by Shen Kuo, the Chinese encyclopaedist. There he describes the use of suspended magnetic compasses a hundred years earlier.

Alexander Neckam or Nequam (1157–1217) was a British polymathic scholar who, during his scholarly travels, perhaps in Paris, learned about a vessel that had a compass-like mechanism with a needle fixed on a magnetic axis that would rotate until it pointed to the north. This mechanism helped medieval seafarers' orientation, especially when cloudy skies and bad weather would prevent the view of constellations and other landmarks used by travelers.

5 Thermal Energy (Heat Energy)

Thermal energy is produced through the motion of atoms, or molecules, or small particles, as the friction produces heat. Although the atomic motion makes thermal energy resemble kinetic energy, the former pertains to a holistic motion of an object, whereas the latter pertains to the atomic movement only. Furthermore, kinetic energy is measured by temperature, hence a high temperature object possesses more energy than a low temperature object. Thermal energy is generated when burning oil, gas, coal, wood, and other forms of energy. The British thermal unit (Btu) is the unit of measure for thermal energy.

Table 1.8 demonstrates the conversion factors of different energy sources translated into Btus.

Table 1.8 British thermal units conversion factors.

Source of energy	*Physical units and Btu*	*Giga Jule (GJ)* (*1 billion or* 10^9 *joules*)
Crude oil	1 ton = 39.68 Btu	41.87
	1 barrel (42 gallons) = 5,729,000 Btu	6.12
	1 gallon = 136,404.76 Btu	0.146
Electricity	1 kilowatthour = 3,412 Btu	0.0034
Natural gas	1 cubic foot = 1,032 Btu	0.00108965
	1 therm = 100,000 Btu	0.105587
Motor gasoline	1 gallon = 120,405 Btu	0.12713
Diesel fuel (distillate fuel with less than 15 parts per million sulfur content)	1 gallon = 137,381 Btu	0.145056
Heating oil (distillate fuel with 15 to 500 parts per million sulfur content)	1 gallon = 138,500 Btu	0.146237
Residual fuel oil	1 barrel (42 gallons) = 6,287,000 Btu	6,633
	1 gallon = 149,690.48 Btu	0.15805
Propane	1 gallon = 91,333 Btu	0.09643
Wood	1 cord (1.25 tons) = 20,000,000 Btu[3]	21.1174
Coal	1 short ton (2,000 pounds) = 19,882,000 Btu	20.9928

Sources: compiled and converted by Author, based on:

- EIA, *British Thermal Units*. Last updated: December 15, 2014. Available at: www.eia.gov/energyexplained/index.cfm/index.cfm?page=about_btu, last accessed: December 25, 2016.
- EIA, *Units and Calculators*. Last updated: 2016. Available at: www.eia.gov/energyexplained/index.cfm/index.cfm?page=about_energy_units, last accessed: December 25, 2016.
- American Physical Society, *Energy Units*. Last updated: 2016. Available at: www.aps.org/policy/reports/popa-reports/energy/units.cfm, last accessed: December 25, 2016.

a James Watt (1736–1819), a Scottish chemist, instrument maker and inventor. In 1764 he commenced working on the steam condensing engine, and over the years attained greater energy efficiency by converting larger ratios of thermal energy obtained from steam into mechanical energy. He conducted experiments on the energy exerted by a horse, subject to feet x pounds/minute.

b James Prescott Joule (1818–1889) was a physicist whose research verified the association between heat (thermal energy), electrical and mechanical energy. His findings established the basic principles of the law of energy conservation, which later formed The First Law of Thermodynamics.

The earliest estimation formula of the gas molecule velocity is attributed to him. (*Famous Scientists*, 2016a).

In 1847 he presented a lecture titled “On matter, living force, and heat,” describing the behavior of heat and the characteristics of thermal energy and heat. He identified kinetic energy as “sensible heat,” whereas potential energy is “latent heat,” interactive energy in a specified arrangement of particles.

Upon examining sensible heat, or such as is signified by the thermometer, he observed that heat will likely comprise of the living force (i.e., thermal energy) of the particles of the objects in which it is evoked, whereas, in others, especially in the case of latent heat, the occurrences are generated when detaching one particle from another, with the intention to induce their attraction by way of increased, more extensive space (Joule, 1884).

In 1852 Joule and Lord Kelvin (then called William Thomson) observed that any time a gas is permitted to expand without producing exterior energy, its temperature declines. During the 19th century the "Joule-Thomson effect" was widely used, facilitating the construction of a sizeable refrigeration industry (*Encyclopedia Britannica*, 2002).

6 Radiant Energy (Electromagnetic Waves, Solar, Lighting and Heating Energy, Radiation)

While ancient and medieval civilizations used natural benefits of radiant energy, it was after the 17th century that scientific advancements and technological inventions elevated the use of this energy type.

a William Gilbert (1544–1603), a British physicist and philosopher, and Court Physician to Her Majesty Elizabeth I, is considered the father of magnetism, electricity and electrical engineering. He used the Ancient Greek word "electron" for amber, to generate the word electricity. He carried out numerous experiments demonstrating how the power of electric charge is generated through the abrasion of different materials.

In 1600 he published the book "On Magnets," under the title *De Magnete, Magneticisque Corporibus, et de Magno Magnete Tellure* (On the Magnet and Magnetic Bodies, and on the Great Magnet the Earth).

b Benjamin Franklin (1706–1790), U.S. Founding Father and 6th President of Pennsylvania, studied Gilbert's works and in 1747 proclaimed his theory that electric charge consists of two types of electric forces, i.e., the attractive and repulsive forces.

c In the 1920s the scientific works of André-Marie Ampère, Hans Christian Ørsted, Joseph Henry and Michael Faraday suggested the interconnectivity between magnetic and electric energy. Their findings shaped the theory of electromagnetic energy, which was later extended by James Maxwell.

d James Clerk Maxwell (1831–1879) was a Scottish mathematician and physicist who in 1864 published his work *Dynamical Theory of the Electromagnetic Field.*

His research associated three types of energy: Electricity, light and magnetism, thus creating his revolutionary theory on electromagnetic radiation.

In 1867 his mathematical equations proved the interconnectivity between light and electricity, where he explained how light waves together with radio waves form electromagnetic waves that move in space, diffusing through an energized, accelerated particle. Maxwell's work strongly suggested that light is an electric manifestation and hence inspired the works of Hertz, leading to the discovery of radio waves and Einstein's theory of relativity.

e Sir William Crookes (1832–1919) was a British physicist, chemist and master experimenter, whose work on radiant energy and cathode-rays was pivotal in the evolution of atomic physics. He invented the Crookes radiometer, and discovered the vacuum, electrical phenomena and rare earth elements. He also discovered thallium.

Crookes was a mentor to Nikola Tesla at a scientific and spiritual level. Crookes' work on radiant matter inspired Tesla, who, upon visiting Crookes' home and laboratory, constructed a Tesla coil, i.e., a critical component used in vacuum technology to discover leakages in equipment.

Crooke's research has also inspired modern scientists, who seek to apply Crookes' works in solar energy technologies (*Encyclopedia Britannica*, 2016a; Crookes, 1879; Brock, 2008).

f Nikola Tesla (1856–1943) admired Crooke's radiometer and passionately dedicated the rest of his life to exploring the possibilities of radiant energy and frequencies, considering them as products of the universe.

At the age of 76 he patented the "Apparatus for the Utilization of Radiant Energy". His scientific foundation was based upon the principle that the sun is the source of radiant energy, which is available in abundance throughout the universe. Upon investigating the consequences of electrostatic charges, his experimental research on radiant energy confirmed how it is effortlessly moveable from one location to another. He also proved that light and sound both move via sine waves.

By 1932 he announced the findings of his investigations on cosmic rays, and highlighted that their abundance and sustainable supply makes them an extremely attractive alternative to conventional energy.

Tesla's ingenuity was better understood and appreciated in the 1990s, i.e., half a century after his death.

The foundations of his research findings will be used to harness solar energy in the 21st century, and his work will inspire scientists through centuries to come.

Tesla invented, envisioned or inspired the development of many hundreds of innovations that have a significant part in modern civilization. All of these innovations are widely used in the energy industry: Wireless transmission, computers, smartphones, laser beams, x-rays, remote controls, neon lights, robotics.

In "The War of the Currents" era, the U.S. government and scientific community had to decide if the nation's electrical system should be designed to use alternating current (AC) or direct current (DC). Tesla's work was focused on alternating current (AC), whereas Thomas Edison promoted direct current (DC). Edison brought out a campaign in opposition to AC, declaring it was harmful and could potentially cause mass fatalities. To prove AC's reliability, Tesla responded by publicly exposing himself to shocks of 250,000 volts. Eventually, AC prevailed triumphantly: In the 21st century alternating current serves as the foundation of our power system (Tesla, 1901; DOE, 2013).

g Thomas Henry Moray (1892–1974). Dr. Moray was a pioneer of Radiant Energy (RE) research. He dedicated 30 years of his life to developing the "T. Henry Moray's 'R.E.' (Radiant Energy) device." The Moray valve was invented in 1925, over two decades before transistor valves were patented.

From an early age, Thomas Henry Moray was attracted by Tesla's scientific work on electrical energy. Ironically, just like Tesla, Moray's scholarly achievements and theories were too advanced for his times. As a consequence, his entire life was dogged by financial hurdles, doubts from the scientific community, and opposition from competitors.

He developed the current concept of "zero-point" energy of a vacuum, which he then called "cosmic ray." Zero-point energy (ZPE) or ground-state energy is defined as the least possible energy that can be employed in a quantum technology. Zero-point energy may refer to atomic ground state, sub-atomic ground state, or even the entire quantum vacuum.

His radiant energy device, also called a solid-state detector, was a 50,000-watt case, power-driven by radiant energy. Subsequent experiments use the radiant energy principles in space exploration.

The works of Nikola Tesla and Thomas Henry Moray were based on tapping infinite volumes of cosmic energy. It is certain that their works will be revisited by future scientific communities, and their efforts will be both vindicated and studied as prototypes of unique scientific ingenuity (Moray, 1978).

Risk Management: An Econometric Approach

While the first section of this chapter defined energy risk and stipulated the ideal conditions from a sociopolitical, economic, commercial and global transport perspective, this section will encompass the elements of risk and the path to preparedness and recovery. The stages of risk management are demonstrated in Figure 1.13 and are duly analyzed below.

Risk Management (RM)

Risk management is defined as the continuous process that aims to monitor and control company procedures.

It consists of techniques for risk management systems, identification, evaluation, supervision and command. Among these systems, many are upgraded all through the lifecycle of the company or a particular project since new risks can be diagnosed. The purpose of risk management is to eliminate the likelihood and impact of risks, however this process can present the great opportunity of identifying a company's strengths, or facilitating visibility within the supply chain. Visibility will enable energy corporations to make optimum strategic decisions, as now they have a clear picture of the company's strengths and weaknesses, and a better idea of how budget and resources should be allocated.

Figure 1.13 The stages of risk management.

Source: the Author.

The Risk Formula

The Risk Formula is used to estimate each individual incident and the pertinent consequences. A cumulative formula will be required to assess the overall impact in the risk assessment process that is company-specific, region-specific, supply-chain specific, and for a particular energy segment.

The Risk Formula:

$$\mathbf{R} = \sum_{i,c} L(i)\ x\ Lc|i\ x\ I(i,c)$$

Where:

R = Risk

Σ = Sum

i, c = Cumulative incidents and consequences

L(*i*) = Likelihood of an incident

L(*c**i*) = Likelihood of consequence for a specific event

I(i, c) = Impact of incident and consequence

The Stages of Risk Management

Risk entails threat or exposure to hazard, injury or loss, and can be measured by the degree of probability of such loss. When an incident occurs, both the authorities and the industry seek to verify that the company has exercised due diligence, i.e., acted in a proactive manner, and made every possible effort to prevent or alleviate the risk. Due diligence goes beyond proactiveness, as companies need to prove that they have been alert and prepared throughout every stage of their operations. Namely, energy professionals need to cope with the aspects of risk through preventive action, contingency planning, emergency response, and corrective action.

Stage 1: Risk Identification and Assessment

Risk identification and risk assessment typically commence prior to an energy project's preliminary stages. Vulnerabilities will need to be identified and assessed. Throughout the different stages of a project management plan, the type of risks can greatly vary but also the cumulative amount of hazards expands.

When a threat is recognized, risk assessment is the preliminary stage to evaluate the likelihood of manifestation, its overall company- or project-related impact, potential delays, loss of resources, and so on. The stages of risk assessment can also be seen in Figures 1.14 and 1.15.

Throughout each stage of a company's risk management (RM) process, in particular as part of risk planning, it is important to identify the critical success factor (CSF) per each crucial RM stage, as well as to critically evaluate how a CSF can influence risk. Table 1.9 below provides a sample analysis as part of corporate risk planning.

During the risk assessment (RA) process, the RA matrix, as seen in Table 1.10, will be used in order to evaluate the risk impact and likelihood. Furthermore, it is necessary to

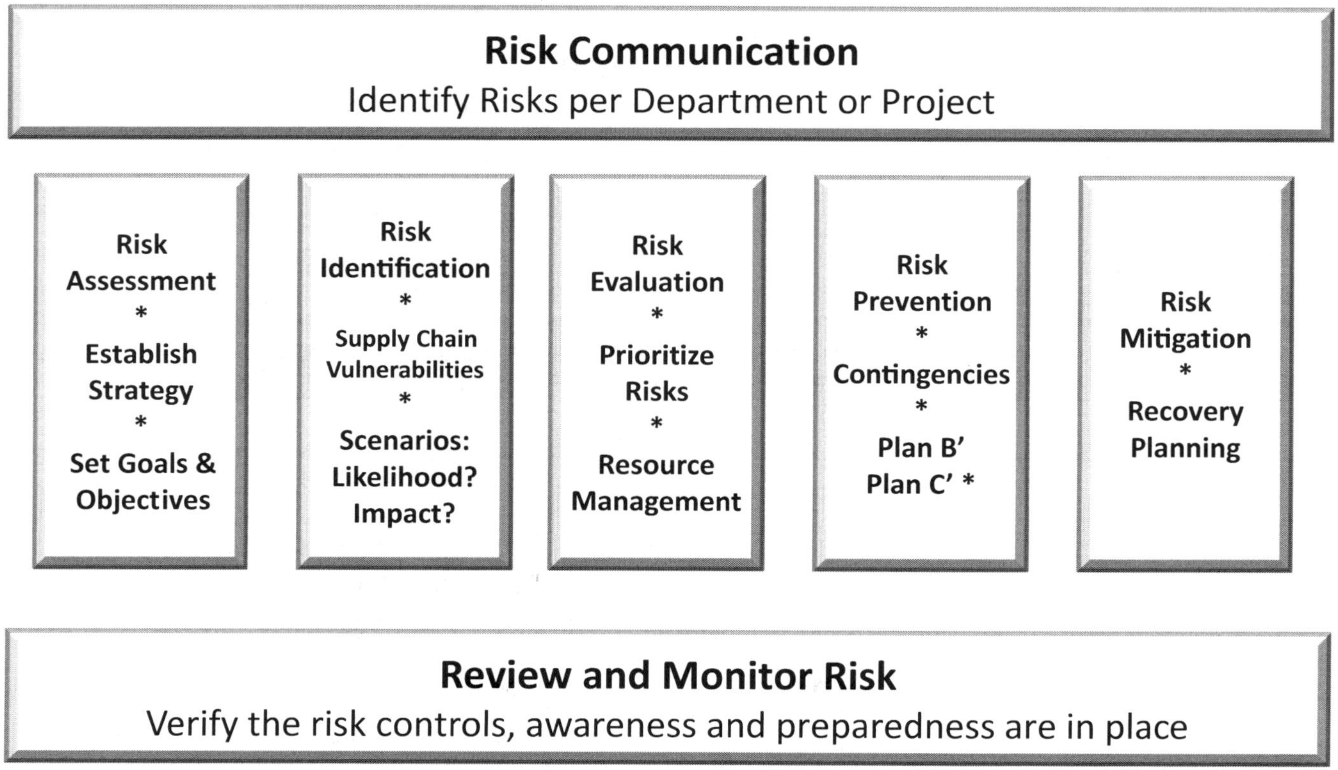

Risk Communication

Identify Risks per Department or Project

Risk Assessment
*
Establish Strategy
*
Set Goals & Objectives

Risk Identification
*
Supply Chain Vulnerabilities
*
Scenarios: Likelihood? Impact?

Risk Evaluation
*
Prioritize Risks
*
Resource Management

Risk Prevention
*
Contingencies
*
Plan B'
Plan C' *

Risk Mitigation
*
Recovery Planning

Review and Monitor Risk

Verify the risk controls, awareness and preparedness are in place

Figure 1.14 Risk communication through all risk management stages.

Table 1.9 Risk planning.

Examples of critical success factors applied to risk planning	
Critical success factor (CSF)	*How CSF can influence risk*
Strategic plan efficiency	Strategic plan outcome, impact, input/resources available, corporate support
Stakeholders' involvement	Participation or intervention of stakeholders in risk management process and desired outcomes
Project management	Identify and communicate project-specific requirements without acting to the detriment of performance
Business risk objectives	Accurate risk identification and prioritization of objectives
Appropriate scope	Specific, unambiguous project objectives for each phase
Resource management and utilization of existing infrastructure	Leveraging of resource management tools and utilization of existing infrastructure
Corporate, industry and/or regulatory requirements	Define unambiguous, realistic goals at a corporate, industry and/or regulatory level
Consistent methodology	Consistency in corporate practices until project management completion
Realistic expectations and evaluations	Realistic expectations and evaluations based upon stakeholders' engagement
Interdependencies	Interdependent relationships based on reliable internal and external key players. Knowledge sharing among participants, information exchange throughout the supply chain
Capital investment	Allocation of sufficient funding to implement scope
Implementation	Maintain development to assigned phases
Organizational culture	Organizational culture and paradigms enable the company to successfully carry out the project, and implement the risk management plan
Supportability	Ability of the company, stakeholders and supply chain partners to support the new process or project
Flexibility	Ability to react to changes in project requirements to ensure successful completion

Source: the Author, modified version from: *Implementation of Electronic Right-of-Way Management Systems Versus Paper Systems. 5 Risks and Barriers to Implementation*, 2016. Federal Highway Administration (FHWA), U.S. Department of Transportation. Available at: www.fhwa.dot.gov/real_estate/publications/e-row_management/page05.cfm, last accessed December 25, 2016.

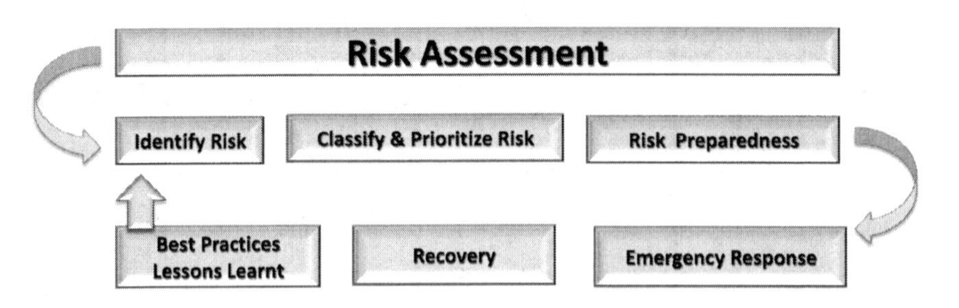

Figure 1.15 Risk assessment matrix.

Table 1.10 Risk assessment matrix. Risk = Probability x Impact.

Probability (1–5) 1=min, 5=max	*Impact (1–5) 1=min, 5=max*				
	(5) Catastrophic 81–100%	*(4) Substantial 61–80%*	*(3) Moderate 41–60%*	*(2) Minor 21–40%*	*(1) Negligible 1–20%*
(5) Frequent 81–100%	Extreme Risk	Extreme Risk	High Risk	Moderate Risk	Low Risk
(4) Likely 61–80%	Extreme Risk	High Risk	Moderate Risk	Low Risk	Low Risk
(3) Moderate 41–60%	High Risk	High Risk	Moderate Risk	Low Risk	Low Risk
(2) Unlikely 21–40%	High Risk	Moderate Risk	Low Risk	Low Risk	Improbable
(1) Improbable 1–20%	High Risk	Moderate Risk	Low Risk	Improbable	Improbable

Table 1.11 Stakeholders' contribution to risk management.

Stakeholders' Contribution				
Stakeholders #	*Stakeholders' Requirements Significance X Urgency*	*Stakeholders' Commitment*	*Assets / risks*	*Action Plan*
#1. Government (Upstream) (Midstream) (Downstream)				
#2. Industry (Upstream) (Midstream) (Downstream)				

obtain the stakeholders' contribution in upstream, midstream and downstream operations. Contractual stipulations coupled with verbal validations can help the risk management team have a clear idea on stakeholders' requirements and commitment according to their role on the energy project. A final list with action plan items will help both stakeholders and risk managers strategize for the next steps from an operational and risk management perspective. This is reflected in Table 1.11.

Stage 2: Risk Evaluation and Prioritization

The process pursuant to the risk evaluation is the risk prevention or elimination, where a strategy requires allocation of resources. Finally, the risk is prioritized according to likelihood and impact.

The Risk Prioritization Formula:

$$\textbf{Risk} = \frac{\textbf{Hazard} \times \textbf{Vulnerability}}{\textbf{Capacity to Cope}}$$

Ideally, a company's capacity to cope in terms of resources, robust preventive action plan, emergency response capabilities and recovery capabilities, exceed the potential hazards and vulnerabilities (PHE, 2016).

Stage 3a: Evaluate Risk Management Tools

Stage 3b: Implement Risk Management Protocols

Stage 3c: Monitor and Control Findings

Risk management analysis is strongly associated with a preventive action culture. Companies should evaluate the available risk management tools, implement the risk management protocols, but also monitor and control findings. Namely:

- Companies adopt a proactive culture and thoroughly evaluate the potential risks before initiating a new project. Professionals strategically plan ahead and exercise due diligence to make sure that potential risks can be prevented, or managed.
- Professionals implement project management and procedure evaluation to verify how to construct a safe and robust project throughout all of its stages, without exposing its people, the environment, stakeholders or any aspect of the company's tangible and intangible assets in jeopardy. Professionals identify potential risks and systemic vulnerabilities, and recommend a preventive course of action.
- Familiarization of the site, technologies and emergency procedures through training and drills are a part of the preventive action, contingency planning and emergency response.

  Risk reduction (RR)

  Risk reduction pertains to the corporate process of proactively eliminating any potential risks, be it internal (company-related), or external. The goal of risk reduction is loss control (LC), i.e., by eliminating the frequency of incidents and/or severity of damages (IRMI, 2016).

  Pursuant to the risk reduction process, the risk management team can make a comparison between risk before risk reduction and risk after risk reduction.

The Risk Reduction (RR) Formula:

$$\textbf{Reduced Risk} = \frac{\textbf{Risk after RR implementation}}{\textbf{Risk before RR implementation}}$$

The outcome of this formula will signify the effectiveness of the current risk reduction process. Most importantly, a periodic comparison of these outcomes will demonstrate to the risk management team the impact of their implementation efforts.

Stage 4: Contingency Planning

The purpose of producing a powerful contingency plan is to proactively plan and be ready for unforeseen incidents such as geopolitical turmoil, machinery failure, bureaucratic delays, and so on. Contingency planning is designed to prepare a company and its employees to pursue a Plan B (or C, or D), should the initial plan fails for reasons beyond the company's control. Creating a contingency plan entails proactively making project management and resource management variations from the original course of action, to ensure that the company's original goals will still be delivered, even though deviating from the original logistics arrangement.

As energy corporations are conscious of key performance indicators (KPIs), it is worth noting that time, resources and effort invested in contingency planning equates to time, resources and effort preserved when a catastrophic event happens. Contingency planning is a hybrid risk management tool that prepares the company for the next steps of emergency response and business continuity.

Stage 5: Emergency Response

Emergency response is defined as the planning, leading and organizing of readily available tools and assets towards risk mitigation. Emergency responders should be prepared to handle an urgent situation up to its recovery. The purpose of this synchronized response is to safeguard human health, the environment and the company, while reducing the consequences of the incident (NIH, 2016).

When an incident takes place, protecting human and environmental safety are the first priorities. The second priority entails securing the incident scene, while protecting it from the imminent danger. The elements of time and resources available are critical. Certain risks can be foreseen days or hours before their occurrence, in which case emergency responders have ample time to prevent, mitigate and recover. Resource management involves the availability of resources, but also the flexibility to utilize them as needed, depending on the incident requirements. The plan should likewise incorporate a procedure for damage assessment, salvage operations, asset protection, security surveillance, insurance and cleanup immediately after an incident.

An example of the applications of an emergency response plan is the emergency response of several government agencies during the Hurricane Katrina disaster. This is a typical example of a large-scale incident that impacts the lives of millions of people. A Geographic Information System (GIS) was employed to map the geospatial location of oil and gas wells, chemical plants, refineries, and storage units. Similarly, GIS marked the location of emergency response sites, and storage areas of resources, Superfund clean-up locations, and other locations of petrochemical or toxic contamination in areas where flooding occurred.

Stage 6: Recovery and Corrective Action

Disaster recovery (DR) consists of a range of protocols and strategic plans facilitating the speedy recovery and business continuation. While full recovery might take months, even years, the first step towards recovery is to isolate the threat, and preserve a risk-free system upon which authorities and companies can build their corrective action strategy. Contingency planning is hereby fully implemented, with substitute resources and supply chain routes being used. In a capital-intensive industry such as the energy industry, insurance reimbursement and resource management plays a critical role.

Major disasters in the past have shown us that while the recovery of physical assets is a relatively easy task, when the disaster involves loss of human life, environmental pollution, or exercising of unsafe practices, the corporate recovery can become an arduous task. From a legal perspective, companies can prove they have exercised due diligence. However, from an ethical perspective, public opinion, the media, or even the stock market sentiment may not be willing to forgive a company who failed to protect its people, the environment, and also the industry's reputation (NIH, 2016).

Statistically, they say that an incident has the power to crash a company's stock price in the stock market, literally overnight, with average losses estimated to account for 33.33% of the stock price.

Figure 1.16 demonstrates the tiers of strategic and tactical risk, commencing with the transparency of risk-based decisions, risk awareness, company-wide information sharing, and finally a process for continuous improvement based on the feedback received.

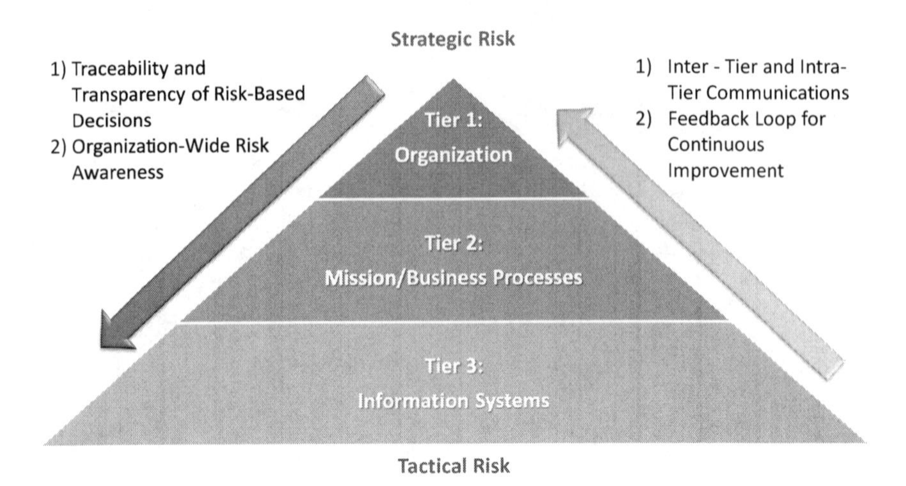

Figure 1.16 Strategic and tactical risk.

Source: NIST Special Publication (SP) 800-161. Title: *Supply Chain Risk Management Practices for Federal Information Systems and Organizations.* Publication date: April 2015. Available at: http://csrc.nist.gov/publications/drafts/800-161/sp800_161_2nd_draft.pdf.

Risk Classifications

Risk factors can be classified as external and internal.

a **External factors** represent influences beyond the company or supply chain operations, including politics, social turmoil, wars, extreme weather, reserves characteristics, geological conditions, etc.

b **Internal factors** represent inherent characteristics, systemic strengths and vulnerabilities, influences beyond the company or supply chain operations, including strategic planning, investment and budget allocation, negotiations and leveraging, operations, technologies, etc.

At the same time, risks are interrelated because:

a when external pressures impact a project, risk mitigation and problem resolution are attained through internal resources;

b when internal weaknesses or vulnerabilities jeopardize a project's success, external factors can provide favorable circumstances under which the internal challenges can be minimized.

Risk management has the capacity to identify risk at any given time, and utilize any and all of the above factors, for problem resolution and risk mitigation purposes.

Key Risks

Here are some of the key risks encountered in the energy industry.

1 GEOPOLITICS AND NATIONAL RISKS: POLITICAL SHIFT OR POLARIZATION

The energy industry is greatly influenced by the political climate, not only at the time a license is granted but also throughout the lifespan of an energy project.

- The risk of polarization: Entails individual voters or the public opinion influenced by a specific political party or belief. Hence, different regimes may favor a different energy agenda.
- Political shift: In domestic projects, the risks pertain to change in the political party, and introduction of diverse energy agendas. For example, the permission to drill in a particular region could be granted by one government, and revoked or scrutinized by the next (or vice versa).

Both community polarization and political changes impose risks for energy companies, as it becomes difficult to meet polarized demands or conflicting agendas over a short period of time. For example: A particular political agenda could interpret energy security as environmental integrity, hence their national security strategy focuses on climate change, global warming, green energy subsidies and investment needed in green technologies. Another political agenda could concentrate on energy security from a perspective of social wellbeing and a nation's need to meet the domestic consumption. Another viewpoint could focus on global competition, economic growth, new jobs, or energy independence.

Again, every time a regime reshuffles the energy security priorities, the energy industry must redefine its scope, and invest its time and resources into this change.

2 GLOBAL POLITICAL CLIMATE

Global projects with multinational energy supply chains entail higher political risks as they need to align their strategy with the political agendas of all countries involved. Just imagine a single energy supply chain passing through nations of diverse regimes: From totalitarian to liberal, from war zones to politically unstable nations with civil unrest. If abiding by a single domestic political agenda is complicated, the adopting of a multiscale energy agenda entails aggregate risks.

Geopolitical risk should also be compared with new regulations and expropriation risk, e.g., government revoking permits, increasing taxation or fees, or ceasing projects.

3 LEGAL RISKS: CONFLICTS AND BREACH OF CONTRACT

Legal conflicts often arise from the interpretation of contractual stipulations, and different entities may have a conflict of perception due to their diverse professional background and terminology, difference in native language or difference in culture.

On certain occasions there is intentional ambiguity in the terminology, when a party's intentions are to divert from the initial agreement and to reach a settlement that is to their own benefit.

4 CONCESSION AGREEMENTS

Long-term concession agreements and licenses are typically preferred for three reasons:

1. **Maximizing profitability and ROI**: There is a high cost of initial investment, planning and development of infrastructure and superstructures. A long-term agreement will enable the return on investment and possible mortgage repayment and amortization.

2 **Reaching a market peak**: The longer the contract, the greater the financial security and potential profitability of the parties involved. Market cycles and economic conditions tend to fluctuate, hence a long-term contract will ensure there will be ample time for market expansion and peak stages.
3 **Loyalty brings income security**: Concession agreements indirectly ensure loyalty among the parties involved. When a particular energy park or mine or other site is secured for a particular purpose, the financial investors are assured that no disruptions or other activities will disrupt the energy production. At the same time, land owners are granted a regular income and will not have to worry about future inability to utilize their land due to modifications such as deforestation, geological alterations, or even diagenesis of the sedimentary rocks or soil.

5 REGULATIONS: HARMONIZATION AND INTERPRETATION

The regulatory framework for the upstream, midstream and downstream of oil and gas has been established at a global, national and regional level. Risk is also present when companies operate at a global level, as many of these regulations may not be consistent, or their interpretation may be perceived differently by different entities. Finally, risk factors relate to the company's regulatory compliance strategy, pertinent to the company resources, investment and methods used to comply with these regulations.

6 HUMAN FACTORS OR PERSONNEL PERFORMANCE

Human factors or personnel performance may cause security or safety-related damage, i.e., intentional or unintentional motives respective:

- Security risks include intentional harm such as terrorism, sabotage, equipment destruction, physical or cyber theft. Such actions are motivated by ideological fanaticism (e.g., political, religious, social, racial rivalry, etc.) or competition, revenge, personal profitability, etc.
- Safety or environmental risks entail unintentional damage and are caused by human error or fatigue, leading to inaccurate situational awareness, or misjudgement, or damage of physical or intellectual property, data, etc.
- Quality risks are a distinct category that results from the taking of shortcuts, leading to poor quality and underperformance. Such risks are present when the performers suffer from time pressure (e.g., deadlines), or resource scarcity, that impacts their operational and technical performance.

7 MISJUDGMENT OF RESERVES CAPACITY

In fossil fuels and energy mining operations, such risks entail safety factors and unintentional threats such as:

- Misjudgment related to field mapping and reservoir description, or volumetric methods used to evaluate the reserves.
- Project management is both useful and successful when the energy leaders are aware of a project's inherent risks at an early stage of upstream activities.

8 MISJUDGMENT OF WEATHER OR GEOLOGICAL RISKS

- Extreme weather, geological morphology, etc.
- Safety risks may lead to accidents or near misses. Delays, disruptions, additional resources are needed.

Professional skills, training, familiarization and efficiency may impact performance and the human factor.

9 TECHNICAL RISKS

The equipment's performance varies according to the manufacturer. Post-purchase it is impacted by age, operational efficiency and maintenance. Technologies must be project-specific or site-specific, while bearing in mind geological and weather-related risks. Obsolete technologies, or the use of the wrong technologies for a particular project, are also risk factors.

10 PRODUCTION AND OPERATIONAL RISKS

These pertain to unsuccessful or inefficient processes, policies or systems, such as cargo handling operations, equipment breakdown, pilferage, sabotage, fraudulent, tampered or missing documentation, etc. Operations are also impacted by external factors such as weather, politics, social turmoil, and wars.

11 COMMERCIAL AND FINANCIAL RISKS

These risks evolve around the ability of a nation, or energy project or individual stakeholder to obtain optimum commercial and financial results throughout the market cycles. This entails the alignment to the supply/demand conditions.

- Market risk, i.e., pertaining to market cycles, unexpected fluctuations, lack of supply–demand equilibrium, and so on;
- Financial risk, i.e., unanticipated variations in rates of interest, rates of exchange, stock, values, futures trading, trade or operational costs, and so on;
- Consumer credit/delinquency risk;
- Assets liquidity risk, i.e., incapacity to sell assets at reasonable prices, or inability to purchase or substitute goods due to scarcity or other hindrances.

12 INVESTMENT RISK

Price fluctuations increase risks related to energy investment strategies pertinent to energy preservation and development. All categories of stakeholders, regardless of whether they pertain to businesses evaluating new energy production possibilities, or fuel-efficient, environmentally friendly engines, or experience challenges evaluating if latest price ranges suggest long-term values. Timing will determine the effectiveness of the investment, as day-to-day market fluctuations can greatly impact the project's success and market sentiment.

Return on investment (ROI) can be attained when the resource input has delivered the optimum financial and commercial output. Investment risks pertain to misjudgment of the potential feasibility and/or profitability of a project, and may be classified as follows:

- over-investment, which is the allocation of funds and resources in a project that did not achieve its ROI goals;
- under-investment, i.e., unmaterialized profits due to the diminished allocation of funds in a project with great potential; and
- mis-investment, i.e., the inadvertent allocation onto the wrong project (e.g., exploration or market distribution in the wrong country, the wrong energy field or land, the wrong technology, the wrong products, etc.) leading to unmaterialized profits.

13 PRICE VOLATILITY RISK

Price volatility risk is attributable to supply and demand alterations, which in turn affect the products' and assets' prices.

Furthermore, electrical power and gas frequently cannot be transported to locations with unplanned or uncontrolled demand fluctuations. Especially in cases where growth in demand occurs and low-cost storage is not available in the region. This is particularly challenging for electric power. Public strategy initiatives to diminish price level volatility have concentrated on boosting both reserve input capacity as well as storage and transportation infrastructure potential.

Companies doing business in the energy industry, such as oil and gas and electricity companies, are especially vulnerable to market risk or, to be precise, value risk, as a result of the unpredictable fluctuations of energy commodities and stock markets. Especially the electricity prices are noticeably more erratic than alternative product costs (EIA, 2002).

Conclusions

The energy industry is truly global by nature, with multinational collaborations and a myriad of stakeholders. Since high risk is traditionally associated with complex supply chain networks, risks are present in the entire process from upstream, midstream, downstream and retail.

Energy corporations embrace risks that range from price fluctuations, losing the supply and demand equilibrium, geopolitical factors rendering a region physically unsafe, and also factors related to both physical and cyber threats, combined with regulatory compliance, especially when linked with HSSQE (i.e., Health, Safety, Security, Quality, Environment).

Since the energy industry is heavily regulated, every risk may impose physical (tangible) or cyber (intangible) threat, but concurrently holds the company non-compliant with serious legal, commercial and financial repercussions.

Hence a company's attempt to identify, mitigate and recover from any risk is also related to:

a regulatory compliance, e.g., HSSQE (Health, Safety, Security, Quality, Environment) and so on; but also
b social responsibilities, towards the community, the state, the nation and other industries.

The above aspects of regulatory compliance and social responsibilities will be duly analyzed in Chapter 2 of this book.

References

Aristotle (1933). "Aristotle Metaphysics". *Loeb Classic Library*. Available at: www.loebclassics.com/view/aristotle-metaphysics/1933/pb_LCL271.459.xml, last accessed: October 7, 2016.

Ball, R. B. B. (1908). *A Short Account of the History of Mathematics* (4th edition). Available at: www.maths.tcd.ie/pub/HistMath/People/Lagrange/RouseBall/RB_Lagrange.html, last accessed: October 12, 2016.

BBC (2016). "Venezuela leader Nicolas Maduro lifts power rationing", BBC News, July 4. Available at: www.bbc.com/news/world-latin-america-36702341, last accessed: December 5, 2016.

Brock, W. (2008). *William Crookes (1832–1919) and the Commercialization of Science*. Abingdon: Routledge.

Bunch, B. and Hellemans, A. (2004). *The Book of Science and Technology*. Boston: Houghton Mifflin Company.

Burns, M. (2015). *Logistics and Transportation Security. A Strategic, Tactical, and Operational Guide to Resilience*. Boca Raton: CRC Press.

Burns, M. (2016). "The paradox of energy security". 10th Mare Forum U.S.A. 2016, October 13, Hilton Post Oak, Houston, Texas, U.S.A. Available at: http://usa2016.mareforum.com, last accessed: October 21, 2016.

Castello, R. (2016). "Business SA calculates September's massive blackout cost the state 367 million". *Adelaide Now*, Business SA, December 9. Available at: www.adelaidenow.com.au/news/south-australia/business-sa-calculates-septembers-massive-blackout-cost-the-state-367-million/news-story/363e7e31a85ff2ddd11a036dd9134e90, last accessed: January 12, 2017.

Central Intelligence Agency (CIA) (2017). *The World Factbook*. Central Intelligence Agency. Available at: www.cia.gov/library/publications/the-world-factbook/geos/xx.html, last accessed: January 3, 2017.

Chemistry Explained (2016). Henry Cavendish. *Chemistry Explained*. Available at: www.chemistryexplained.com/Bo-Ce/Cavendish-Henry.html, last accessed: October 7, 2016.

Crookes, W. (1879). "On Radiant Matter. A Lecture Delivered to the British Association for the Advancement of Science. Sheffield, August 22, 1879". *American Journal of Science*. October, Series 3 Vol. *18*: 241–262. Available at: www.ajsonline.org/content/s3-18/106/241.full.pdf+html, last accessed: October 5, 2016.

Department of Energy, Energy Information Administration (DOE, EIA) (2002). *Derivatives and Risk Management in the Petroleum, Natural Gas, and Electricity Industries*. Energy Information Administration, U.S. Department of Energy, Washington, DC 20585, October. Available at: www.hks.harvard.edu/hepg/Papers/DOE_Derivatives.risk.manage.electric_10-02.pdf, last accessed: December 3, 2016.

Department of Energy (DOE) (2013). *Top 11 Things You Didn't Know About Nikola Tesla*. U.S. Department of Energy, November 18. Available at: https://energy.gov/articles/top-11-things-you-didnt-know-about-nikola-tesla, last accessed: October 1, 2016.

Department of Energy (DOE) (2016a). *U.S. Energy Department's Quadrennial Energy Review (First Installment)*. U.S. Department of Energy. Available at: https://energy.gov/epsa/quadrennial-energy-review-first-installment, last accessed: December 2, 2016.

Department of Energy (DOE) (2016b). *U.S. Energy Department's Quadrennial Energy Review*. U.S. Department of Energy. Available at: https://energy.gov/epsa/quadrennial-energy-review-second-installment, last accessed: December 2, 2016.

Department of Energy (DOE) (2016c). *Strategic Petroleum Reserve*. U.S. Department of Energy. Available at: https://energy.gov/fe/services/petroleum-reserves/strategic-petroleum-reserve, last accessed: December 5, 2016.

Department of Homeland Security (DHS) (2011). *ICE/HSI, DHS training-using-armored-vehicles*. USCG, Department of Homeland Security. Available at: www.dhs.gov/photo/hsi-using-armored-vehicles-training-ice, last accessed: December 2, 2016.

Department of Homeland Security (DHS) (2016). *Energy Sector*. U.S. Department of Homeland Security. Available at: www.dhs.gov/energy-sector, last accessed: December 2, 2016.

Dupuy, M. and Xuan, W. (2016). "China's string of new policies addressing renewable energy curtailment: update April 18, 2016". *Renewable Energy World*. Available at:

www.renewableenergyworld.com/articles/2016/04/china-s-string-of-new-policies-addressing-renewable-energy-curtailment-an-update.html, last accessed: December 5, 2016.

EagleFord Texas (2015). "Oil and gas industry benefits from magnetic technology". *EagleFord Texas*, October 6. Available at: http://eaglefordtexas.com/news/id/158201/oil-and-gas-industry-benefits-from-magnetic-technology, last accessed: October 16, 2016.

EFM Leeds (2016). *Bernoulli Equation*. EFM Leeds University. Available at: www.efm.leeds.ac.uk/CIVE/CIVE1400/Section3/bernoulli.htm, last accessed: October 12, 2016.

Encyclopedia Britannica (2002). "James Prescott Joule". *Encyclopedia Britannica*. Available at: www.britannica.com/biography/James-Prescott-Joule, last accessed: December 21, 2016.

Encyclopedia Britannica (2016a). "William Crookes". *Encyclopedia Britannica*. Available at: www.britannica.com/biography/William-Crookes, last accessed: October 5, 2016.

Encyclopedia Britannica (2016b). "Michael Faraday". *Encyclopedia Britannica*. Available at: www.britannica.com/biography/Michael-Faraday, last accessed: October 1, 2016.

Encyclopedia Britannica (2016c). "George Ohm". *Encyclopedia Britannica*. Available at: www.britannica.com/biography/Georg-Ohm, last accessed: November 1, 2016.

Encyclopedia Britannica (2016d). "Gottfried Wilhelm Von Leibniz". *Encyclopedia Britannica*. Available at: www.britannica.com/biography/Gottfried-Wilhelm-Leibniz, last accessed: October 12, 2016.

Encyclopedia Britannica (2016e). "Sebastian Ziani De Ferranti". *Encyclopedia Britannica*. Available at: www.britannica.com/biography/Sebastian-Ziani-de-Ferranti, last accessed: October 3, 2016.

Energy Information Administration (EIA) (2002). *Derivatives and Risk Management in the Petroleum, Natural Gas, and Electricity Industries*. Energy Information Administration, U.S. Department of Energy, Washington, DC 20585, October. Available at: www.hks.harvard.edu/hepg/Papers/DOE_Derivatives.risk.manage.electric_10-02.pdf, last accessed: December 3, 2016.

Energy Information Administration (EIA) (2015). *Nonrenewable Energy Sources – Energy Explained*. Last reviewed: October 27, 2015. Available at: www.eia.gov, last accessed: December 5, 2016.

Energy Information Administration (EIA) (2016a). *EIA International Energy Outlook 2016*. U.S. Energy Information Administration. Available at: www.eia.gov/outlooks/ieo/pdf/0484(2016).pdf, last accessed: January 1, 2017.

Energy Information Administration (EIA) (2016b). *Annual Energy Outlook, 2016*. U.S. Energy Information Administration. Available at: www.eia.gov/pressroom/presentations/sieminski_06282016.pdf, last accessed: December 5, 2016.

Energy Information Administration (EIA) (2016c). *U.S. Energy Industry*. U.S. Energy Information Administration. Available at: www.eia.gov/energyexplained/index.cfm/data/index.cfm?page=us_energy_industry, last accessed: November 1, 2016.

Energy Information Administration (EIA) (2017a). *Energy Explained*. U.S. Energy Information Administration. Available at: www.eia.gov/energyexplained/?page=us_energy_home, last accessed: January 1, 2017.

Energy Information Administration (EIA) (2017b). *Coal Imports and Exports*. U.S. Energy Information Administration. Available at: www.eia.gov/energyexplained/index.cfm?page=coal_imports, last accessed: January 3, 2017.

Energy Quest (2016). "Scientists, Ferranti". *Energy Quest*. Available at: www.energyquest.ca.gov/scientists/ferranti.html, last accessed: October 3, 2016.

ExxonMobil (2016). *The Outlook for Energy: A View to 2040*. ExxonMobil. Available at: http://cdn.exxonmobil.com/~/media/global/files/outlook-for-energy/2016/2016-outlook-for-energy.pdf, last accessed: January 1, 2017.

Famous Scientists (2016a). "James Prescott Joule". *Famous Scientists*. Available at: www.famousscientists.org/james-prescott-joule, last accessed: October 5, 2016.

Famous Scientists (2016b). "Michael Faraday – biography, facts and pictures". *Famous Scientists*. Available at: www.famousscientists.org/michael-faraday, last accessed: October 1, 2016.

Felix Marine (2016). *FC Spec Sheet*. Felix Marine. Available at: http://felixmarine.net/pdf/FC_Spec_Sheet.pdf, last accessed: October 12, 2016.

Flo-Rite Fluids (2016). *Flo-Rite Fluids, Company Profile*. Available at: https://flo-ritefluids.leadpages.co/free-trial, last accessed: October 4, 2016.

Forbes (2017). The World's Top 25 Companies. *Forbes*. Available at: www.forbes.com/pictures/edjl45efeik/the-worlds-top-25-companies/#615546dc340c, last accessed: January 1, 2017.

FoxNews (2016). "Governor: State House saved by 'Ben Franklin' lightning rod". *FoxNews*, July 2. Available at: www.foxnews.com/us/2016/07/02/governor-state-house-saved-by-ben-franklin-lightning-rod.html, last accessed: October 1, 2016.

Fox, W. (1909). "Luigi Galvani". *The Catholic Encyclopedia*. New York: Robert Appleton Company, New Advent. Available at: www.newadvent.org/cathen/06371c.htm, last accessed: 21 November 2016.

Geosoft (2016). *Gravity and Magnetic Modelling*. Geosoft. Available at: www.geosoft.com/gravity-and-magnetic-modelling#sthash.eYWxliLZ.dpuf, last accessed: October 12, 2016.

Gilbert, W. (1600). *De Magnete, Magneticisque Corporibus, et de Magno Magnete Tellure* (On the Magnet and Magnetic Bodies, and on the Great Magnet the Earth). Digital copy, available at: www.new-science-theory.com/william-gilbert-de-magnete.pdf, last accessed: October 5, 2016.

Heaviside, O. (1894). *Electrical Papers*. 1. London: Macmillan and Co.

History MCS (2016). Biographies, Johann Bernoulli. *MacTutor History of Mathematics archive*. Available at: www-history.mcs.st-andrews.ac.uk/Biographies/Bernoulli_Johann.html, last accessed: September 9, 2016.

Hubbert, M. K. (1956). "Nuclear energy and the fossil fuels", in *American Petroleum Institute Spring Meeting: San Antonio, Texas, p. 40, by M. King Hubbert Chief Consultant (General Geology), Presented Before The Spring Meeting Of The Southern District Division Of Production. American Petroleum Institute Plaza Hotel, San Antonio, Texas March 7–8–9.* Available at: www.hubbert-peak.com/hubbert/1956/1956.pdf, last accessed: December 14, 2016.

International Energy Agency (IEA) (2016a). *Renewable Energy*. IEA. Available at: www.iea.org/about/faqs/renewableenergy, last accessed: December 5, 2016.

International Energy Agency (IEA) (2016b). *About IEA*. Available at: www.iea.org/about, last accessed: December 6, 2016.

International Energy Agency (IEA) (2016c). *IEA International*. Available at: www.iea.gov/beta/international, last accessed: December 6, 2016.

International Energy Agency (IEA) (2016d). *Energy Security*. Available at: www.iea.org/topics/energysecurity, last accessed: January 3, 2017.

International Energy Association (IEA) WEO (2016). *World Energy Outlook* (WEO). Available at: www.iea.org/media/publications/weo/WEO2016Chapter1.pdf, last accessed: December 3, 2016.

International Risk Management Institute (IRMI) (2016). *Risk Reduction*. International Risk Management Institute. Available at: www.irmi.com/online/insurance-glossary/terms/r/risk-reduction.aspx, last accessed: December 2, 2016.

Joule, J. P. (1884). *Matter, Living Force, and Heat: The Scientific Papers of James Prescott Joule*. London: The Physical Society of London.

Karlsruhe Institute of Technology (KIT) (2016). *Karlsruhe Institute of Technology (KIT) Electrotechnisches Institut (ETI) Timetable 1856–1873: From the Invention of the Dynamo to the DC Motor (1856)*. Available at: www.eti.kit.edu/english/1390.php, last accessed: October 3, 2016.

Math Pages (2016). "Christiaan Huygens' Principle". *Math Pages*. Available at: www.mathpages.com/home/kmath242/kmath242.htm, last accessed: October 12, 2016.

Merriam Webster (2016). Definition: "Infrastructure". *Merriam Webster*. Available at: www.merriam-webster.com/dictionary/infrastructure, last accessed: December 2, 2016.

Millikan, R. A. and Bishop, E. S. (1917). *Elements of Electricity*. Chicago: American Technical Society.

Moray, T. H. (1978). *The Sea of Energy in Which the Earth Floats*. Salt Lake City: Cosray Research Institute.

MPower UK (2016). "Battery and energy technologies. history of technology". *The Electropaedia*, Chester: Woodbank Communications Ltd. Available at: www.mpoweruk.com/history.htm, last accessed: October 3, 2016.

National Institute of Environmental Health Sciences (NIH) (2016). *Emergency Response*. National Institute of Environmental Health Sciences. Available at: www.ready.gov/business/implementation/emergency, last accessed: October 1, 2016.

Neidhöfer, G. (2008). *Michael von Dolivo-Dobrowolsky und der Drehstrom. Geschichte der Elektrotechnik VDE-Buchreihe*, Volume 9, Berlin Offenbach: VDE Verlag.

New Science Theory (2016). "René Descartes, mechanical push universe theory". *New Science Theory*. Available at: www.new-science-theory.com/rene-descartes.php, last accessed: October 7, 2016.

Newton, Isaac (2016). *Philosophiæ Naturalis Principia Mathematica*. Available at: https://docs.lib.noaa.gov/rescue/Rarebook_treasures/QA803A451846.PDF, last accessed: October 3, 2016.

New York Times (2016). "Oil prices, energy environment". *New York Times*, June 14. Available at: www.nytimes.com/interactive/2016/business/energy-environment/oil-prices.html?_r=0, last accessed: December 3, 2016.

NobelPrize (2016). *Albert Einstein – Biographical*. NobelPrize.org. Available at: www.nobelprize.org/nobel_prizes/physics/laureates/1921/einstein-bio.html, last accessed: October 7, 2016.

Prometheas (2016). *Prometheas, Mythology*. The Hellenic Society Prometheas, Inc. Available at: www.prometheas.org/mythology.html, last accessed: November 1, 2016.

Public Health Emergency (PHE) (2016). *Risk Management Plan*. U.S. Department of Health and Human Services, Office of the Assistant Secretary for Preparedness and Response. Available at: www.phe.gov/about/amcg/contracts/Documents/risk-management.pdf, last accessed: October 1, 2016.

Russell, J. S. (2003). *Perspectives in Civil Engineering: Commemorating the 150th Anniversary of the American Society of Civil Engineers*. Reston: The American Society of Civil Engineers.

Science World (2016). "Christiaan Huygens, Science". *Science World*. Available at: http://scienceworld.wolfram.com/biography/Huygens.html, last accessed: October 15, 2016.

Stanley, D. (1998). *Magnetic Fluid Conditioner*, U.S. Patent U.S. 5783074 A, Inventor David Stanley, 1995–1998. Available at: www.google.com/patents/US5783074; http://patft.uspto.gov/netacgi/nph-Parser?Sect2=PTO1&Sect2=HITOFF&p=1&u=/netahtml/PTO/search-bool.html&r=1&f=G&l=50&d=PALL&RefSrch=yes&Query=PN/5783074, USPTO Patent full-text and image database, July 21, last accessed: October 12, 2016.

Suzuki, D. and Moola, F. (2010). *Let's Dare to Consider a National Energy Plan for Canada*. Available at: www.davidsuzuki.org/blogs/science-matters/2010/09/lets-dare-to-consider-a-national-energy-plan-for-canada, last accessed: December 5, 2016.

Tesla, N. (1901). *Tesla Patent 685,957 Apparatus For The Utilization Of Radiant Energy*. Application filed March 21, serial no. 52,153, patented November 5. Available at: www.google.com/patents/US685957, last accessed: October 5, 2016.

Texas Renewable Energy Industries Alliance (TREIA) (2016) *Renewable Energy Defined*. Texas Renewable Energy Industries Alliance. Available at: www.treia.org/renewable-energy-defined, last accessed: November 1, 2016.

ThinkGeoenergy (2016). "IRENA opens renewable energy cost analysis portal". *ThinkGeoenergy*. Available at: www.thinkgeoenergy.com/irena-opens-renewable-energy-cost-analysis-portal, last accessed: in October 2, 2016.

Trading Economics (2016). "Commodity, crude oil". *Trading Economics*. Available at: www.tradingeconomics.com/commodity/crude-oil, last accessed: December 5, 2016.

UK Oil & Gas PLC (UKOG) (2016). *Energy for Britain*. UK Oil & Gas PLC. Available at: www.ukogplc.com/page.php?pID=74, last accessed: December 2, 2016.

United Nations (UN) (2017). *Sustainable Development Goals: Ensure Access to Affordable, Reliable, Sustainable and Modern Energy*. United Nations. Available at: www.un.org/sustainabledevelopment/energy, last accessed: January 12, 2017.

United Nations Conference on Trade and Development (UNCTAD) (2016). *Review of Maritime Transport 2016*. Available at: http://unctad.org/en/PublicationsLibrary/rmt2016_en.pdf, last accessed: January 2, 2017.

UN Chronicle (2014–2015). Goal 7—Ensure Access to Affordable, Reliable, Sustainable and Modern Energy for All. *UN Chronicle*, the magazine of the United Nations. Vol. LI, No. 4. Available at: https://unchronicle.un.org/article/goal-7-ensure-access-affordable-reliable-sustainable-and-modern-energy-all, last accessed: January 12, 2017.

United Nations, Department of Economic and Social Affairs (UNESA) (2016). *World Population Prospects, The 2015 Revision*. United Nations, Department of Economic and Social Affairs. Available at: https://esa.un.org/unpd/wpp, last accessed: December 3, 2016.

UPS Battery Center (2016). *Thales Miletus, Father of Electricity*. UPS Battery Center. Available at: www.upsbatterycenter.com/blog/thales-miletus-father-of-electricity, last accessed: October 5, 2016.

US Coastguard (USCG) (2010). *History, The Piracy Mission*. U.S. Coastguard, DHS. Available at: http://coastguard.dodlive.mil/files/2010/03/Pirates-1024x545.jpg, last accessed: December 2, 2016.

U.S. Coastguard (USCG) (2010). *Human Factor, Deepwater Horizon, April 21, 2010*. U.S. Coastguard. Available at: www.uscg.mil/npfc/img/Slideshow/ltimg/DWH.jpg, last accessed: December 2, 2016.

U.S. Geological Survey (USGS) (2016). *USGS Estimates 20 Billion Barrels Oil In Texas Wolfcamp Shale Formation*. U.S. Geological Survey. Available at: www.usgs.gov/news/usgs-estimates-20-billion-barrels-oil-texas-wolfcamp-shale-formation, last accessed: December 2, 2016.

U.S. History (2016). "The electric. Ben Franklin". *U.S. History*. Available at: www.ushistory.org/franklin/science, last accessed: October 1, 2016.

World History Project (2016). "Thomas Edison". *World History Project*. Available at: https://worldhistoryproject.org/topics/thomas-edison, last accessed: October 1, 2016.

Your Dictionary (2016). Definition: "Infrastructure". *Your Dictionary*. Available at: www.yourdictionary.com/infrastructure, last accessed: December 2, 2016.

2 Major Security and Safety Threat Types

My message for companies that think they haven't been attacked is:

"You're not looking hard enough."

James Snook, Deputy Director in the Office for Cyber Security,
UK Government Cabinet Office, 2016

This chapter analyzes and contrasts different security and safety threats in terms of risk types, identification, mitigation, and recovery techniques. The role of energy is pivotal within the perspective, as energy sustains a nation's defense, population health and wellbeing, transportation systems, trade and storage of foods, medicines and other temperature-specific commodities. Waste management, temperature and pressure controlling equipment, air purifiers, and all household equipment depend on energy security infrastructure.

Energy is needed for water management, equipment to biologically treat invasive species and other microbial organisms that impact human health, animals, the environment and the global food chain. A nation's critical infrastructure is the backbone of socioeconomic growth, health and wellbeing for its citizens, and national security and territory sovereignty.

Over the past decades the United Nations and their affiliated organizations, in collaboration with governments and industries, have been working on alleviating any threats pertaining to health, security, safety, quality and environment (HSQE). For the sake of precision, let us rename the HSQE as HSSQE, to incorporate both security and safety protection. This chapter will encompass the different threat types in the areas of Security, Safety, Health and Environment, which are both interrelated and yet possess distinct particularities (Figure 2.1).

HSSQE Harmonization

The interrelation of the HSSQE disciplines led to harmonization, i.e., a common strategy pertaining to risk assessment, mitigation and recovery, that helps address interrelated issues from a geopolitical, regulatory, emergency response, economic, operational, or technical viewpoint.

Table 2.1 demonstrates the interrelation among these risks. It is observed that, regardless of the threat type, i.e., man-made or natural disaster, cyber or physical threat, and

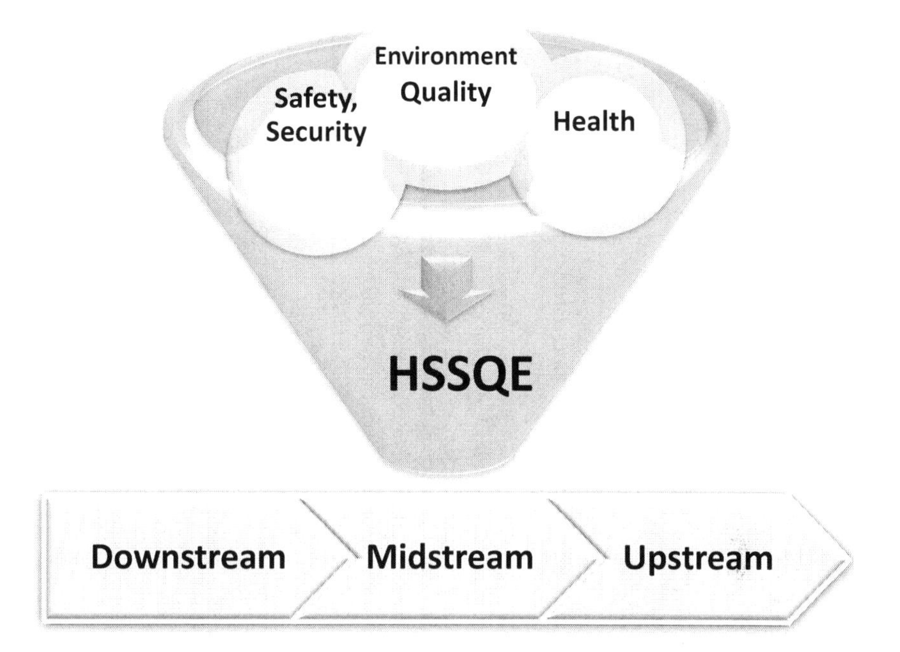

Figure 2.1 Integrated Health, Safety, Security, Quality and Environmental (HSSQE) systems.
Source: the Author.

intentional or unintentional attack, every single incident impacts all HSSQE aspects and has the same consequences:

- Safety impact: Damages physical assets;
- Security impact: Weakens or damages security systems;
- Environmental impact: Damages ecosystem, including human health, flora and fauna. Impacts water and food quality. Consequent health impact.
- Health impact affects human health, flora and fauna: May cause fatalities or injuries;
- Fear or distress may affect emotional health and the national or community wellbeing;
- Quality impact is present in every type of threat, as quality is an all-encompassing process;
- Cyber threats are also interrelated with the potential to damage cyber and physical systems.

Defining Safety vs Security

Based on the premise that homeland security officers as well as HSSQE (Health, Security, Safety, Quality, Environment) personnel need to deal with both intentional (security) and unintentional (safety) threats, this chapter discuss both of these areas.

Table 2.1 Threat types and the interrelation of safety, security, health and environmental incidents.

Threat Types	*Examples*	*HSSQE Impact*			
		Safety	*Energy Security*	*Environmental and Biological Threats*	*Health Threats*
Natural disasters: Geophysical phenomena ***(Unintentional)***	• Hurricane • Earthquake	Damage of physical assets	Weakening of physical assets causes security vulnerabilities	Ecosystem damage Pollution with impact on water, food, etc.	Health impact to: • humans, • animals, • ecosystem *** • fatalities, • injuries *** • emotional health *** • fear affecting national and community wellbeing.
Man-made disaster: ***(Intentional)***	• Terrorist attack • Sabotage	Damage of physical assets	Security systems attacked	Environmental degradation with impact on water, food, etc.	
Man-made accident ***(Unintentional)***	• Fatigue or lack of situational awareness	Damage of physical assets	Physical threats cause security vulnerabilities	Ecosystem damage Pollution	
Physical threat on network: Infrastructure/ superstructure Supply chain ***(Unintentional)***	• Structural deficiencies	Damage of physical assets	Damage of physical assets Causes security vulnerabilities	Ecosystem damage Pollution with impact on water, food, etc.	
Cyber threats ***(Intentional)***	• Hacking of energy companies	Damage of physical assets	Breach of security systems, physical and cyber	Cyber attack may target water, food and environmental integrity	
Environmental, pollution & biological hazards ***(Intentional & Unintentional)***	• Pipeline leakage • Vessel grounding & oil leakage • Invasive species carried by offshore platforms	Chemical or biological degradation of physical assets	Damage of physical assets causes security vulnerabilities	Ecosystem damage Pollution with impact on water, food, etc.	

Source: the Author.

The definitions of security and safety deserve our undivided attention, as these two terms are being used interchangeably by some government and corporate entities, whereas others distinguish their use and meaning.

For the sake of good order, we hereby make a distinction between the intentional planned attacks and unintentional randomly occurring disasters, by identifying the former as "security threats," and the latter as "safety threats."

Both safety and security risks have the ability to trigger damages of similar magnitudes and ranges such as loss of life, personal injuries, environmental damage, or harm to tangible and intangible assets. Nonetheless, considering the element of intention, security threats are more subtle and possibly more harmful as the attacker plans to attack a system by studying its vulnerabilities. They may frequently plan multiple consecutive or synchronized attacks, and will not stop until their mission is accomplished (Burns, 2016a).

Based on the above definitions, it becomes clear that:

Security threats deal with intentional damage, which is a result of a deliberate, premeditated act. Security threats include terrorist attacks, hijacking/piracy, identity theft, cyber-security, and so on.

Safety threats pertain to naturally occurring threats, such as a) natural disasters and weather conditions, e.g., hurricane Katrina; b) natural deterioration of structures, e.g., corrosion, structural fatigue.

The human factor also impacts safety by failing to protect a system through lack of action, lack of intervention, or human error – for example, fatigue or lack of training.

Finally, safety entails the element of coincidence, or a combination of accidental factors. For example, the heavy rain combined with structural fatigue imposed serious damages to the structure.

When it comes to energy supply, a balance should be attained between all the different components of the energy supply chain:

- fast operational processes should not compromise situational awareness, security or safety checks;
- technological innovation should not overshadow crew empowerment or human ingenuity;
- energy affordability should not compromise health or environmental compliance.

Synergistic initiatives and forward thinking will help us achieve optimum productivity, increase the flow of energy sources, and grow financially, without compromising human health, environmental integrity, or sociopolitical stability.

Energy Security Threat Types

Defining Energy Security

Energy represents the powerhouse of each nation's economic activities, and, as such, governments consider energy as one of the few critical infrastructure segments. Security is defined as "freedom from measures taken to guard against espionage or sabotage, crime, attack or escape" and this is generally the definition we are using when we refer to intentional attacks (*Merriam-Webster* Dictionary, 2016a).

The most common energy security threats pertain to the global antagonism over power resources, sociopolitical instability of numerous energy-producing nations, and acts of sabotage against energy supply chains, which are frequently expressed as physical or cyber-attacks against energy production or storage units, or refineries, or energy transportation and infrastructure.

Energy security, therefore, is defined as the act of safeguarding a country's energy supply chains and electric grid, while making certain that a dependable uninterrupted energy availability are a nation's and a corporation's primary concerns.

Energy Security and the Levels of Defense

Energy security is multi-dimensional as it has many faces and risk types, which we hereby call the levels of defense. It is classified into three main categories, in accordance to its attack types: Supply chain security (see Chapters 1 and 3); physical security (see Chapters 4–10), and cyber security (see Chapter 11). The three levels of defense, as presented in Figure 2.2, encompass the upstream, midstream and downstream of security throughout their global networks.

Physical and Cyber Security

In this increasingly complex and dynamic environment, any incidents involving physical or cyber security involve soaring costs for governments and companies alike, as they cause damages and supply chain delays while depriving citizens of energy sources.

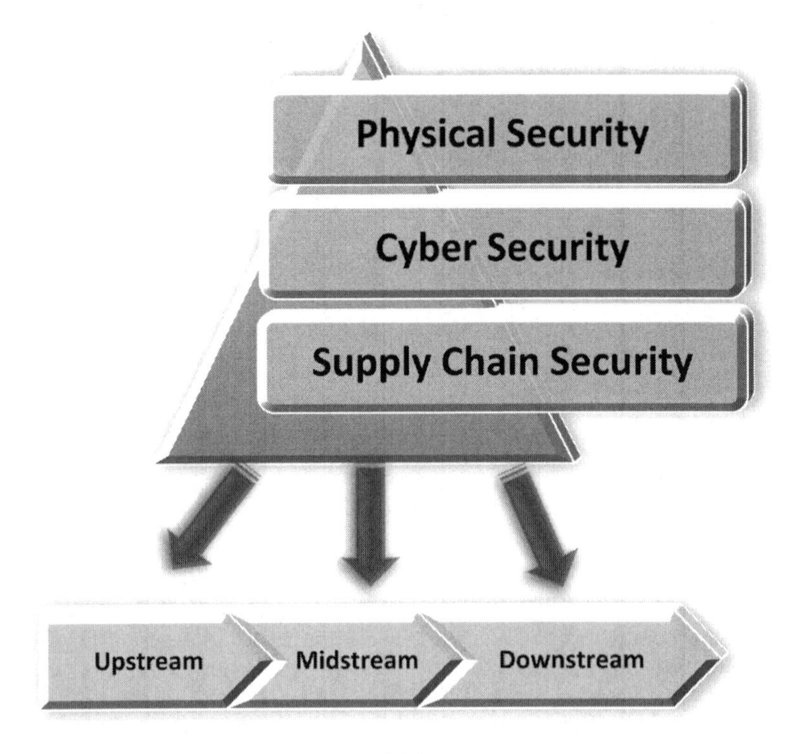

Figure 2.2 Energy security and the levels of defense.

Source: the Author.

Energy security is becoming increasingly more challenging to control as technological know-how accelerates with intricacy. This creates more systemic vulnerabilities, and demands the utilization of more government and corporate resources.

1 Physical Security

Physical security is the term for precautionary strategies that are developed to secure the uninterrupted supply of energy, ensuring the smooth function of all government, industrial and domestic processes.

Energy supply challenges or disruptions may be caused by reasons including but not limited to the uninterrupted supply to deny unlawful or unauthorized access to facilities, technologies, electronic devices, storage units and working areas for input or output. Their goal is to defend nations, civilians or professionals and the natural or physical environment from any damage or harm.

At a national (homeland) level, physical security is intended to protect our homeland, its citizens and residents, innovation, patents, private data and public domain data, superstructure, infrastructure, including energy generators, grid systems, pipelines, any and all supply chain components that pertain to energy.

At a corporate level, physical energy security entails the protection of workforce, and deterrence of physical theft, espionage or damage to energy technologies, equipment, superstructure components, IT software and hardware, and data related to physical activities. It safeguards our system from any physical access that can potentially result in significant loss, injury or harm to an organization, government agency or corporation.

This comprises a) protection from natural disasters, such as earthquakes, tsunamis, fire, hurricanes, flooding; but also b) protection from intentional attacks to our system, such as robbery, pilferage, criminal damage, terrorism, or acts of sabotage and espionage.

Physical security requires the employment of a security strategy that incorporates several layers of interconnected security professionals and systems.

Physical security measures are typically experienced, utilized, or connected with government officials, energy professionals or civilians, and entail:

1. biometric devices
2. hard tokens
3. professional access cards such as TWIC cards in the U.S.
4. ID badges
5. building lights
6. security fencing and gates, razor wires, security guards
7. alarm system and intrusion detection systems (IDS)
8. closed-circuit television (CCTV) monitoring systems
9. protective barriers, locking mechanisms
10. access control protocols, i.e., technology used to authenticate entry into an area or logging into a system, and several other systems.

In a computerized framework, security pertains to both physical security and cyber security: While cyber security entails the remote theft of data, manipulation of information or identity theft – all remotely accessed – physical security includes the actual pilferage (theft) of electronics, cellular devices such as laptop computers, smart phones, tablets, and so on. These items are convenient targets due to portability and mobile access (Hutter, 2016).

2 Cyber Security

Cyber security is defined as the state of being protected against the criminal or unauthorized use of electronic data, or the measures taken to achieve this. Cyber security is the body of technologies, processes and practices designed to protect networks, computers, programs and data from attack, damage or unauthorized access.

Cyber security entails defending data, critical information and programs from different cyber threat types, which include cyber espionage, identity theft, cyber terrorism and cyber warfare.

Cyber security threats attack governments and their people. The particularity of cyber security is the very thin line between breach of government, industry or private systems: Once the cyber attacks penetrate a government database, they frequently obtain access to industrial data.

Case Study: NATO and Energy Security

The hindrance of energy supply may affect regional and national security and have an effect on NATO's military procedures. Although these challenges are mainly under the jurisdiction and accountability of national authorities, NATO is constantly involved with matters of energy security and additionally expands the ability to enhance energy security, while focusing on topics where it can be meaningful. In this respect, NATO aims to: Boost its strategic understanding of energy trends with security effects; improve its expertise in promoting the security of critical energy infrastructure; and focus on considerably enhancing the military energy capabilities (NATO, 2016).

NATO's position in the domain of energy security was initially recognized in 2008 during the Bucharest Summit. Energy security is a crucial component of global sustainability and has obtained growing significance over the past decade or so, as a result of the new security framework.

Energy efficacy is significant from a supply chain, economic, commercial and environmental perspective.

- **Energy security**, from a consumer's perspective, is defined as accessibility to sustainable, cost-effective and reliable energy flow, and this has turned into a topic of increasing interest in Europe and North America over the past years. Several nations in the western hemisphere have not attained total energy self-sufficiency, hence depend on foreign production flows and global supply chains, frequently in distant and volatile locations. As a result, their energy dependence extends to a number of exterior factors and trends, many of which are areas for concern.
- One crucial area of concern is the development of the international energy industry. The **energy supply–demand equilibrium** is anticipated to change due to a rising growth in the energy demand, primarily in developing economies where a change of over 74% is estimated from 2005 to 2030. The International Energy Agency (IEA) reports that the international energy demand will expand by more than 50% by the year 2030 in comparison with 2005 (IEA, 2015; IEA, 2016). Furthermore, the industry's efforts to cut down CO_2 emissions and promote greener energy may strongly recommend a change in the present industry patterns.
- Concerns of a **political accessibility of energy sources** by energy-producing nations have been amplified by a number of occurrences.

- There are warning signs of impending threats due to **terrorist attacks** upon energy infrastructures, with international repercussions. Terrorists may find it easier to penetrate fragile international supply chains in afflicted regions.

Energy security infrastructure is a domain of utmost geopolitical significance that defines a nation's stability and existence. For all of these reasons, the domain of energy security is of critical significance to NATO Allies (NATO Parliamentary Assembly, 2016).

Energy Safety Threat Types

> Safety trumps everything we do, including production and profit.
>
> J. Brett Harvey, at Consol Energy Inc.

Energy Safety: Threat Types

The primary definition of safety is "the condition of being free from harm or risk; being safe from undergoing or causing hurt, injury, or loss." It also includes a safety "device (such as a weapon or an appliance), designed to prevent inadvertent or hazardous operation" (*Merriam-Webster* Dictionary, 2016b).

Energy safety pertains to the safe supply and use of electricity and gas. As discussed in the introductory part of this chapter, safety risks pertain to naturally occurring energy threats such as extreme geophysical or geological phenomena, extreme weather conditions, and structural deterioration, i.e., corrosion, structural fatigue, and so on.

Table 2.2 depicts the various safety threat types, and provides key guidelines as to the best practices implemented at a national and industrial level related to the preventive action and contingencies.

A secure energy supply chain entails the reliable, safeguarded and protected upstream, midstream and downstream activities of energy sources towards its global markets. The energy flow stages entailed include the exploration, production, storage, transportation, refineries and storage, transportation, distribution and consumption. Each nation has a homeland security strategy, matching with energy security regulations and guidelines on how to act upon security disruptions caused by geopolitical, warfare and territory disputes, terrorism, sabotage, economic, commercial, technological, and other factors.

Energy and affordability have become necessary to the survival of contemporary nations. Nevertheless, the unequal availability and supply of energy flows around the world, combined with the significant priority of exploring and extracting conventional and unconventional resources, has resulted in considerable weaknesses. Risks to international energy security consist of political uncertainty of energy-producing nations, exerting financial, political or commercial influence on global supplies, corruption, antagonism over sources of energy, assaults against commercial facilities, but also geophysical phenomena, extreme weather, incidents, mechanical failures and human error.

Energy risks related to safety may include technological or mechanical malfunctions and threats of chemical, thermal, mechanical, electrical, pneumatic, hydraulic, and other hazard types, and impose threat to nations, industries, and populations. Occupational incidents may occur during the maintenance and repair of machinery. The improper storage and transportation, and the unanticipated leakage or ignition of energy may lead to severe personal injuries and death, some times combined with environmental pollution (OSHA, 2017).

Table 2.2 Safety threat types, preventive action and contingencies.

Safety Threat	*Preventive Action and Contingencies*
Aging supply chain infrastructure, limited investment • It is worth noting that during or after a financial crisis, there is limited cash flow or resources available for asset replacement, retrofitting, maintenance or repairs. • At a global level, aging supply chain infrastructure has been observed, and high levels of investment are needed.	• Increase input by generating funds, resources and allocating sufficient time for system restoration. • Adequate budget allocation in enhancing systemic safety and reliability. • Efficient maintenance and repairs. • Insurance policy stipulations.
Human factor • Human error. • Complacency. • Fatigue. • Lack of training. • Strikes.	• Familiarization of facilities and equipment. • Training and use of safety equipment. • Drills. • Recruitment of highly capable, well-trained personnel. • Crew recycling programs, ensuring that training and familiarization time and resources will be utilized in a long-term recruitment scheme. • Impact on long-term performance and the building of energy security culture. • Back-to-back contracts with recruiters, sharing liability. • Insurance policy stipulations.
Extreme weather and geophysical phenomena (Act of God; Force Majeure) Weather-generated system failures exclusively in the U.S. range from $18 billion to $33 billion per annum.	• System's structural integrity. • De-activation, system shut-down. • Evacuation. • Deviation. • Insurance policy stipulations.

Energy cargo theft, energy equipment theft International energy pilferage is on an upswing, most of which derives from unlawful pipeline hot tapping or pressure tapping. The industry's overall damages exceeding $37 billion per annum. The major energy-producing nations that are mainly afflicted by energy theft include the U.S., Nigeria, Iraq, and Mexico. Theft of energy machinery is also very common, especially during periods of diminishing energy production operations. In the region of West Texas alone, drilling companies lose about $20 million per annum. The cost of primary theft of energy commodities in Nigeria alone exceeds $1.5 billion, and in Mexico several millions of oil drums vanish on an annual basis.	• Security and safety technologies; Closed Circuit Cameras (CCTVs). • Selection of optimum supply chain partners and trade routes. • Insurance policy stipulations.
Assets and systems' failure • Unintended release/leakage of energy. • Equipment failure. • Structural fatigue. • Machinery failure. • Poor maintenance and repairs. • Corrosion, oxidization.	• Increase input by generating funds, resources, and allocating sufficient time for system restoration. • Adequate budget allocation in enhancing systemic safety and reliability. • Efficient maintenance and repairs. • Insurance policy stipulations.

Source: the Author.

Energy Safety: The Regulatory Framework

Most developed economies share similar definitions of energy safety, and their regulatory frameworks appear to be harmonized.

According to the U.S. Department of Labor, accidents caused by the inability to monitor and control dangerous energy during transportation, storage, repairs and/or maintenance operations, may cause explosions, fires, nuclear incidents, electrocution, falling objects leading to fracturing body parts or crushing, sharp objects leading to cutting, lacerating, or even amputating, and so on.

Accidents may be caused during day-to-day operations, including but not limited to:

- Repairing the wiring network of a manufacturing assembly line may lead to the workers' electric shocks.
- During maintenance or repairs of a downstream piping system connection can accidentally switch on a steam valve, causing serious injuries.
- The effort to clear a jammed conveyor may lead to the equipment's unexpected discharge, smashing employees who may be working on the conveyor.

As a rule of thumb, high profit, high-growth, technology-intensive industries entail high levels of risk. Furthermore, occupations related to dangerous goods (i.e., hazardous materials) are also likely to impose high risk levels. For this reason, the energy industry and pertinent occupations within the global supply chain have a long tradition of proactive versus reactive working culture in order to eliminate all types of risks.

Inability to handle dangerous energy sources exceeds 10% of the significant incidents in several industrial and business sectors. Appropriate lockout/tagout (LOTO) procedures and a safe working culture protect employees from harmful energy discharges. The LOTO regulations determine the company's liability to safeguard employees from harmful energy. Furthermore, the OSHA regulations, *The Control of Hazardous Energy (Lockout/Tagout) (29 CFR 1910.147)*, offer guidelines for monitoring and handling various kinds of harmful energy sources (OSHA, 2017).

According to the Canadian Standards Association (CSA), hazardous energy is described as: "any electrical, mechanical, pneumatic, chemical, nuclear, thermal, gravitational, or other energy that can harm people" (CSA Z460: "Control of Hazardous Energy – Lockout and Other Methods" (CCOHS, 2018)). Certain energy resources such as electricity have apparent presence and risks, whereas other energy types like high pressure within a system may entail concealed threats (CCOHS, 2016).

According to the Government of New Zealand, "safe" is described as a procedure of immaterial, inconsequential threat or risk whereas "unsafe" is considered to be a considerable threat of considerable impact or serious damage.

Electricity Regulation #20 of the electricity regulations specifies electrical operations and/or installations as hazardous, if:

- they contain damaged properties as part of their fittings;
- there are inadequately recognized conductors in terms of color, design and material;
- there are unreliable connections of poor structure;
- there are damaged thermostatic or pressure systems; inability to retain appropriate temperatures and pressures;
- there is a probability of combustion, ignition, blasting;

- there are poor wiring systems and cable connections, deteriorated, pulled or twisted in an unsafe manner;
- an inadequate working area disables safe working movements.

(New Zealand Government, 2016)

Gas Regulation #11 of the gas regulations defines unsafe distribution systems or gas installations. It is worth noting that preventive action and company procedures are not in position to ascertain gas leakage and the following activities.

Networks and installations are unsafe if:

- gas leakage is noticeable, prevented or mitigated;
- gas leakage or release is effectively managed;
- individuals and the environment are not subjected to dangerous goods prone to or igniting gas combustion;
- individuals, the environment and assets are not in contact with dangerous goods or circumstances;
- mechanical breakdown or safety equipment failure will not render energy production or storage-related technologies vulnerable to abnormal pressure higher than the ranked strain of any energy upstream, midstream and downstream technologies;
- the reliable seclusion and disconnect of energy supply sources;
- damaged protection capabilities and safety functions of equipment;
- damaged, corroded, leaking piping systems, outlets and connectors.

(New Zealand Government, 2016)

The aforementioned regulatory provisions represent only a small portion of the safety risks related to both industrial and domestic use of energy. And while an extensive analysis will be provided in Chapters 4 through 11 of this book, it is worth noting the numerous facets of safety and security, and the fact that risk prevention is becoming as complex as our national, industrial and domestic energy networks.

Case Study: International Energy Forum (IEF)

The IEF is the globe's premier ongoing meeting of energy ministers. Encompassing all six continents and comprising approximately 90% of world demand and supply for oil and gas, the IEF is distinct for the reason that it consists of not merely energy- consuming and producing nations of OPEC and the IEA, but also Transit States and leading participants beyond their memberships, such as China, India, Brazil, Mexico, Russia and South Africa. The IEF is supported by an appointed permanent Secretariat located in the Diplomatic Quarter of Riyadh, Saudi Arabia.

As the IEF has celebrated its 25th anniversary since its inception, it considers as its principal accomplishment, deriving from its past discussions, its success in promoting the awareness of the significance of energy interdependence among nations and key players. As an alternative to handling it as a cause for rigidity and conflict, the IEF has been promoting interdependence among producers and consumers. Its role has been extremely timely and significant. The industry has come a long way beyond the 1970s and 1980s communication challenges and the IEF indicates how much the energy industry has improved since then (IEF, 2017).

Environmental and Biological Threat Types

Defining Environmental Pollution

Pollution is defined as the defilement of air, water, soil (including agricultural areas), but also of the atmosphere through noise and vibration, through the release of an impurity or waste into the ecosystem. Pollution may occur naturally, for instance by means of natural petroleum seepage, or as the result of man-made activities (anthropogenic pollution) including industrial pollution, the release of industrial waste, or the accidental leakage of hazardous cargoes.

As a consequence of a series of technological innovations and environmental policy protocols, air pollution levels have decreased considerably throughout the past 25 years. However, the rapid growth in population and global trade, combined with the improved standards of living of less developed and developing countries, requires the constant monitoring and controlling of the environmental impact deriving from diverse energy sources.

Hazardous Air Pollutants

Hazardous air pollutants, or toxic air pollutants or air toxics, are pollutants that may cause cancer or other serious health effects, including reproductive effects, birth defects, and adverse environmental and ecological effects. Examples of toxic air pollutants include benzene, ethylbenzene, toluene, mixed xylenes, n-hexane, carbonyl sulfide, ethylene glycol, and 2,2,4-trimethylpentane, perchlorethlyene, etc. (EPA, 2014).

One of the primary pollution sources is known as "nonpoint source pollution," which takes place as a consequence of runoff. Nonpoint source pollution consists of several minor sources, such as septic tanks, transportation vehicles (trucks, ships, rail, cars) as well as large-scale activities occurring in industrial plants, forests, farms, livestock areas, and so on. Eighty percent of marine pollution derives from land-based activities that end up in the sea (EPA, 2016a; Burns, 2014; Burns, 2012).

Fluorinated and Greenhouse Gases

There are seven main categories of greenhouse gases that are released as a consequence of human actions and that are contained in global levels of greenhouse gas emissions.

Four of these greenhouse gases are considered as fluorinated gases, hence their environmental impact is more severe. As demonstrated in Table 2.3, four out of the seven most common greenhouse gases belong to the fluorinated gases category. Their chemical composition, natural availability, utility and behavior, are duly discussed in this section.

Fluorinated Gases (FGs or F-Gases)

Fluorinated gases are extremely potent greenhouse gases that are generated for a number of manufacturing processes. These are often utilized as alternatives for stratospheric ozone-burning ingredients. The European Commission's goal is to eliminate F-gas emissions by two-thirds by the year 2030 (EC, 2017).

Table 2.3 Greenhouse and fluorinated gases.

	List of greenhouse and fluorinated gases	*Greenhouse gases*	*Fluorinated gases*
1	Hydrofluorocarbons (HFCs)	X	X
2	Perfluorocarbons (PFCs)	X	X
3	Sulfur hexafluoride (SF_6)	X	X
4	Nitrogen trifluoride (NF_3)	X	X
5	Carbon dioxide (CO_2)	X	
6	Methane (CH_4)	X	
7	Nitrogen oxide; Nitrous oxide (NOx)	X	

Source: the Author.

There are four main categories of fluorinated gases, which include:

1 **Hydrofluorocarbons (HFCs) and hydrochlorofluorocarbons (HCFCs)**: These types of gases have a similar chemical composition to the CFCs (see below), and are widely used in manufacturing, refrigeration and for air conditioning uses. They are typically released in reduced volumes, and are considered as less harmful to the ozone stratum. They are used in diverse industries and technologies, for example air conditioning, heat pump or refrigeration equipment, fire extinguishers and aerosols, blowing agents for foams, solvents, and so on.

   Hydrofluorocarbons (HFCs) and hydrochlorofluorocarbons (HCFCs) have substituted the following fluorinated gases, as they combine maximum utility from an industrial and household viewpoint, while having an improved environmental performance compared with the following.

2 **Chlorofluorocarbons (CFCs) and halons, halocarbons**: Mixtures of carbon, fluorine, hydrogen and chlorine, which are vapors utilized in refrigerants and propellants in aerosol and pressurized gas products. Since the 1980s, the Montreal Protocol eliminated their use due their ozone-depleting nature. They have been gradually substituted by HFCs, which are considered as more environmentally friendly.

3 **Perfluorocarbons (PFCs)** are utilized in the electronics industries, i.e., for plasma cleaning of silicon wafers; but also in the pharmaceutical and cosmetic industries. They are also found in obsolete firefighting systems.

4 **Sulfur hexafluoride (SF_6)** is an inert gas, primarily used in medicine (respiratory physiology) as a test gas; also in ophthalmology (eye/vitreoretinal surgeries) it is injected to reinstate the vitreous chamber. It is also widely utilized in industry to generate magnesium and aluminum, in high voltage switches and also as insulating gas (NCBI, 2016a).

5 **Sulfur dioxide** is an invisible gas with an unpleasant, distinct smell. It is a colorless gas generated primarily from fossil fuel burning, especially coal and petroleum. It reacts quickly with alternative elements to shape dangerous substances, including sulfate particles, sulfuric acid, sulfurous acid, and so on.

   Approximately 99% of the sulfur dioxide in the environment is man-made as its main origin is in industrial operations, in particular the handling of sulfur-related compounds such as the creation of electricity from fossil fuels, i.e., coal, oil or gas. Certain mineral ores additionally contain sulfur, hence sulfur dioxide is discharged when they are refined or processed. Furthermore, commercial operations that require the ignition of sulfur-rich fossil fuels may be significant resources for sulfur dioxide. Transport-related fuel combustion may also generate sulfur dioxide in the atmosphere.

Sulfur dioxide is produced as part of a photochemical reaction, i.e., when light is absorbed by a compound. It also reacts with oxygen to generate sulfur trioxide, sulfuric acid and sulfates. It is considered to be one of the conventional types of pollution, being responsible for sulfurous smog in global metropolitan areas.

6 **Nitrogen trifluoride (NF_3)** is a colorless, nonflammable, odorless, gaseous compound. It is extremely toxic and imposes inhalation hazards. It is corrosive to tissue. It is widely used in generating other chemical substances, and is a compound of rocket fuels. In case of extended placement in the vicinity of fire or high temperatures, there is a threat for storage containers to break abruptly, and rocket/blast (NCBI, 2016b; NJ.Gov, 2017).

Greenhouse Gases

Several of the chemical substances in the earth's atmosphere behave as greenhouse gases. As soon as sunlight reaches the earth's surface area, a portion of it is reflected back, directed towards space in the form of heat, i.e., infrared rays. Greenhouse gases take in this infrared emission and capture its high temperature in the environment, resulting in a greenhouse phenomenon.

Four greenhouse gas types – 1) hydrofluorocarbons (HFCs); 2) perfluorocarbons (PFCs); 3) sulfur hexafluoride (SF_6); and 4) nitrogen trifluoride (NF_3) – also belong to the fluorinated gases category, and have been covered above.

CARBON DIOXIDE (CO_2)

Carbon dioxide, CO_2, is an unscented, colorless, non-combustible gas, contained in the atmosphere and shaped during inhaling and exhaling, typically obtained through coal, natural gas, coke, or through burning, from carbs through fermentation, by the acidic reaction with limestone or additional carbonates, or from springs through natural means.

Carbon dioxide is widely used in industry in fire-fighting systems, or in a frozen solid form like dry ice in carbonated refreshments, etc. It is used in soft drinks to generate the bubbles, but also used in the movies and theater in order to generate special effects, e.g., fog.

It penetrates the atmosphere by means of consuming fossil fuels (coal, oil, natural gas), solid wastes, and timber, as well as a consequence of specific chemical reactions (EPA, 2017c).

Diverse fuels discharge distinctive levels of CO_2 in relation to the energy generated when consumed. To assess pollutants throughout fuels, evaluate the quantity of CO_2 released per unit of energy production or temperature content (EIA, 2016c).

The quantity of CO_2 generated when an energy segment is consumed is an attribute of the carbon chemical composition on behalf of the fuel. Temperature is generated when carbon (C) and hydrogen (H) merge with oxygen (O) during ignition. The heat level material, or the energy quantity produced when an energy source is ingested, is mainly relying on the carbon and hydrogen components of the fuel. Natural gas mainly consists of methane (CH_4), which includes an increased energy content pertinent to other energy sources. Therefore, it has a comparatively reduced CO_2-to-energy material (EIA, 2016d).

Table 2.4 demonstrates the estimations of world energy-related carbon dioxide emissions from 1990 to 2035.

Table 2.5 shows the amount of CO_2 emissions released in the atmosphere through the most commonly used fuels.

Table 2.4 World energy-related carbon dioxide emissions, 1990, 2005, 2007 and 2035.

	1990	*2005*	*2007*	*2035*
Estimated emissions (million metric tons)	21,537	28,329	29,728	42,386
Change from 1990 (mil. metric tons)		6,793	8.191	20,849
Change from 1990 (percent)		31.50%	38%	96.80%
Average annual change from 1990 (percent)		1.80%	1.90%	1.70%
Change from 2005 to 2035 (mil. metric tons)				14,057
Change from 2005 to 2035 (percent)				49.60%
Annual average change from 2005 to 2035 (percent)				1.40%

Source: U.S. Congressional Budget Office. *The Renewable Fuel Standard: Issues for 2014 and Beyond.* June 26, 2014. Available at: www.cbo.gov/sites/default/files/cbofiles/images/pubs-images/45xxx/45477-land-figure1b.png, last accessed: January 3, 2017.

Table 2.5 Pounds of CO_2 released per million British thermal units (Btu) of energy for several fuels.

Coal (anthracite)	228.6
Coal (bituminous)	205.7
Coal (lignite)	215.4
Coal (subbituminous)	214.3
Diesel fuel and heating oil	161.3
Gasoline	157.2
Propane	139.0
Natural gas	117.0

Sources:

EIA (2016a). *EIA International Energy Outlook 2016.* U.S. Energy Information Administration.

EIA (2016b). *How Much Carbon Dioxide is Produced when Different Fuels are Burned?* U.S. Energy Information Administration. Available at: www.eia.gov/tools/faqs/faq.cfm?id=73&t=11, last accessed: January 3, 2017.

EIA (2016c). *Energy Emissions.* U.S. Energy Information Administration. Available at: www.eia.gov/environment/emissions/co2_vol_mass.cfm, last accessed: January 3, 2017.

EIA (2016d). *EIA – Greenhouse Gas Emissions Overview.* U.S. Energy Information Administration. Available at: www.eia.gov/environment/emissions/ghg_report/ghg_overview.php, last accessed: January 3, 2017.

EIA (2016e). *Energy Explained.* U.S. Energy Information Administration. Available at: www.eia.gov/energy explained/index.cfm/data/index.cfm?page=environment_home, last accessed: January 3, 2017.

EIA (2016f). *International Energy Outlook 2016, DOE/EIA-0484(2016).* U.S. Energy Information Administration. Available at: www.eia.gov/outlooks/ieo/pdf/0484(2016).pdf, last accessed: January 1, 2017.

Global CO_2 emissions that are energy-pertinent surge from 32.2 billion metric tons in 2012 to 35.6 billion metric tons in 2020, as well as to a predicted 43.2 billion metric tons in 2040, a rise of 34% above the projection timeframe. A large part of the increase in emissions is caused by developing non-OECD nations, several of which depend greatly on fossil fuels to fulfill the rapid growth of energy demand (EIA, 2016b).

METHANE (CH_4)

Methane is a chemical compound, the primary ingredient of natural gas, and widely distributed in nature. It is released throughout the production and haulage of fossil fuels including coal, oil, and natural gas. Methane emissions also derive from agricultural techniques such as livestock farming and through the decomposition of organically generated waste in municipal solid waste landfills.

Colorless, unscented, combustible hydrocarbon gas that is the primary ingredient, i.e., approximately 75% to 99%, of natural gas. It is non-toxic but operates as an asphyxiant and is one of the leading greenhouses gases. Methane is transformed into methanol by catalytic oxidation.

NITROGEN OXIDE (NOX)

Nitrous oxide is a non-flammable gas used as an oxidizer in rocket propellants, and in surgery and dentistry as anesthesia and as a bacteriostatic (kills bacteria). It is used for fast speed engines in transportation, especially rallies. It is released in the course of farming and manufacturing activities, in addition to fossil fuels and solid waste combustion.

These are man-made emissions developed through combustion of vehicle exhausts or smog and include nitric oxide, nitrogen dioxide, nitric acid, nitrous oxide, nitrates, etc.

When non-renewable fuels are consumed, they discharge nitrogen oxides into the environment, thus giving rise to the generation of air pollution, smog, acid rain and so on. The most frequent nitrogen-related substances released into the atmosphere by man-made activities are jointly known as nitrogen oxides.

The majority of the nitrogen oxides launched in the U.S. due to man-made activity are generated by the consumption of fossil fuels connected with logistics and industrial activities. These are discharged into the atmosphere from motorized vehicle exhausts or the internal combustion of engines when using fossil fuels, i.e., coal, oil, natural gas and diesel fuel, as well as in electrical power plants.

They are furthermore discharged during manufacturing or industrial operations including dynamite blasting, welding, electroplating, engraving, and so on.

The natural compounds NO and NO_2 are free radicals, i.e., have unpaired valence electrons making them fairly reactive. Free radicals act as magnetic dipole hence react with magnetic fields.

Some of the major types of nitrogen oxide sources of pollution include:

- land transportation (cars and trucks);
- coal-burning power plants;
- sizeable manufacturing and commercial activities;
- land and air transportation.

The existence of surplus nitrogen in the environment as nitrogen oxides or ammonia is transferred back to terrain, whereby it flushes into the vicinity of sea or rivers. These surplus nutrients give rise to smog and oxygen-deficient aquatic zones.

NITROGEN

Nitrogen is a vital element to flora and fauna (plants and animals), found plentiful in the atmosphere. Nitrogen is generated from man-made activities, including electricity generation, manufacturing, farming and transportation, and can disrupt the natural equilibrium of nitrogen in the atmosphere.

AMMONIA

Ammonia is yet another nitrogen compound released to the atmosphere, mainly from farming activities, and secondarily from non-renewables (fossil fuels). Excessive ammonia and reduced pH impact the environment through toxicity.

OZONE

Ozone is formally a greenhouse gas given it has an impact on international heat levels. It is at increased environmental quantities in the stratosphere, i.e., the second significant layer of Earth's ambiance, which contains approximately 20% of the atmosphere's volume. In cases where ozone is generated naturally, ozone obstructs ultraviolet (UV) light from attaining the earth's layer, which is damaging to flora and fauna. Authorities all across the globe prohibit and manage the generation of numerous industrial gases that damage atmospheric ozone and generate a gap in the ozone layer. At reduced levels of the troposphere, ozone is hazardous to human health. There are global rules to eliminate the development of this ground level ozone (EIA, 2016e; Australian Government, 2005).

Ground-level ozone (smog): Tropospheric or ground-level ozone is not emitted directly into the air, but is created by chemical reactions between oxides of nitrogen (NOx) and volatile organic compounds (VOC).

Ground-level ozone is a colorless, largely irritating gas generated over the earth's surface area. It is known as a "secondary" pollutant since it is generated when two principal pollutants, i.e., nitrogen oxides and volatile organic compounds, respond in natural light and oxygen. These two pollutants derive from man-made activities. Approximately 95% of man-made NOx emissions derive from the combustion of coal, oil, and gasoline in automobiles, households, manufacturing and energy plants. Volatile organic compounds from man-made emissions derive primarily from gasoline burning, upstream fossil fuel production, household timber combustion, as well as from the evaporation of fluid fuels and substances. Considerable amounts of VOCs furthermore derive from biogenic resources, for instance pine trees (Pinophytous, coniferous forests).

Although it is not released straight into the air, ozone is generated by chemical reactions (the re-structuring of a compound's molecular or ionic composition) among nitrogen oxides and volatile organic compounds in the existence of natural daylight. Nitrogen oxides and VOC pollutants are mainly generated by manufacturing facilities, power plants, electricity and vehicle emissions, gasoline fumes, and chemical substances (EPA, 2015).

Suspended Particulate Matter (SPM)

Suspended particulate matter encompasses a mixture of finely divided solid or liquid compounds disseminated into the environment by processes at the earth's surface. The SPM chemical configuration consists of carbon, higher hydrocarbons formed by incomplete combustion of hydrocarbon fuels, and sulfur (OECD, 1997).

Particulate matter is generated by industrial activities, traffic and internal combustion processes.

Suspended particulate matter is a collective name for fine solid or liquid particles added to the atmosphere by processes at the earth's surface. Particulate matter includes dust, smoke, soot, pollen and soil particles (EEA, 2017).

Table 2.6 demonstrates the Air Quality Index for particle pollution.

Table 2.6 Air Quality Index for particle pollution.

Air Quality Index	*Air quality*	*Health advisory*
0 to 50	Good	None
51 to 100	Moderate	Unusually sensitive people should consider reducing prolonged or heavy exertion
101 to 150	Unhealthy for sensitive groups	People with heart or lung disease, older adults, and children should reduce prolonged or heavy exertion
151 to 200	Unhealthy	People with heart or lung disease, older adults and children should avoid prolonged or heavy exertion. Everyone else should reduce prolonged or heavy exertion
201 to 300	Very unhealthy	People with heart or lung disease, older adults and children should avoid all physical activity outdoors. Everyone else should avoid prolonged or heavy exertion

Source: EPA (2015). *Air Quality Guide for Particle Pollution.* August 2015, EPA-456/F-15-005. Available at: www3.epa.gov/airnow/air-quality-guide_pm_2015.pdf, last accessed: December 14, 2016.

Volatile Organic Compounds in Consumer and Commercial Products

The use of non-renewables contributes to the emission of volatile organic compounds (VOCs) in two ways: a) directly as energy products such as gasoline, or b) indirectly as by-products, such as vehicle gas emissions.

Volatile organic compound (VOC) pollutants bring about the development of ground-level ozone and fine particulate matter (PM2.5), which generate smog (Canada Government, 2017). There are various other greenhouse gases that are not included in U.S. or global greenhouse gas stocks.

Table 2.7 demonstrates the emission limits for existing and new power plants in selected countries/regions (mg/m3).

Non-Anthropogenic Oil and Gas Pollution: Natural Marine Seepage

There are four main classifications of energy pollution; the first type of pollution occurring naturally (non-anthropogenic), whereas the other three being man-made (anthropogenic):

1. Natural marine seeps pertain to non-anthropogenic environmental pollution sources that signify more than 60% of petroleum released within the aquatic environment in North America. As a result of progressive and decreased seepage volumes of petroleum into the water, the environment manages to recuperate at a swift rate, with the least ecological deterioration.
2. Petroleum and gas exploration and production comes down to 3% of all the man-made or 2% of overall pollution of fossil fuels discharged into the North American shoreline environment.
3. Refining, transport, storage and supply chain distribution extends to 9% of all the man-made and 4% of all cumulative pollution sources; and
4. Ingestion by public and private transportation companies, i.e., mainly non-energy industries, such as terrestrial, public and commercial waste bring about 85% of all the man-made or 33% of all oil inputs.

(BOEM, 2009)

Table 2.7 Emission limits for existing and new power plants in selected countries/regions (mg/m3).

Region	*Policy*	*SO2*		*NOx*		*PM*	
		Existing	*New*	*Existing*	*New*	*Existing*	*New*
China	Emission standard of air pollutants for thermal power plants	200–400	100	200	100	30	30
European Union	Industrial Emissions Directive	200–400	150–400	200–450	150–400	20–30	10–20
USA	New Source Performance Standards	160–640	160	117–640	117	23	23
India	Environment (Protection) Amendment Rules, 2015	200–600	100	300–600	100	50–100	30
Indonesia	MOE decree no. 21, 2008	750	750	850	750	150	100
Japan**	Air Pollution Control Law	NA	NA	123–513	123–513	30–100	30–100
Mexico***	Mexican Official Standard NOM-085-ECOL-1994 (in PPMV for SO2 and Nox	550–2200	30–2200	110–375	25–375	60–450	60–450
Philippines	National Emission Standards for Particulate Matter for Stationary Sources	1,000–1,500	200–700	1,000–1,500	500–1,000	150–200	150–200
South Africa	The Minimum Emissions Standards published by the Government	3,500	500	1,100	750	100	50
Korea	Special Measures for Metropolitan Air Quality Improvement	286	229	308	164	40	20–30
Thailand	Royal Thai Government Gazette	700–1,300	180–360	400	200	80–320	80
Vietnam	Industrial Emission standards for dust and inorganic substances	1,500	500	1,000	650–1,000	400	200

Sources:

IEA (2016). *World Energy Outlook Special Report (2016), Energy and Air Pollution*. International Energy Agency. Available at: www.iea.org/publications/freepublications/publication/WorldEnergyOutlookSpecialReport2016EnergyandAirPollution.pdf, last accessed: January 2, 2017.

IEA (2015). *Energy and Climate Change: World Energy Outlook Special Report*. International Energy Agency, OECD/IEA, Paris, France. Available at: www.iea.org/publications/freepublications/publication/WEO2015SpecialReportonEnergyandClimateChange.pdf, last accessed: January 2, 2017.

Notes: “Existing” refers to the emission limit for currently operating power plants. “New” refers to the limit for planned or proposed plants.

*U.S. emission limits were converted from lb/MBtu to mg/m3 assuming an F-factor of 1 800 standard cubic feet of CO_2 and a CO_2 content of 12% in the flue gas.

**Japan’s Air Pollution Control Law (APCL) specifies emission limits for SOX, NOX and PM that differ depending on the scale of facilities, technologies and regions. Prior to the construction of new plants in Japan, local authorities and power generation companies usually arrive at bilateral pollution prevention agreements more stringent than those mandated in the APCL.

***For Mexico, SO_2 and NOX are expressed in parts per million by volume (PPMV).

While the sources of man-made environmental contamination acquire substantial exposure, very little is published on the naturally leaking oil and gas that the oceans and shorelines have been exposed to and ingesting for 400 million years (NRC, 2012).

A scientific investigation carried out by NASA and the Smithsonian Institution determined that the contamination from natural seepage from marine oil wells add up to 62 million gallons per annum, therefore outnumbering contamination from conventional, unconventional, sea and land drilling combined, which amount to 15 million gallons a year. These significant conclusions concur with an earlier investigation from the National Research Council (NRC, 2012, 2003; NASA, 2016).

Environmental Regulations for the Energy Industry

In this book chapter we duly examined the different types of chemical compounds found in the energy industry, and what became apparent is that global pollution is the outcome of collective activities from households, public utilities and industries alike.

Global organizations, governments and the energy industry continuously aim to monitor, control and eliminate pollutants by employing several methods and initiatives, as shown in Table 2.8.

What is evident in Figures 2.3, 2.4, 2.5 and Tables 2.9 and 2.10 is that new technologies, regulations and processes eliminate environmental threats while improving energy efficiency. Furthermore, a lot of progress has been observed, as governments, global agencies and the industry adopt these systems and comply with regulations that advance the safe, secure and greener industry.

Chapters 4 to 10 of this book further verify these assumptions, as they cover the regulations for each energy sector in both renewable and non-renewable energy sources.

Table 2.8 Initiatives to improve energy efficiency and environmental integrity.

	Methods & initiatives	*Global organizations*	*Governments*	*Energy & manufacturing Industries*
1	Monitoring & controlling of energy impact upon the environment	X	X	X
2	Ratification of new regulations pertinent to emission limits of pollutants	X	X	X
3	Compliance of new regulations pertinent to emission limits of pollutants	X	X	X
4	Research & investment in green technologies to limit emissions	X	X	X
5	Research & investment in new, greener energy grades		X	X
6	Establishing environmental & health impact assessments	X	X	X
7	Promoting public awareness & data sharing	X	X	X

Source: the Author.

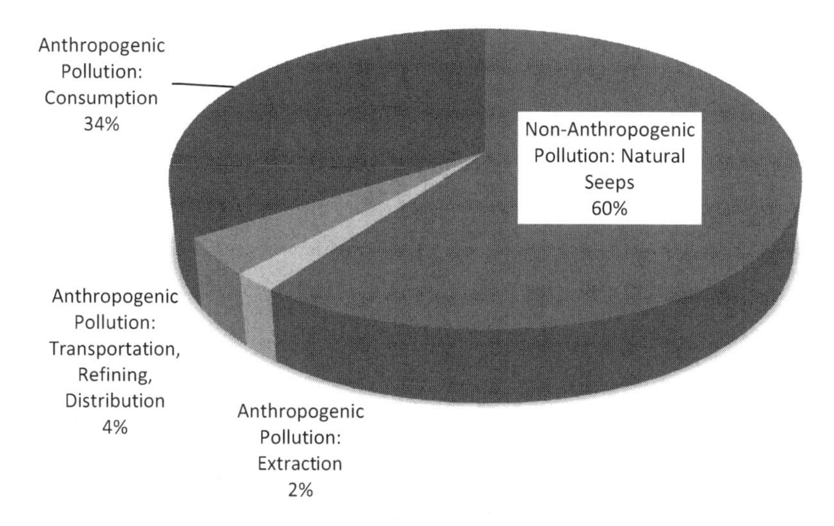

Figure 2.3 North America, marine pollution from oil spills.

Source: compiled by the Author, based on data in BOEM (2009), *Oil in the Sea 2003*. Bureau of Ocean Energy Management, and National Research Council. Available at: www.boem.gov, last accessed: December 12, 2016.

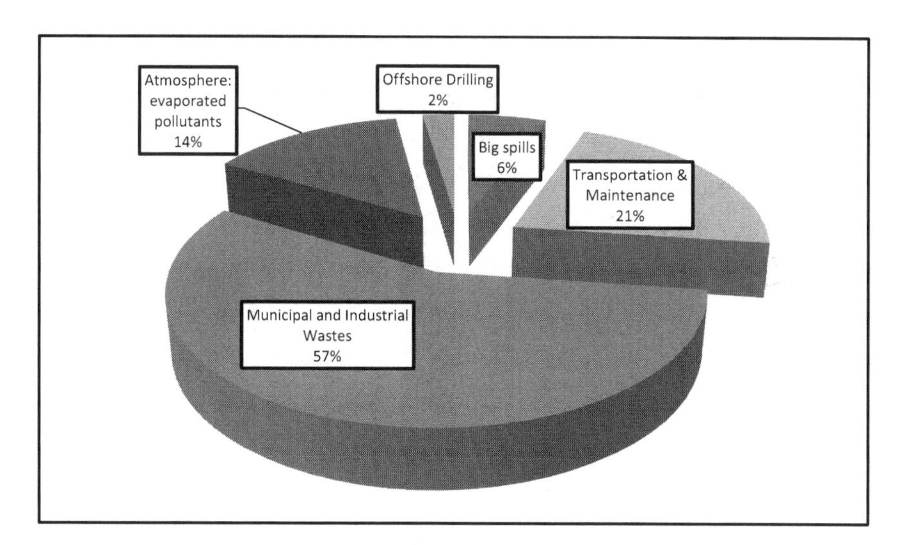

Figure 2.4 Sources of pollution in the marine environment (1985).

The Energy Environmental Paradox

Our modern economic driving forces are characterized by irrational market cycles with violent fluctuations. Each chapter of this book underlines the significance of the energy industry, and appreciates the industry's contribution to civilization. The energy industry is a substantial economic driver as it promotes opportunities for global development, market equilibrium and advancement.

Table 2.9 Sources of pollution in the marine environment (1985).

Global pollution source	*Million gallons*
Big spills	37
Routine maintenance	137
Municipal and industrial wastes	363
Atmosphere: evaporated pollutants	92
Offshore drilling	15
Million gallons of pollutants per year	**644**

Sources:
National Research Council (2003). *Oil in the Sea*. Washington, DC: National Academy Press.
U.S. Coast Guard (1990d). *Update of Inputs of Petroleum Hydrocarbons into the Oceans due to Marine Transportation Activities*. Washington, DC: National Research Council. National Academy Press.

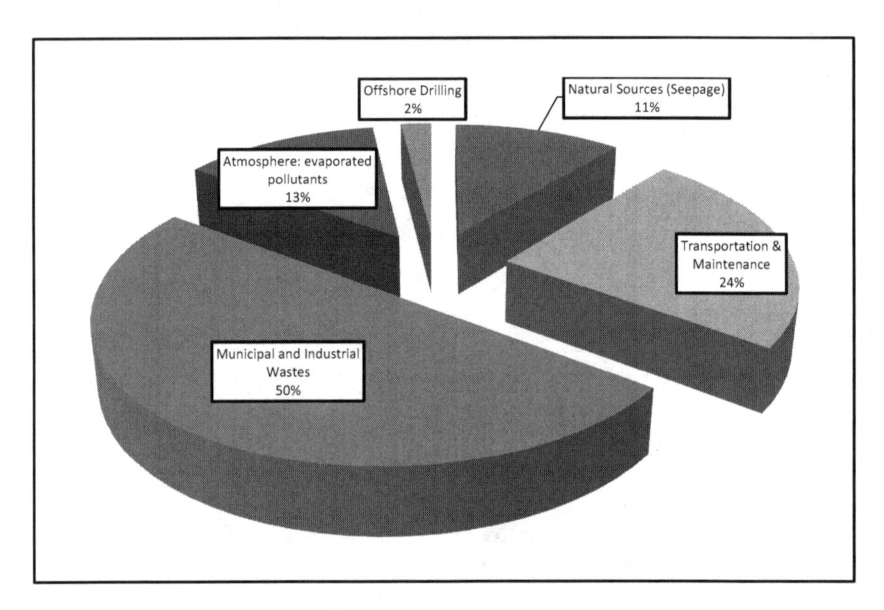

Figure 2.5 Sources of pollution in the marine environment (1993).

Sources:

UN (1993). *Impact of Oil and Related Chemicals and Wastes on the Marine Environment*, GESAMP Report No. 50. London: IMO, Joint Group of Experts on the Scientific Aspects of Marine Pollution (GESAMP). Available at: www.un.org/earthwatch/about/docs/scpGESAMP.htm, last accessed: January 2, 2017.

And while the industry generates millions of jobs, government taxation, innovative research and education opportunities for the underprivileged, there is an ongoing debate as to whether the economic benefits are more or less significant than the environmental impact.

Incidents like the Deepwater Horizon disaster in the Gulf of Mexico seem to polarize mass media and the society as a whole. On one hand, the proponents of fossil fuel exploration purport that the energy sector greatly sustains the nation's social and financial stability. On the other hand, environmental proponents consider that the economic benefits are overshadowed by environmental considerations and prospective health impact. Admittedly, scholarly efforts to verify the environmental impact of energy sources

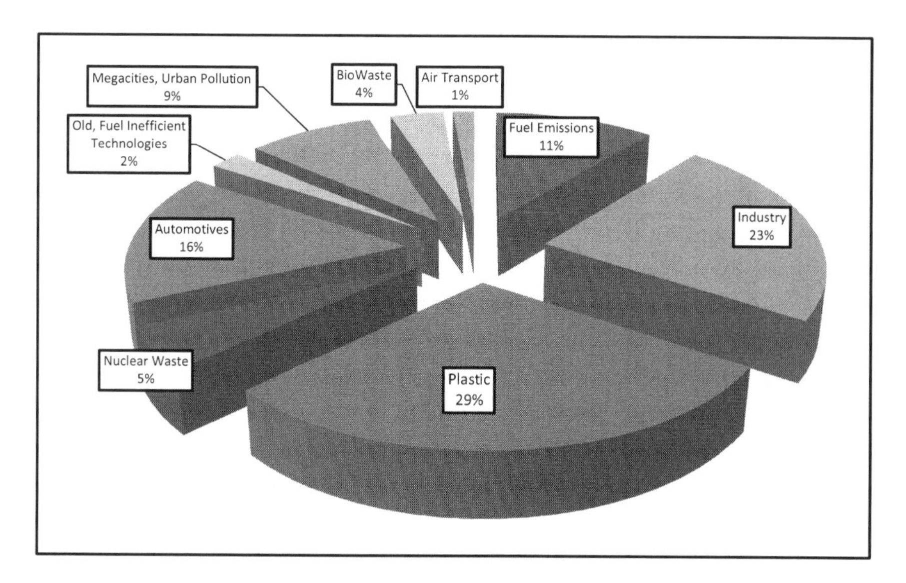

Figure 2.6 Global solution sources (2014).

Source: IEA (2016). *World Energy Outlook Special Report (2016) Energy and Air Pollution*. International Energy Agency. Available at: www.iea.org/publications/freepublications/publication/WorldEnergyOutlookSpecialReport2016EnergyandAirPollution.pdf, last accessed: January 2, 2017.

Table 2.10 Selected primary air pollutants and their sources (2015).

Pollutants	*Sources*	
Sulfur dioxide	99%	Combustion of coal, oil, gas, bioenergy, waste
Nitrogen oxides	99%	Fuel and process emissions
Particulate matter 2.5	85%	Transportation emissions and evaporation
Carbon monoxide	92%	Municipal and building utilities: lighting, cooking, heating
Volatile organic compounds	66%	Energy extraction, storage, transport and refineries

Source: IEA (2016). *World Energy Outlook Special Report, Energy and Air Pollution*. International Energy Agency. Available at: www.iea.org/publications/freepublications/publication/WorldEnergyOutlookSpecialReport2016EnergyandAirPollution.pdf, last accessed: January 2, 2017.

generates more questions than answers. Energy and transportation professionals who have served the industry for several decades find it common to encounter companies with "strong overall safety records," with significant investments in innovative, green, fuel-efficient technologies. In the case of Deepwater Horizon, upon examining the legal case and official reports, BP's record was so exemplary, as confirmed by the Officials of Mineral Management Service, that the rig was never targeted or included in the inspectors' "watch list" for problem rigs. (BSEE, 2012; BOEM, 2012). Nonetheless, this is an instance where an incident could almost swipe over a century of exemplary performance, compliance, innovation, financial investment, taxation, generation of jobs, etc. Irrespective of the outstanding contribution to the economy and the society, the overall penalty and fines paid by BP have exceeded $21 billion. For an industry that is the backbone of global economy, the impact upon a company's brand can be most severe.

Carbon: The Cornerstone of Life

Carbon is the cornerstone of life on Earth and the Universe as a whole. It is an organic compound that is inherent in any living organism, flora, fauna or microbe on earth.

Just like any other natural compound, carbon is a primary ingredient in the atomic structure, i.e., it has the capability to form countless natural compounds that are essential for the creation of all living organisms. And while carbon, oxygen and nitrogen are the three most vital constituents for organic life, growth and reproduction, carbon is the most impactful compound due to its unique capability to form distinct chemical bonds with other elements, with enduring footprint. Consequently, carbon is the source of all fossil fuels, and represents the prevailing energy forms that are used throughout the history of humanity, for survival, nutrition, construction, transportation, manufacturing. Most important, carbon (in particular carbon fiber) is used in nanotechnology, space travel, and deep space exploration (Riebeek, 2011; Burns, 2016b).

Earth's Balancing Act and the Greenhouse Gases Hypothesis

Global climate fluctuations are the outcome of a balance between energy penetrating and exiting the earth's system. Global energy balance can be affected by a number of different factors, such as:

- fluctuations in solar energy absorbed by the earth's atmosphere and surface;
- fluctuations in solar energy reflected or deflected by the earth;
- alterations in the greenhouse phenomenon, which impacts the level of heat absorbed by Earth's atmosphere and subsequently retained (EPA, 2017c).

Hence, warmer temperatures occur when solar energy is assimilated in the Earth's atmosphere. Conversely, Earth's temperature drops when solar energy is deflected from the earth's surface towards the outer space.

Global Warming and the Environmental Footprint of Large-Scale Energy Sources

Official scientific findings and government bodies ascertain that "Warmer temperatures occur when solar energy is assimilated and retained in the Earth's atmosphere" (EPA, 2017c).

And while numerous research projects have focused on fossil fuels, due to their leading role in energy consumption, the assimilation of solar energy does not only occur with fossil fuels, but will also occur with any energy source that captures solar energy or results from photosynthetic processes.

Carbon-based energy segments have been the prevailing energy source for more than two centuries, hence their environmental impact has been under the spotlight. One of the benefits of researching a widely used energy source, such as fossil fuels, is the scientists' ability to accurately interpret the findings, since they are based on a large-scale sample. During the past decades, both governments and the fossil fuel industry have funded numerous research projects on a variety of scientific testbeds to verify the environmental conditions, processes and footprint of carbon-based energy.

On the other hand, renewable energy sources still represent a small portion of the global energy consumption (less than 10% globally), and very little research has been done in order to thoroughly examine their environmental impact.

A significant research gap lays on the fact that most renewable energy production units are small to medium scale, hence existing research projects can only reflect findings based on the small to medium impact of such installations, without taking into consideration the exponential growth of environmental pollution, which comes together with the large-scale use of any and all energy segments, both renewables and non-renewables.

As a number of global initiatives promote the growth of renewable energy sources, there is a compelling need to continuously monitor the exponential environmental impact as a result of their growing capacity.

Greenhouse Gases and the Environmental Footprint of Water Vapor (H_2O), Methane (CH_4) and Carbon Dioxide (CO_2)

The greenhouse effect pertains to several greenhouse gases (GHG) generated from water vapor (H_2O), methane (CH_4) and carbon dioxide (CO_2) that have the capacity to absorb energy. Again, the impact of fossil fuels, i.e., carbon dioxide and methane, has been duly studied, and the use of fossil fuels has almost been demonized, despite its immense contribution to human evolution, survival, security and wellbeing. At the same time, little has been said about water vapor, which is the most vital component of greenhouse gases representing over 60% of the global warming effect.

The Impact of Water Vapor (H_2O), on Global Warming, Greenhouse Gases and Sea Level Rising

Global warming is caused by the balance between water vapor and the other greenhouse gases, hence as higher volumes of carbon dioxide are released into the environment, higher volumes of water vapor are also generated (NASA, 2008). And while atmospheric humidity triggers the effect of carbon dioxide, this process is not exclusively the outcome of fossil-fuel processing.

As a number of government policies aim to reduce environmental pollution such as greenhouse gases and to promote alternative energy, there is a need for extensive research to verify the impact of water vapor (H_2O) on the environment, and to prove, or disprove, the aforementioned assumptions.

As a subsequent step, scientists will need to identify the technological advancements needed in order to mitigate the elevated temperatures and greenhouse gases deriving from large-scale installation of alternative energy sources.

Reduction of Air Pollutants in Europe

According to the European Union Emission Inventory Report (EEA, 1990–2014), for the past 25 years (1990 to 2014) the EU-28 documented the elimination of emissions of air pollutants. First, the most significant reduction was encountered in sulfur oxides (SOx) which dropped by approximately 90%. The second most significant reduction entails the non-methane volatile organic compounds (NMVOC) which decreased by about 60%. Consequently, pollution levels of nitrous oxides (NOx) were almost reduced by 50% and pollution levels of fine particulate matter (PM2.5) dropped by almost 33%. The least reduction was documented for ammonia (NH_3), pollution levels of which dropped by approximately 25% (Eurostat, 2017; EEA, 1990–2014).

Case Study: The Kyoto Protocol

United Nations Framework Convention on Climate Change

The Kyoto Protocol is a global agreement connected to the United Nations Framework Convention on Climate Change, which is committed to the establishing of globally binding emission decrease goals. It was implemented in Kyoto, Japan, on December 11, 1997 and enforced on February 16, 2005. The comprehensive rules for the execution of the Protocol were implemented at COP 7 in Marrakesh, Morocco, in 2001, and are known as the "Marrakesh Accords."

Acknowledging that developed economies are primarily accountable for the existing high amounts of GHG polluting gases in the environment on account of over 150 years of manufacturing activity, the Protocol imposes a weightier pressure on developed nations as part of the principle of "common but differentiated responsibilities."

The Kyoto Protocol closely monitors specific types of air pollutants known as fluorinated greenhouse gases. These, as the hydrofluorocarbons (HFCs), perfluorocarbons (PFCs) and sulfur hexafluoride (SF6) are collectively known, are further controlled by specific EU legislation (UNFCCC, 2017).

Health Threat Types

Health Impact Assessment (HIA)

Health impact assessment (HIA) is a tool that can help communities, decision makers, and practitioners make choices that improve public health through community design.

Health impact assessment is a process that helps evaluate the potential health effects of a plan, project, or policy before it is built or implemented. An HIA project:

- offers tangible guidelines to optimize health benefits, and eliminate health risks or harmful impacts;
- decides the possible outcomes pertinent to the health status of a selected demographic segment, and analyzes the patterns of specific health influences inside the population;
- takes into account feedback from decision makers, entities affected by the decision, or stakeholders;
- employs various kinds of evidence and diagnostic methods;

- has the flexibility to work within budget and time frames;
- offers proof and suggestions for improvement to decision-makers within a reasonable time frame, i.e., at a stage where decisions can still make an impact (EPA, 2017d).

The HIA assessment comprises of the following major tasks:

- screening, i.e., determining strategy, task, or policy conclusions where an HIA could be applicable;
- scoping, i.e., preparing the HIA and distinguishing project-specific health impacts that should be included;
- assessing, i.e., ascertaining impacted population segments based on geospatial and demographic data, and evaluating the pertinent health impacts;
- recommendations, i.e., advising useful and realistic measures to enhance health benefits and eliminate health risks;
- reporting, i.e., delivering outcomes to leaders, stakeholders, impacted communities, and other interested parties;
- monitoring and evaluation, i.e., ascertaining the HIA's effect on the conclusions and health condition.

(CDC, 2017)

Health impact assessment (HIA) offers a solution, or improved corporate paradigm. As opposed to risk assessment that focuses on people and events through a process of risk identification, prevention and elimination, HIA focuses on policy improvement or alternative solutions and strategic processes. In addition, it regards the ecosystem as a natural capital with great benefits and rewards, and not merely as a risk-prone environment (UNEP, 2016).

Integrated Environmental and Health Impact Assessment

- As the energy industry evolves with innovative technologies and pioneering methods, the pertinent risks become more intricate, and a harmonized approach is needed to assess the environmental and health impacts.
- Impact assessment methodologies are also comparative processes that incorporate the examination and contrasting of diverse possible risk-based strategies.
- Stakeholders' participation is required for the environmental and health-related risks.
- The interrelation between health and environmental impact should also be examined.

Identifying Sources of Pollution

Upon examining global statistics on the world's most polluted nations, the world's most health-afflicted regions and the world's most energy-intensive economies, it was observed that the process of developing comparative analyses and drawing conclusions is extremely complex.

1 A wide variety of **"environmental pollution measuring instruments"** are available that detect, measure and analyze sea, land and air pollution. However, when examining environmental pollution indices in a particular region, it becomes difficult to accurately pinpoint the true cause of pollution.

As an example, the world's most populated nations such as China, India, Russia, the U.S. and so on, are involved in heavy manufacturing, complex transportation networks, and consume large volumes of energy, chemicals, cosmetics, and household substances. Thus, an environmental impact assessment should include statistical analyses such as multi-linear regression models, encompassing all parameters that may contribute to pollution.

2 **Dynamic environmental assessment**: Using less, cleaner fuels for high-performance technologies. Environmental assessment studies over the past decades or so included a series of measurements and recommendations on how to eliminate pollution.

Nevertheless, innovative technologies and the capacity of fuel-efficient, emission-efficient engineering allow a more dynamic intervention, where scientists and energy industries can eliminate environmental pollution in an imminent, highly effective manner.

Health impact functions either as a result of individual subjection to ecological hazards or through harnessing of nature's energy resources and benefits. These two options are equally conciliated through actions and beliefs, determined by social, cultural or professional morals and ideals.

Table 2.11 demonstrates the greenhouse and fluorinated gases whose exposure presents a health impact.

Sulfur dioxide impacts human health when inhaled or comes into contact with the throat and eyes. It may induce dizziness, nausea, headaches, breathing problems, wheezing and coughing, a suffocating feeling, or chest pain and pressure. The side effects of sulfur dioxide are experienced rapidly, with a symptom peak around 10 to 15 minutes after inhalation (EIA, 2016e).

Volatile organic compounds (VOCs) consist of several carbon substances, eliminating carbon monoxide, carbon dioxide, metallic carbides or carbonates, carbonic acid and ammonium carbonate, which take part in atmospheric photochemical reactions (40 CFR 51.100, 2009) and play a role in the creation of smog, i.e., ground-level ozone. They are identified as organic as their molecular components include the element of carbon. Volatile organic compounds are frequently toxic, and as they evaporate they may interact with other contaminants to develop ground-level ozone, i.e., smog. Typical VOCs encompass benzene, formaldehyde, 1,3-butadiene and toluene.

Particulate Matter (PM) or Particle Pollution

Particulate matter represents a mixture of liquid and solid pollutants, such as airborne smog, dirt and dust, plant pollen, and liquid compounds like diesel exhausts. It is considered as a high-risk compound as, due to its small size and dual (liquid/solid) nature, it can easily penetrate the human body.

Moreover, the mixture of different dangerous particles results in a more dangerous cocktail of chemicals. Once it penetrates the body's immune system it permeates lung and cardiopulmonary tissue, causing lung and heart irritation, and it interferes with cellular processes and leads to chronic diseases or even death (EPA, 2017b).

Fine particulate matter (PM2.5) found in smoke and exhaust fumes is connected with more serious health impacts compared with coarse particulate matters (PM10) such as dust (Alaska Government, 2017).

Nitrogen oxide exposure is slightly more mild, yet it causes respiration system damage such as difficulty in inhalation, throat muscle spasms, and liquid build-up in the respiratory system. It can obstruct the blood's capacity to carry oxygen via the human body, resulting in headache, exhaustion, or vertigo (NLM, NIH, 2017).

Table 2.11 Health impact of greenhouse and fluorinated gases.

	List of Gaseous Compounds: Greenhouse Gases "G" and Fluorinated Gases "F"	*Type: ("F", "G")*	*Effect on Human Health *Low concentrations are harmless or may cause mild side effects. Acute or prolonged contact may generate more severe impact.*
1	Hydrofluorocarbons (HFCs)	G, F	Widespread use in household and industrial products, especially cleaners, sprays and glues. Due to its toxic nature, acute or prolonged exposure to some hydrofluorocarbons may cause brain or heart damage.
2	Perfluorocarbons (PFCs)	G, F	Widespread use in household and industrial products, but also found in water and food. • Impact on brain and heart. • May affect the developing fetus and child, including possible changes in growth, cognition, learning, and behavior patterns. • May impact fertility and reproductive hormones. • Cholesterol surge. • Immune system and cancer instances. • Further research needed to evaluate mixtures of PFAS. • Long atmospheric lifespan due to chemical stability.
3	Sulfur hexafluoride (SF6)	G, F	• Impact on brain and heart. • Long atmospheric lifespan due to chemical stability.

(continued)

Table 2.11 (continued)

	List of Gaseous Compounds: Greenhouse Gases "G" and Fluorinated Gases "F"	*Type: ("F", "G")*	*Effect on Human Health* **Low concentrations are harmless or may cause mild side effects. Acute or prolonged contact may generate more severe impact.*
4	Nitrogen trifluoride (NF3)	G, F	• Inhalation and lung problems. • Eyes and skin irritation. • Inability of the blood to move oxygen may cause dizziness, fatigue, headaches. • Seizures. • Damage to liver and kidneys. • Fluorosis, i.e., impact on bones and teeth. • Acute or large scale exposure may cause collapse, or even death.
5	Carbon dioxide (CO_2)	G	• Brain and suffocation threat. • Inhalation and lung problems. • Rapid heart rate. • Cardiopulmonary problems. • Eyes and skin irritation; frostbites.
6	Methane (CH_4)	G	• Brain and suffocation threat. • Explosion, blasting hazard.
7	Nitrogen oxide; nitrous oxide (NOx) Ready: Industrial exposure to nitrogen dioxide may cause genetic mutations, damage a developing fetus, decrease fertility in women. In the case of acute exposure it may lead to death.	G	• Greenhouse gas, atmosphere impact, i.e., air quality. • Inhalation and lung problems. • Cardiopulmonary problems.

Source: the Author.

Hazardous Air Pollutants and the Three Types of Toxicity

Hazardous air pollutants, or toxic air pollutants or air toxics, are pollutants that may cause cancer or other serious health effects, including reproductive effects, birth defects, and adverse environmental and ecological effects. Examples of toxic air pollutants include benzene, ethylbenzene, toluene, mixed xylenes, *n*-hexane, carbonyl sulfide, ethylene glycol, and 2,2,4-trimethylpentane, perchlorethlyene, etc. (EPA 2014).

Toxicity is classified in three main categories:

1. **Acute toxicity** pertains to an organism's exposure to an intense, severe exposure to toxic substances.
2. **Sub-chronic toxicity** pertains to the impact of a toxic compound for over a year yet less than a life span.
3. **Chronic toxicity** pertains to the prolonged or gradual deterioration of tissue and/or organs combined, caused by the chronic (long-term) subjection to a chemical compound (NOAA, 2017).

Ozone

Health hazards may derive from inhalation or optical contact, as ozone is a significant part of photochemical pollutants. Inhaling ozone can induce respiratory diseases including symptoms of asthma, inhalation problems, ocular, nasal and esophagus discomfort, especially in sensitive age groups such as the elderly, infants and pregnant women. Ground-level ozone may also damage vulnerable plant life and ecosystems.

Methane

Methane's life span in the environment is roughly 12 years, i.e., shorter than carbon dioxide (CO_2), yet it can easily trap higher radiation levels compared with CO_2 (EPA 2015; EPA, 2010; U.S. DoS, 2007).

Disciplines pertaining to environmental health research examine the wellbeing and ecological impact of contact with environmental pollutants, including unprocessed energy sources, chemical substances and natural pollutants. When energy security also includes the intentional damage through terrorism or sabotage, identical health and environmental impacts are encountered.

Case Study – U.S. Office of Health and Safety

The U.S. Office of Health and Safety secures employee health and safety criteria and objectives for the department to ascertain safety of workers from the risks connected with occupational functions. The office carries out health research to ascertain employee and general public health impact from contact with dangerous goods (hazmat) and facilitates global health research and initiatives.

Furthermore, it deploys medical monitoring and testing packages for existing and previous employees, while also advocating for the Department of Labor in the execution of the Energy Employees Occupational Illness Compensation Program Act (EEOICPA). Finally, the office offers help to the headquarters and industry components in execution of policy and handling employee health and safety practices (Energy.gov, 2017a).

Traditional methods of risk assessment have provided good service in support of policy, mainly in relation to standard setting and regulation of hazardous chemicals or practices.

Health Recommendations: Clean Energy Strategies for Governments, Industries and Communities

There are several methods for governments, industries and communities to eliminate pollution, such as:

- establishing an energy conservation strategy;
- investing in energy-efficiency technologies;
- monitoring, controlling and eliminating pollutants;
- establishing long-term goals to minimize pollution, i.e., developing annual reports on energy consumption, fuel efficiency measures and estimated greenhouse emissions;
- implementing an energy management plan that enables businesses to handle energy with comparable expertise utilized to handle other corporate performance areas;
- exploring contingency planning, such as fuel-efficient technologies, dual-fuel technologies or a combination of renewable and non-renewable energy (EPA, 2016a).

Energy Security: Government Agencies and Regulations

This section demonstrates selected regulatory initiatives for all energy sources, both renewables and non-renewables, as established by the United Nations, and major global economies.

1 **United Nations (UN)**
The United Nations has established a set of rules and protocols to support a strategy for global energy from diverse viewpoints, including but not limited to: Security and safety, but also commercial, technological and geopolitical issues. It also embraces energy security from a conservation and environmental perspective (UN, 2017).
2 **ITER Fusion (35 Nations)**
Fusion and its nuclear reaction can be a prospective supply of safe, green and unlimited power. ITER, which in Latin means “the way,” is an ambitious energy program in collaboration with 35 nations. Country members of ITER include the U.S., China, Russia, India, the European Atomic Energy Community (Euratom), Japan, and South Korea. Their goal is to utilize fusion’s energy, and employ today’s research that will generate tomorrow’s fusion energy plants. Most important, they work to establish the globe’s greatest magnetic fusion device: “Tokamak,” or “toroidal chamber with magnetic coils” is an experimental device that will be designed to generate fusion energy. It aims to capture atomic fusion in the form of elevated temperature in the bulwarks of the superstructure. If the ITER scientific team succeeds, Tokamak will manage to power our planet, just as the universe powers our solar system (ITER, 2017).

Top Energy Key Players (in Alphabetical Order)

1 **Australia**
Australia’s Department of the Environment and Energy undertakes accountability for energy and environmental policies. The Administrative Arrangement Order of 2016 stipulates the new environmental framework.

2 **Brazil**

In Brazil, the Ministry of Mines and Energy (MME) (Ministério de Minas e Energia) is a federal department founded in 1960. It promotes investment strategies in energy and mining operations, invests in research and generates federal policies. Brazil's MME consists of two divisions:

a energy (encompassing the electricity and petroleum industry divisions); and
b mining and metallurgy.

The Ministry of Mines and Energy is in charge of petroleum policies (ANP) and electric power policies (ANEEL). In addition, it is the stakeholder of the federal energy corporations Petrobras and Eletrobras. It is also connected with the Ministry of the Environment, which is a separate department (MME, 2017).

3 **Canada**

Canada is a major energy producer, with diverse natural resources including oil, gas, coal, renewables, and minerals. The Ministry of National Resources Canada (NRCan) oversees the nation's energy resources, as well as minerals, metals, forests, mapping and earth sciences.

The Canadian constitution stipulates that provincial authorities are in charge of the natural resources on land, whereas the federal government is responsible for the offshore resources. The government is also in charge of global relations and boundaries, trade agreements, commercial synergies, economics and statistics (NRCan, 2017).

4 **China, National Energy Administration (NEA)**

China's Twelfth National People's Congress and the Circular of the State Council on the Administration of the Ministries and Management founded the national energy bureau to coordinate the advancement of energy regulations for coal, oil, natural gas, electricity, new energy and renewable energy, in addition to commercial guidelines and specifications for oil refining, coal fuel, and fuel ethanol.

Its purpose is to construct energy advancement plans and regulations for the above energy segments, and to coordinate the execution and establishment of energy emergency plans and incident response (NEA, 2014).

China's National Center for Climate Change Strategy and International Cooperation is in charge of China's climate and energy policy. China has attempted to mitigate greenhouse pollution. Its National Action Plan entails the enhancement of electricity generation from alternative energy sources and nuclear energy, enhancing the productivity of coal plants, and the continuing development of coal-bed and coal-mine methane.

5 **England (United Kingdom)**

The UK Department of Energy and Climate Change (DECC) was established in 2008 in order to undertake the national energy agenda, that was previously allocated to the Department for Business, Enterprise and Regulatory Reform, but also functions pertinent to the environment and climate change, which were previously handled by the Department for Environment, Food and Rural Affairs (UK Government, 2017a; UK Government, 2017b).

The Oil and Gas Authority (OGA) was established on April 1, 2015 as the oil and gas regulating instrument, when an Energy Bill in the House of Lords stipulated the transfer of the Secretary of State for Energy and Climate Change's regulatory powers on oil and gas to the OGA.

The OGA was founded to regulate, impact and boost the UK oil and gas activities together with various other regulatory government bodies. It handles offshore CO_2

storage and offshore gas storage and transportation. Furthermore, it was given several powers to ensure it will precipitate the economic growth of UK fossil fuels through a number of acts, all stipulated in the Infrastructure Act (2015), the Energy Act (2016), and the Energy Act (2011) that gave the OGA dispute-resolution controls pertinent to third-party admission to upstream oil infrastructure (OGA UK, 2017a; OGA UK, 2017b; OGA UK, 2017c).

The Petroleum Act 1998 (the Petroleum Act) determines the regulatory framework encompassing the oil and gas discovery and generation in the UK (while excluding the onshore activities in Northern Ireland) (UK Government, 2017a).

6 European Union

In the European Union it is the European Commission and the European Parliament that represent the principal governing bodies that have established a set of rules for both the renewable and non-renewable energy segment. A set of governing rules stipulate the granting of licenses for fossil fuel upstream and midstream; offshore safety; shale gas extraction; carbon capture and storage; as well as coal and other solid fuels. Some of these rules include:

- The Safety of Offshore Oil and Gas Operations Directive 2013/30/EU, enforced in 12 June 2013, which set up guidelines and regulations pertinent to licensing requirements, exploration, safety, security, etc.
- The 1994 Offshore Protocol concentrates on the elimination and mitigation of environmental issues from offshore actions and offers basic rules for everyday operations. The Energy Efficiency Directive 2012/27/EU (EED) incorporates suggestions on energy efficiency improvements within the European Union and the safety of offshore operations by creating an extensive framework for the prevention of major occurrences (Europa EU, 2017).
- EU, atomic, hydrogen, fusion: Fusion for Energy (F4E). Fusion entails the energizing process of the solar system and all living things. It joins hydrogen and other atoms. Plasma pertains to an electrically-charged gas, or the transformation of any gas into the fourth phase of matter (i.e., the solid, liquid and gaseous state), due to high temperature. While it is seldom available on earth, over 99% of the universe is in a plasma phase (Europa EU, 2017).

7 India

The energy policy of India is generally determined by the nation's growing energy deficit and amplified focus on creating substitute power sources, especially nuclear, solar and wind power. Since 2013, its primary power usage has reached the 3rd highest global rate, following China and the U.S.

The Ministry of New and Renewable Energy (MNRE) is developing a database pertinent to the renewable energy segment growth. It was initially founded in 1992 as the Ministry of Non-Conventional Energy Sources, and was renamed as MNRE in 2006. It is primarily in charge of energy research and advancement, intellectual property (IP) safeguard, and global collaboration, marketing, and control in alternative energy segments including solar, wind, hydro, and biogas. Its broad goal is to produce and utilize new and alternative energy for meeting the country's increasing energy needs while relieving the developing concern for energy security (MNRE, 2017).

OPEC is India's major energy partner, as its geopolitical strategy entails the strengthening of relations with other Asian and Middle Eastern nations. India purchases and imports most of its oil from Iran, Kuwait, and Saudi Arabia.

8 Iran

The Ministry of Petroleum of the Islamic Republic of Iran is the nation's energy agency, in command of the nation's vast oil and gas reserves. Iran is a major global oil and gas exporter within OPEC, and its economy strongly relies on energy exports. China, Japan, India, and South Korea are its top trading partners for oil, and a trade agreement with the European Union signified a significant natural gas coalition with an ambitious pipeline project.

9 Russia

Minenergo is the Ministry of Energy of the Russian Federation, which is in charge of the executive and legislative bodies for energy and mineral reserves. The ministry coordinates the nationwide energy extraction projects in collaboration with national and global energy majors, and oversees the pipeline and land agreements, as well as the trade agreements conducted at a diplomatic level. Minenergo was established in 2008 after a restructuring of the Cabinet divisions: The Ministry of Industry and Energy became the Ministry of Industry, with increased economic and trade-related responsibilities (MinEnergo, 2017).

Russia and China have a long history of geopolitical alliances, and the energy factor strengthens this bond. China's insatiable need for energy, combined with Russia's abundant reserves, have shaped a major, multi-faceted bond that is sealed with an ongoing petroleum agreement.

10 Saudi Arabia

The Kingdom of Saudi Arabia is the nation with the world's greatest crude oil reserves. The Ministry of Energy, Industry and Mineral Resources is the entity in charge of the policies and legal framework overseeing oil, gas, and minerals. Vision 2030 is a government strategic plan issued in 2016 with the purpose of modernizing the national economy and society.

Growth through diversification is the nation's vision for the future, based on the following energy pillars:

a reduction of the country's dependence on petroleum revenue;
b growth of the renewable energy segment;
c liberalization of national energy costs.

Saudi Arabia and the U.S. energy departments have an impactful cooperation, and an annual bilateral energy dialogue (Energy.gov, 2017b).

11 U.S.A.

In the United States, the Department of Energy (DOE) is in charge of the nation's security and prosperity through mitigating the energy, nuclear and environmental issues by the use of science and technology. It is also responsible for the U.S. Navy's production of nuclear reactor energy, radioactive waste disposal, energy conservation and domestic production. The department runs 17 national laboratories (DOE, 2017).

The Energy Information Administration is a division of the Department of Energy under the U.S. Federal Statistical System, whose task is to gather, evaluate and publicize data pertinent to energy policies, economics, environmental and commercial issues (EIA, 2017).

NAICS 21 is the classification of the North American Industry Classification System that consists of organizations that produce mineral solids such as liquid minerals (i.e., crude petroleum), gases (i.e., natural gas), and metals (i.e., coal and other construction mineral deposits) (EPA, 2017e).

The Clean Air Act Amendment (1990) supported the utilization of natural gas and gasoline as a consequence of the shift towards cleaner sources of energy and improved safety during extraction and hauling.

The Energy Policy and Conservation Act of 1975 (EPACA) is a United States Act of Congress enacted in December 22, 1975 in order to meet the country's energy needs and enhance energy efficiency strategies in response to the 1973 oil crisis by developing an extensive solution to federal government energy policy. In 2016 it was amended through Public Law 94–163, and reinforced in the same year (EPACA, 2016).

The U.S. Code of Federal Regulations (CFRs) is the administrative law to stipulate rules and regulations issued by the executive branches and federal government. Some of the CFRs pertaining to the energy industry include:

a 49 CFR 1.53; 101-172-185 on hazmat transportation guides, hazmat response and hazmat segregation tables, and pipeline safety.
b The U.S. DOT Hazardous Materials Regulations (HMR).
c 29 CFR Part 1910, hazmat response; industry requirements; occupational safety and health standards.
d 33 CFR Part 155, oil or hazardous material pollution prevention; internal combustion engines.
e 40 CFR Hazardous Wastes, offshore exploration and transportation, etc.

The Federal Energy Regulatory Commission (FERC) is the U.S. federal agency that handles the diffusion and wholesale distribution of natural gas and electricity, the hauling of petroleum and its products via pipeline in domestic trade, in addition to licensing agreements for hydropower projects (FERC, 2017). The Energy Policy Act of 2005 and its enhanced strategic plan empowered FERC.

Finally the National Nuclear Security Administration, Office of Nuclear Energy, is in charge of the nation's nuclear strategy.

Key Terms on Energy Security Regulations

This section covers the main terms pertaining to energy security regulations and government strategies, mainly energy protectionism, subsidies, anti-dumping and Dual Prices. These terms apply to different energy segments, on both renewable and non-renewable energy, hence this concise definition guide will help the readers better understand these rules and how they play out in a global energy security context.

1 **Free trade**

Free trade refers to the non-intervention practices at a global or national level, where the supply–demand equilibrium is not affected by quotas, tariffs, selective taxation and other practices in the realm of protectionism. In the energy sector, this strategy allows consumers to select the energy options of their choice, be it local or imported from abroad.

2 **Energy protectionism**

Protectionism is defined as a government's strategy to protect and support local industry segments by imposing taxation on foreign imports. In an energy context, protectionism rules apply when government taxation affects imported energy segments, thus indirectly encouraging local consumers to shift into local energy sources. Protectionism can positively impact a nation's economy, yet, from a theoretical economic perspective, it disrupts free trade and the normal supply–demand equilibrium. The local demand for energy may be dictated by financial, and not quality or efficiency, benchmarks.

3 **Energy subsidies**

Subsidies are grants or financial incentives given by government or regional authorities in order to support selected energy or industrial segments. Most nations support renewable energy segments, such as wind, solar, biomass, etc. Subsidies allow selected energy products to become more popular, yet, again, they intervene with the market supply–demand equilibrium. Energy preference may be based on financial, and not quality or efficiency-related criteria.

4 **Dumping practices, anti-dumping and countervailing duties**

a **Dumping** pertains to national energy products being more expensive locally and cheaper overseas. This brings in the foreign products at a disadvantage to home products, hence **anti-dumping and countervailing duties** are tools that governments can use in order to better protect their domestic products.

b **Anti-dumping tariffs** are government bills to levy tariffs on imported energy products when their cost is cheaper than the fair market prices.

c **Anti-dumping laws**, when enforced, have the power to prohibit the import and sale of products that derive from discriminating or manipulating foreign trade practices.

d **Countervailing duties** refer to import taxation enforced on particular products in order to safeguard foreign dumping strategies. Countervailing measures in the energy industry support a nation's strategy for energy independence.

Trade agreements among nations are ruled by laws of reciprocity, hence governments can decide on a case-by-case basis when free trade should be exercised, or whether certain strategies, be it subsidies, protectionism, or anti-dumping policies, should apply to specific products and specific global partners. Deviating from free trade eventually becomes a reciprocal decision: When nations exercise trade or currency manipulation techniques their global partners adopt similar policies. And while such interventions can impact the supply–demand equilibrium, they also serve as strategic tools for nations to reach energy security and energy independence while boosting the local economy.

From a political science perspective, an examination of national strategies on currency manipulation or trade protectionism practices can reveal the national strategy of different political regimes, as well as the leveraging power of nations at any moment in history.

References

Agency for Toxic Substances and Disease Registry: Centers for Disease Control and Prevention (ATSDR.CDC) (2017a). *Health Effects of HFCs*. U.S. Department of Health and Human Services, Public Health Service Agency for Toxic Substances and Disease Registry. Available at: www.atsdr.cdc.gov/toxguides/toxguide-70.pdf, last accessed: January 2, 2017.

Agency for Toxic Substances and Disease Registry: Centers for Disease Control and Prevention (ATSDR.CDC) (2017b). *Health Effects of PFAS*. U.S. Department of Health and Human Services, Public Health Service Agency for Toxic Substances and Disease Registry. Available at: www.atsdr.cdc.gov/pfc/health_effects_pfcs.html, last accessed: January 2, 2017.

Alaska Government (2017). *Particulate Matter Health Impacts*. State of Alaska, Division of Air Quality. Available at: https://dec.alaska.gov/air/anpms/pm/pm_health.htm, last accessed: January 2, 2017.

Australian Government (2005). *Air Quality Fact Sheet*. Australian Government, Department of the Environment and Heritage. Available at: www.environment.gov.au/protection/publications/factsheet-sulfur-dioxide-so2, last accessed: January 3, 2017.

Bureau of Ocean Energy Management (BOEM) (2009). *Oil in the Sea 2003*. Bureau of Ocean Energy Management, and National Research Council. Available at: www.boem.gov, last accessed: December 12, 2016.

Bureau of Ocean Energy Management (BOEM) (2012). *Deep Water Horizon Report*. Bureau of Ocean Energy Management. Available at: www.boem.gov, last accessed: December 14, 2016.

Bureau of Safety and Environmental Enforcement (BSEE) (2012). *Deep Water Horizon Report*. Bureau of Safety and Environmental Enforcement. Available at: www.bsee.gov, last accessed: December 14, 2016.

Burns, M. (2012). "Strategic environmental assessment in the offshore industry: an economic and empirical study". *Proceedings of the 18th Offshore Symposium, February 7, 2013, Houston, Texas*. Alexandria: Texas Section of the Society of Naval Architects and Marine Engineers (SNAME).

Burns, M. (2014). *Assessing the Environmental Vulnerability Risk at Major Seaports. A Comparative Analysis between the Major Ports of the US Gulf*. SNAME Annual Maritime Convention, October 23, Houston, Texas. Available at: https://sname.eventsential.org/Speakers/Details/48180, last accessed: May 5, 2017.

Burns, M. (2016a). *Transportation and Logistics Security: A Strategic, Tactical and Operational Guide to Resilience*. London: CRC Press, Taylor & Francis.

Burns, M. (2016b). *Energy Independence in North America and the Energy Markets: The Impact on Shipping, the Energy Industry and Regulatory Development*. Maritime Transportation of Energy. 10th Mare Forum, USA, October 13, Hilton Post Oak, Houston, Texas, USA.

Canadian Centre for Occupational Health and Safety (CCOHS) (2016). *Canada's Definition of Hazardous Energy*. Hazardous Energy Control Programs, Canadian Centre for Occupational Health and Safety. Available at: www.ccohs.ca/oshanswers/hsprograms/hazardous_energy.html, last accessed: December 6, 2016.

Canadian Centre for Occupational Health and Safety (CCOHS) (2017). *OSH Fact Sheets. Carbon Dioxide*. Canadian Centre for Occupational Health and Safety. Available at: www.ccohs.ca/oshanswers/hsprograms/hazardous_energy.html, last accessed: January 1, 2017.

Canadian Centre for Occupational Health and Safety (CCOHS) (2018). *OSH Answers Fact Sheets. Hazardous Energy Control Programs*. Canadian Centre for Occupational Health and Safety. Available at: www.ccohs.ca/oshanswers/hsprograms/hazardous_energy.html, last accessed: March 25, 2018.

Canada Government (2017). *Environment and Climate Change, Canada. Volatile Organic Compounds in Consumer and Commercial Products*. Environment and Climate Change Canada. Available at: www.ec.gc.ca/cov-voc, last accessed: December 14, 2016.

Centers for Disease Control and Prevention (CDC) (2017). *Health Impact Assessment*. U.S. Centers for Disease Control and Prevention. Available at: www.cdc.gov/healthyplaces/hia.htm, last accessed: May 5, 2017.

Department of Energy (DOE) (2017). *About Us*. U.S. Department of Energy. Available at: www.energy.gov/about-us, last accessed: January 12, 2017.

Department of Health Services (DHS) Wisconsin (2017). *Carbon Dioxide*. Wisconsin Department of Health Services. Available at: www.dhs.wisconsin.gov/chemical/carbondioxide.htm, last accessed: January 3, 2017.

Department of State (2007). *United States Report on Implementation of the Declaration on Security in the Americas*, June 22. Available at: www.state.gov/p/wha/rls/112844.htm, last accessed: May 12, 2018.

Energy.gov (2016). *Intelligence and Counterintelligence*. U.S. Office of Health and Safety. Available at: https://energy.gov/office-intelligence-and-counterintelligence, last accessed: December 14, 2016.

Energy.gov (2017a). *Organizational Chart*. U.S. Office of Health and Safety. Available at: http://Energy.Gov/Ehss/Organizational-Chart/Office-Health-And-Safety, last accessed: January 3, 2017.

Energy.gov (2017b). *U.S.–Saudi Arabia Energy Cooperation. International Affairs Initiatives*. Washington, DC: U.S. Department of Energy.

Energy Information Administration (EIA) (2016a). *EIA International Energy Outlook 2016. U.S.* Energy Information Administration. Available at: www.eia.gov/outlooks/ieo/pdf/0484(2016).pdf, last accessed: January 1, 2017.

Energy Information Administration (EIA) (2016b). *How Much Carbon Dioxide is Produced when Different Fuels are Burned?* U.S. Energy Information Administration. Available at: www.eia.gov/tools/faqs/faq.cfm?id=73&t=11, last accessed: January 3, 2017.

Energy Information Administration (EIA) (2016c). *Energy Emissions*. U.S. Energy Information Administration. Available at: www.eia.gov/environment/emissions/co2_vol_mass.cfm, last accessed: January 3, 2017.

Energy Information Administration (EIA) (2016d). *EIA – Greenhouse Gas Emissions Overview*. U.S. Energy Information Administration. www.eia.gov/environment/emissions/ghg_report/ghg_overview.php, last accessed: January 2, 2017.

Energy Information Administration (EIA) (2016e). *Energy Explained*. U.S. Energy Information Administration. Available at: www.eia.gov/energyexplained/index.cfm/data/index.cfm?page=environment_home, last accessed: January 3, 2017.

Energy Information Administration (EIA) (2017). *About Us*. U.S. Energy Information Administration. Available at: www.eia.gov/about, last accessed: January 12, 2017.

Environmental Protection Agency (EPA) (2010). *Methane and Nitrous Oxide Emissions*. Available at: https://nepis.epa.gov/Exe/ZyNET.exe/P100717T.TXT?ZyActionD=ZyDocument&Client=EPA&Index=2006+Thru+2010&Docs=&Query=&Time=&EndTime=&SearchMethod=1&TocRestrict=n&Toc=&TocEntry=&QField=&QFieldYear=&QFieldMonth=&QFieldDay=&IntQFieldOp=0&ExtQFieldOp=0&XmlQuery=&File=D%3A%5Czyfiles%5CIndex%20Data%5C06thru10%5CTxt%5C00000017%5CP100717T.txt&User=ANONYMOUS&Password=anonymous&SortMethod=h%7C-&MaximumDocuments=1&FuzzyDegree=0&ImageQuality=r75g8/r75g8/x150y150g16/i425&Display=hpfr&DefSeekPage=x&SearchBack=ZyActionL&Back=ZyActionS&BackDesc=Results%20page&MaximumPages=1&ZyEntry=1&SeekPage=x&ZyPURL, last accessed: December 14, 2016.

Environmental Protection Agency (EPA) (2014). *Inventory of US Greenhouse Gas Emissions and Sinks*. Available at: www.epa.gov/ghgemissions/inventory-us-greenhouse-gas-emissions-and-sinks-1990-2012, last accessed: December 14, 2016.

Environmental Protection Agency (EPA) (2015). *Air Quality Guide for Particle Pollution, August 2015, EPA-456/F-15-005*. Available at: www3.epa.gov/airnow/air-quality-guide_pm_2015.pdf, last accessed: December 14, 2016.

Environmental Protection Agency (EPA) (2016a). *The Sources and Solutions: Fossil Fuels*. Available at: www.epa.gov/nutrientpollution/sources-and-solutions-fossil-fuels, last accessed: December 14, 2016.

Environmental Protection Agency (EPA) (2016b). *Basic Information [Online]*, Available at: www.epa.gov/ozonepollution/basic.html, last accessed: December 14, 2016.

Environmental Protection Agency (EPA) (2017a). *Overview Greenhouse Gases*. U.S. Environmental Protection Agency. Available at: www.epa.gov/ghgemissions/overview-greenhouse-gases, last accessed: March 12, 2017.

Environmental Protection Agency (EPA) (2017b). *Health and Environmental Effects of Particulate Matter (PM)*. U.S Environmental Protection Agency. Available at: www.epa.gov/pm.../health-and-environmental-effects-particulate-matter-pm, last accessed: March 12, 2017.

Environmental Protection Agency (EPA) (2017c). *Causes of Climate Change*. U.S. Environmental Protection Agency. Available at: https://19january2017snapshot.epa.gov/climate-change-science/causes-climate-change_.html, last accessed: March 12, 2017.

Environmental Protection Agency (EPA) (2017d). *Health Impact Assessments*. U.S. Environmental Protection Agency. Available at: www.epa.gov/healthresearch/health-impact-assessments, last accessed: June 5, 2017.

Environmental Protection Agency (EPA) (2017e). *Oil and Gas Extraction Sector (NAICS 2011)*. U.S. Environmental Protection Agency. Available at: www.epa.gov/regulatory-information-sector/oil-and-gas-extraction-sector-naics-211, last accessed: June 12, 2017.

Environmental Protection Agency (EPA) (2017f). *Clean Air Act Overview (1990), Amendment Summary*. U.S. Environmental Protection Agency. Available at: www.epa.gov/clean-air-act-overview/1990-clean-air-act-amendment-summary, last accessed: June 12, 2017.

EPACA (2016). Energy Policy And Conservation Act [Public Law 94–163, as Amended]; [As Amended Through P.L. 114–255, Enacted December 13, 2016]. U.S. House of Representatives, Office of the Legislative Counsel. Available at: https://legcounsel.house.gov/Comps/Energy%20Policy%20And%20Conservation%20Act.pdf, last accessed: June 1, 2017.

EUR-Lex (2018). *Access to European Union Law. The Energy Efficiency Directive 2012/27/EU (EED)*. EUR-Lex. Available at: https://eur-lex.europa.eu/legal-content/EN/ALL/?uri=CELEX%3A52011PC0690, last accessed: January 18, 2019.

Europa EU (2017). *Fusion Energy, About Fusion*. Europa, European Union. Available at: http://fusionforenergy.europa.eu/aboutfusion, last accessed: June 12, 2017.

European Commission (EC) (2017). *European Commission's Climate Action. Fluorinated Greenhouse Gases*. Available at: http://ec.europa.eu/clima/policies/f-gas_en, last accessed: January 3, 2017.

European Environmental Agency (EEA) (1990–2014). *European Union Emission Inventory Report 1990–2014 Under the UNECE Convention on Long-Range Transboundary Air Pollution (LRTAP)*. European Environmental Agency. Available at: http://ec.europa.eu/eurostat/statistics-explained/index.php/Glossary:European_Environment_Agency_(EEA), last accessed: December 14, 2016.

European Environmental Agency (EEA) (2017). *Suspended Particulate Matter*. Environmental Terminology and Discovery Service (ETDS), European Environmental Information and Observation Network. Available at: http://glossary.eea.europa.eu/terminology/concept_html?term=suspended%20particulate%20matter, last accessed: January 2, 2017.

European Union (EU) (2014). *European Union Regulation (EU) No. 517 of 2014 on Fluorinated Greenhouse Gases. European Union*. Available at: www.epa.ie/pubs/legislation/air/ods/revisedf-gas regulation.html, last accessed: December 3, 2016.

European Union (EU) (2016). *European Union Fluorinated Greenhouse Gas Regs 2016 S.I. No. 658 of 2016. Irish Regulations to Give Further Effect to Certain Elements of the European Regulation on Fluorinated Greenhouse Gases, EU No. 517/2014*. European Union. Available at: www.epa.ie/pubs/advice/air/fluorinatedgreenhousegases/eufluorinatedgreenhousegas2016.html, last accessed: December 14, 2016.

Eurostat (2017). *Statistics Explained, 2017: Air Pollution Statistics*. Eurostat. Available at: http://ec.europa.eu/eurostat/statistics-explained/index.php/Air_pollution_statistics, last accessed: December 14, 2016.

Federal Energy Regulatory Commission (FERC) (2017). *What FERC Does*. Federal Energy Regulatory Commission, U.S Government. Available at: www.ferc.gov, last accessed: June 12, 2017.

Hutter, David (2016). *Physical Security and Why It Is Important*. SANS Institute. Available at: www.sans.org/reading-room/whitepapers/physical/physical-security-important-37120, last accessed: December 30, 2016.

International Energy Agency (IEA) (2015). *Energy and Climate Change: World Energy Outlook Special Report*. International Energy Agency, OECD/IEA, Paris. Available at: www.iea.org/publications/freepublications/publication/WEO2015SpecialReportonEnergyandClimateChange.pdf, last accessed: January 2, 2017.

International Energy Agency (IEA) (2016). *World Energy Outlook Special Report (2016) Energy and Air Pollution*. International Energy Agency, OECD/IEA, Paris. Available at: www.iea.org/publications/freepublications/publication/WorldEnergyOutlookSpecialReport2016EnergyandAirPollution.pdf, last accessed: January 2, 2017.

International Energy Forum (IEF) (2017). *History*. The International Energy Forum. Available at: www.ief.org/about-ief/history.aspx, last accessed: January 2, 2017.

ITER (2017). *Euratomic ITER Fusion*. Available at: www.iter.org, last accessed: June 12, 2017.

Merriam-Webster Dictionary (2016a). Definition: "Security". Available at: www.merriam-webster.com/dictionary/security, last accessed: December 19, 2016.

Merriam-Webster Dictionary (2016b). Definition: "Safety". Available at: www.merriam-webster.com/dictionary/safety, last accessed: December 19, 2016.

MinEnergo (2017). Russia's Ministry of Energy. Available at: https://minenergo.gov.ru, last accessed: June 12, 2017.

Ministry of Mines and Energy (MME) (2017). *Overview*. Ministry of Mines and Energy (Ministério de Minas e Energia). Brazil Portal of Legislation. Available at: www4.planalto.gov.br/legislacao, last accessed: June 12, 2017.

Ministry of New and Renewable Energy (MNRE) (2017). *About Us*. Ministry of New and Renewable Energy, Government of India. Available at: https://mnre.gov.in, last accessed: January 12, 2017.

Ministry of Non-Conventional Energy Sources (MNES) (2017a). *About Us, Initiatives and Achievements*. Ministry of Non-Conventional Energy Sources, India. Available at: www.mnre.gov.in, last accessed: June 12, 2017.

Ministry of Non-Conventional Energy Sources (MNES) (2017b). *Mission and Vision*. Ministry of Non-Conventional Energy Sources, India. Available at: mnre.gov.in/mission-and-vision-2/mission-and-vision, last accessed: June 16, 2017.

NASA (2008). *Water Vapor Confirmed as Major Player in Climate Change*. NASA. Available at: www.nasa.gov/topics/earth/features/vapor_warming.html, last accessed: July 1, 2017.

NASA (2016). *Perils of Oil Pollution*. Available at: http://seawifs.gsfc.nasa.gov/OCEAN_PLANET/HTML/peril_oil_pollution.html, last accessed: June 1, 2017.

National Center for Biotechnology Information (NCBI) (2016a). *Sulfur Hexafluoride*. National Center for Biotechnology Information. Available at: https://pubchem.ncbi.nlm.nih.gov/compound/sulfur_hexafluoride, last accessed: December 12, 2016.

National Center for Biotechnology Information (NCBI) (2016b). *Nitrogen Trifluoride*. National Center for Biotechnology Information. Available at: https://pubchem.ncbi.nlm.nih.gov/compound/nitrogen_trifluoride#section=Top, last accessed: December 8, 2016.

National Energy Administration (NEA) (2014). *National Energy Administration*. The State Council, The People's Republic of China, updated: Sep 12, 2014, 1:56 pm. Available at: http://english.gov.cn/state_council/2014/10/01/content_281474991089761.htm, last accessed: June 12, 2017.

National Library of Medicine, National Institutes of Health (NLM, NIH) (2017). *Chemicals*. U.S. National Library of Medicine. National Institutes of Health, Department of Health & Human Services, Developed by Specialized Information Services, Environmental Health and Toxicology. Available at: https://toxtown.nlm.nih.gov/text_version/chemicals.php?id=19, last accessed: December 12, 2016.

National Oceanic and Atmospheric Administration (NOAA) (2017). *What is the Biggest Source of Pollution in the Ocean?* National Ocean Service, National Oceanic and Atmospheric Administration. U.S. Department of Commerce. Available at: http://oceanservice.noaa.gov/facts/pollution.html, last accessed: December 12, 2016.

National Research Council (NRC) (2003). *Oil in the Sea III: Inputs, Fates, and Effects*. National Research Council, National Academy Press, Washington D.C. Available at: www.nationalacademies.org/nrc, last accessed: September 21, 2017.

National Research Council (NRC) (2012). *Department of Ocean, Earth and Atmospheric Sciences of the Old Dominion University VA*. SCI.ODU, 2012. Available at: http://sci.odu.edu/oceanography and www.nationalacademies.org/nrc, last accessed: December 14, 2016.

National Resources Canada (NRCan) (2017). Home page. National Resources Canada, Government of Canada. Available at: www.nrcan.gc.ca/home, last accessed: June 12, 2017.

NATO (2016). *NATO's Role in Energy Security*. NATO, 22 June. Available at: www.Nato.Int/Cps/En/Natohq/Topics_49208.htm, last accessed: December 12, 2016.

NATO Parliamentary Assembly (2016). *Energy Security: Co-operating to Enhance the Protection of Critical Energy Infrastructures. Rapporteur: Lord Jopling (United Kingdom)*. NATO Parliamentary Assembly, 157 CDS 08 E REV 1. Available at: www.nato-pa.int/default.asp?SHORTCUT=1478, last accessed: December 8, 2016.

NJ.Gov (2017). *Hazardous Substance Fact Sheet. Nitrogen Trifluoride*. New Jersey Department of Health and Senior Services. Available at: http://nj.gov/health/eoh/rtkweb/documents/fs/1380.pdf, last accessed: January 2, 2017.

New Zealand Government (2016). *Energy Safety*. Available at: www.energysafety.govt.nz/installations-networks/audits-and-enforcement/enforcement/definition-of-safe-and-unsafe, last accessed: December 21, 2016.

Occupational Safety and Health Administration (OSHA) (2017). *Controlling Hazardous Energy*. U.S. Department of Labor, Occupational Safety and Health Administration. Available at: www.osha.gov/SLTC/controlhazardousenergy, last accessed: September 21, 2017.

Oil and Gas Authority (OGA UK) (2017a). *About Us*. UK Oil and Gas Authority. Available at: www.ogauthority.co.uk, last accessed: June 16, 2017.

Oil and Gas Authority (OGA UK) (2017b). *Regulatory Framework. Overview*. UK Oil and Gas Authority. Available at: www.ogauthority.co.uk/regulatory-framework/overview, last accessed: June 16, 2017.

Oil and Gas Authority (OGA UK) (2017c). *Legislative Context*. UK Oil and Gas Authority. Available at: www.ogauthority.co.uk/regulatory-framework/legislative-context, last accessed: June 16, 2017.

Organisation for Economic Co-operation and Development (OECD) (1997). *OECD Glossary of Statistical Terms, Glossary of Environment Statistics, Studies in Methods, Series F, No. 67*. Organisation for Economic Co-operation and Development, United Nations, New York. Available at: https://stats.oecd.org/glossary/detail.asp?ID= 2623, last accessed: December 12, 2016.

Riebeek, H. (2011). *The Carbon Cycle*. NASA Earth Observatory. Available at: https://earthobservatory.nasa.gov/Features/CarbonCycle, last accessed: June 1, 2017.

UK Government (2017a). *Organisations. Oil and Gas Authority*. UK Government. Available at: www.gov.uk/government/organisations/oil-and-gas-authority, last accessed: June 16, 2017.

UK Government (2017b). *Organisations. Topics. Energy*. UK Government. Available at: www.gov.uk/government/topics/energy, last accessed: June 16, 2017.

UN (1993). *Impact of Oil and Related Chemicals and Wastes on the Marine Environment*. GESAMP Report No. 50. London: IMO, Joint Group of Experts on the Scientific Aspects of Marine Pollution (GESAMP). Available at: www.un.org/earthwatch/about/docs/scpGESAMP.htm, last accessed: January 2, 2017.

UN (2017). *Sustainable Development*. United Nations. Available at: https://sustainabledevelopment.un.org/index.php?page=view&type=13&nr=359&menu=1634, last accessed: June 1, 2017.

United Nations Environment Programme (UNEP) (2016). *Urban Environment. Waste Dump*. Report Summary. United Nations Environment Programme. Available at: www.unep.org/urban_environment/PDFs/DandoraWasteDump-ReportSummary.pdf, last accessed: January 2, 2017.

United Nations Framework Convention on Climate Change (UNFCCC) (2017). *Kyoto Protocol*. United Nations Framework Convention on Climate Change. Available at: http://unfccc.int/kyoto_protocol/items/2830.php, last accessed: January 2, 2017.

U.S. Congressional Budget Office (2014). *The Renewable Fuel Standard: Issues for 2014 and Beyond*. U.S. Congressional Budget Office, June 26. Available at: www.cbo.gov/sites/default/files/cbofiles/images/pubs-images/45xxx/45477-land-figure1b.png, last accessed: January 3, 2017.

3 Global Energy

A Geopolitical and Economic Landscape

> Upon examining the individual nations' energy strategies, it seems that strategic preferences that appear to be "part of a nation's culture and daily life," are in fact logical, fact-based decisions, dictated by their current energy reserves, technological and environmental capabilities, and the pressure to meet one's socio-economic needs.
>
> Maria Burns, National Academies Annual Meeting 2017

The energy sector is a high-risk, capital-intensive industry, and, as such, security threats are most likely to occur in the major energy-producing and energy-consuming global players. In fact, the energy industry is influenced by a series of economic, commercial and geopolitical changes. Namely:

1. **Global energy markets**, which for the fossil fuel sectors are defined by the existing reserves, the depletion of old reserves and the discovery of new reserves. As the world moves towards unconventional energy, both for renewables and non-renewable energy, the industry's future depends on new, unconventional technologies, regulations and partnerships.
2. **Geopolitics**, shaped by political trends, alliances, and global areas of conflict (war, terrorism, social turmoil). In fact, the vast majority of fossil fuel reserves are found in nations that face some sort of geopolitical challenges.
3. **Economics**, as the energy financial environment, that is sensitive to supply–demand equilibrium rules, which derive from production and consumption patterns.

As depicted in Figure 3.1, energy security is impacted by geopolitics, energy markets and economics. This chapter will demonstrate the close interrelation of these factors, but also the affinity of global energy key players.

Mapping the Global Energy: Key Production and Consumption Regions

National Energy Security Strategies: The Energy Matrix Revisited

The 21st century has brought about a number of radical changes in geopolitics and alliances, but also competition among energy-producing nations, and occasionally a shift of power in particular energy segments. Governments and corporations mitigate their Triple-E goals of energy efficiency, environmental integrity and economies of scale, and synergies, policies and innovative technologies help them meet those goals.

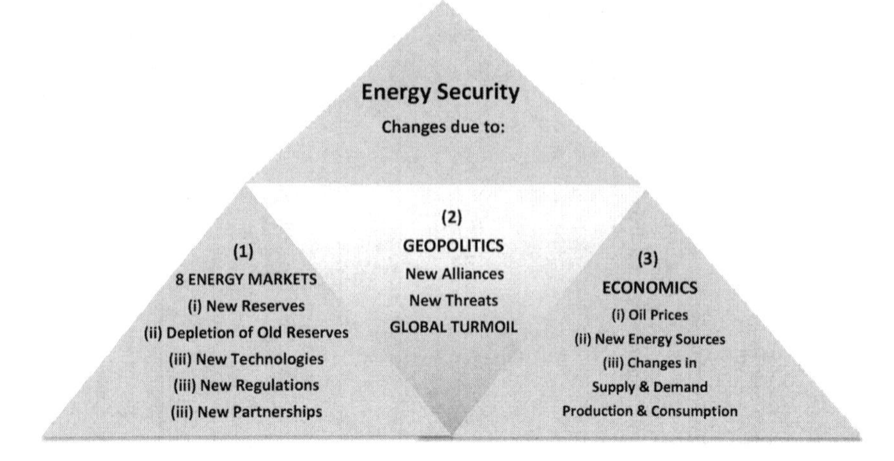

Figure 3.1 National energy security plan and the supply chain perspective.

Source: the Author.

At the same time, polarization exists among the society pertinent to the national energy strategies, and these opposing views are becoming part of equally divided political agendas.

This polarization has one thing in common: Energy security strategies seem to evolve around the five goals of Sustainability, Independence, Efficiency, Affordability and Accessibility, as presented in Chapter 1 of this book, and these are further discussed in this chapter (Figure 3.2).

And while these goals are interrelated, as shown in Figure 3.1, a nation's energy profile will determine how different political agendas will prioritize these goals differently:

a When **Sustainability** is a nation's top priority, there is a need to import, produce and distribute energy at all costs. For nations to attain sustainability, they must alleviate a long list of energy network risks from a physical, cyber and supply chain perspective; resolve all of their diplomatic and sociopolitical conflicts; plan for contingencies related to natural disasters, mechanical failures, and intentional acts of terror, sabotage or industrial espionage.

   Such goals may be hard to attain by energy-dependent nations. For countries with limited natural resources and limited reserves or storage capacity, any disruption in the energy supply chain can seriously harm national security, safety, sovereignty, and the society as a whole. For these nations, importing energy is the only path to sustainability (at least from a short-term perspective), hence they can compromise the strategies for independence, affordability and/or efficiency as long as they find the energy sources that will keep the country in operational mode.

b **Independence** is a top goal for nations that have previously compromised their energy dependence by importing from global partners, and have probably experienced the risks of expensive or disrupted or inefficient energy, price fluctuations due to market instability, supply chain disturbances, etc. For nations to strive for energy independence, they either need to possess adequate energy reserves or advanced technologies for renewables. For nations to attain their independence goals, domestic energy

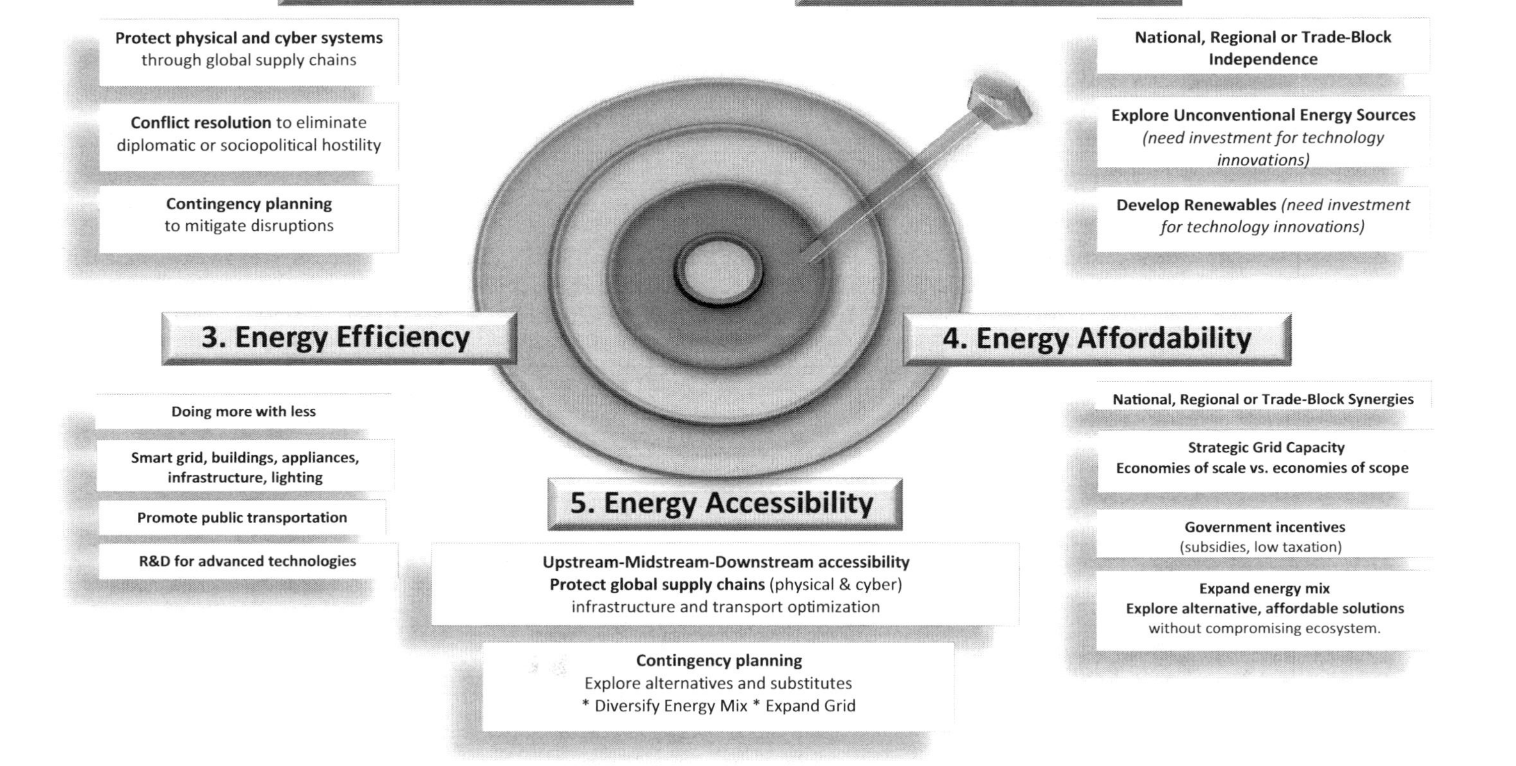

Figure 3.2 National energy security strategies: The energy matrix revisited.

Source: the Author.

reserves must be untapped, and investment in innovative technologies is required to enable the exploration of unconventional energy sources. When national independence is not feasible due to lack of natural resources, investment, grid capabilities or other reasons, nations can form regional or trade-block synergies.

c **Energy efficiency** goals pertain to utilizing energy prudently, i.e., doing more with less. Such energy-saving goals will require the adoption of energy-efficient policies, and the construction of a performance metrics system that encompasses:

   i Construction and conversion guidelines for smart buildings, and incentives to improve existing buildings; guidelines for domestic appliances and lighting.

   ii Efficient use of domestic transportation. For commercial transport this entails the promotion of energy-efficient modes, larger, green vehicles and supply-chain strategies that benefit from economies of scale. For commercial trade, this entails a well-designed public transportation, with usage incentives, i.e., for the use of rail or buses or car-pooling, etc.

   iii Technologies, smart grid systems, smart appliances, engineering and infrastructure networks.

d The **Affordability** goal is another significant priority for nations. The global energy market, just like the economy, is characterized by unpredictable fluctuations that impact prices and people's ability to purchase.

   Population segments in less developed countries (LDCs) or overpopulated nations may pursue cost-effective energy alternatives, thus affecting their national energy mix. Since the grid network expansion seems to be a significant cost consideration that impacts affordability, energy-deprived populations are increasingly inventive, turning to energy sources that do not require access to the grid and combining affordability with independence. For all these reasons, LDCs prefer the use of renewables and certain affordable segments of fossil fuel energy. Bearing in mind the rapid population and economic growth of LDCs, it would be safe to presume that energy companies investing in LDCs and providing solutions now will enjoy a larger market segment in the near future.

e **Accessibility** encompasses the entire supply chain and upstream, midstream and downstream operations, terms that this book uses for all energy sources, and not just for the oil and gas industry, for the sake of harmonization. Accessibility entails the physical and cyber components of energy, and expands through infrastructure, superstructure and transport optimization. Finally, a contingency planning is useful to explore alternatives and substitutes, diversity, the energy mix and expanding the electricity grid.

Upon examining the individual nations' energy strategies, it seems that the preferences that appear to be "part of a nation's culture and daily life," are in fact a logical, fact-based decision based on their current energy reserves, technological capabilities, and the pressure to meet one's socio-economic needs.

National Energy Security From a Production/Consumption Equilibrium

Strategic decisions will inevitably define the tactics and operational protocols to be applied as part of a nation's energy security plan.

The energy planning protocol takes into consideration the resources, infrastructure and networks available locally and determines the energy input, i.e., non-renewables and renewables. The first decision entails the selection of energy segments that will comprise the national energy mix, including the electricity mix. Consequently, once the annual amount of energy needed is estimated, it will be determined whether this energy will be produced locally, or is stored and available locally, or will be imported. Trade agreements will stipulate the partnering nations and global supply chains that will undertake the export/import process.

The energy input comprises of unprocessed or semi-processed energy that is now further processed to meet the final consumers' demand. The output stage includes all energy that will be used for national energy consumption, plus the energy stored as part of the national energy security plan (Figure 3.3), or distributed to local and foreign markets.

Primary and Secondary Energy Sources as part of a National Energy Security Plan

A national energy security plan encompasses primary and secondary energy sources. Primary energy sources entail:

a Non-renewable and/or renewable resources that are generated and preserved naturally. For example, non-renewable energy such as oil, gas, coal, but also renewable energy such as biomass based on wood and agricultural products, etc.
b Renewable energy that is naturally available, such as solar, wind, tidal, thermal, etc.
c Energy that is generated based on such primary sources, such as nuclear energy based on uranium or plutonium mining.

Secondary energy sources are non-natural derivatives generated from primary sources. Hence, both renewables and non-renewable energy sources generate secondary energy segments such as electricity, synthetic and alcohol fuels, and so on.

Measuring the Non-Renewable and Renewable Energy Sources

In order to map the geopolitical landscape it becomes necessary to map the known energy reserves at a global, continental, and national level. Energy security is attained through accessibility and independence, hence a nation's energy reserves can certainly define these boundaries. Leading energy nations are the ones that possess sufficient energy resources for the present and future, at a macro and micro level, but that also have a diverse energy mix containing non-renewable and renewable energy sources, and thus have an energy contingency plan in place.

Non-Renewable Energy

Figure 3.4 depicts the oil and gas resource categories and how these reflect varying degrees of certainty, availability and profitability:

a **Proved reserves** are the reserves whose quality, quantity, mapping, upstream feasibility and overall operations are determined and predictable. A nation's volume of proved reserves may diminish or expand in line with the current market rates and expenditure, driven by the supply/demand equilibrium.
b Some reserves may be proved, yet, in order to verify whether these are **economically recoverable reserves**, a thorough economic study is needed, including estimations such as:

 i a feasibility study;
 ii a cost/benefit analysis;
 iii a return-on-investment analysis.

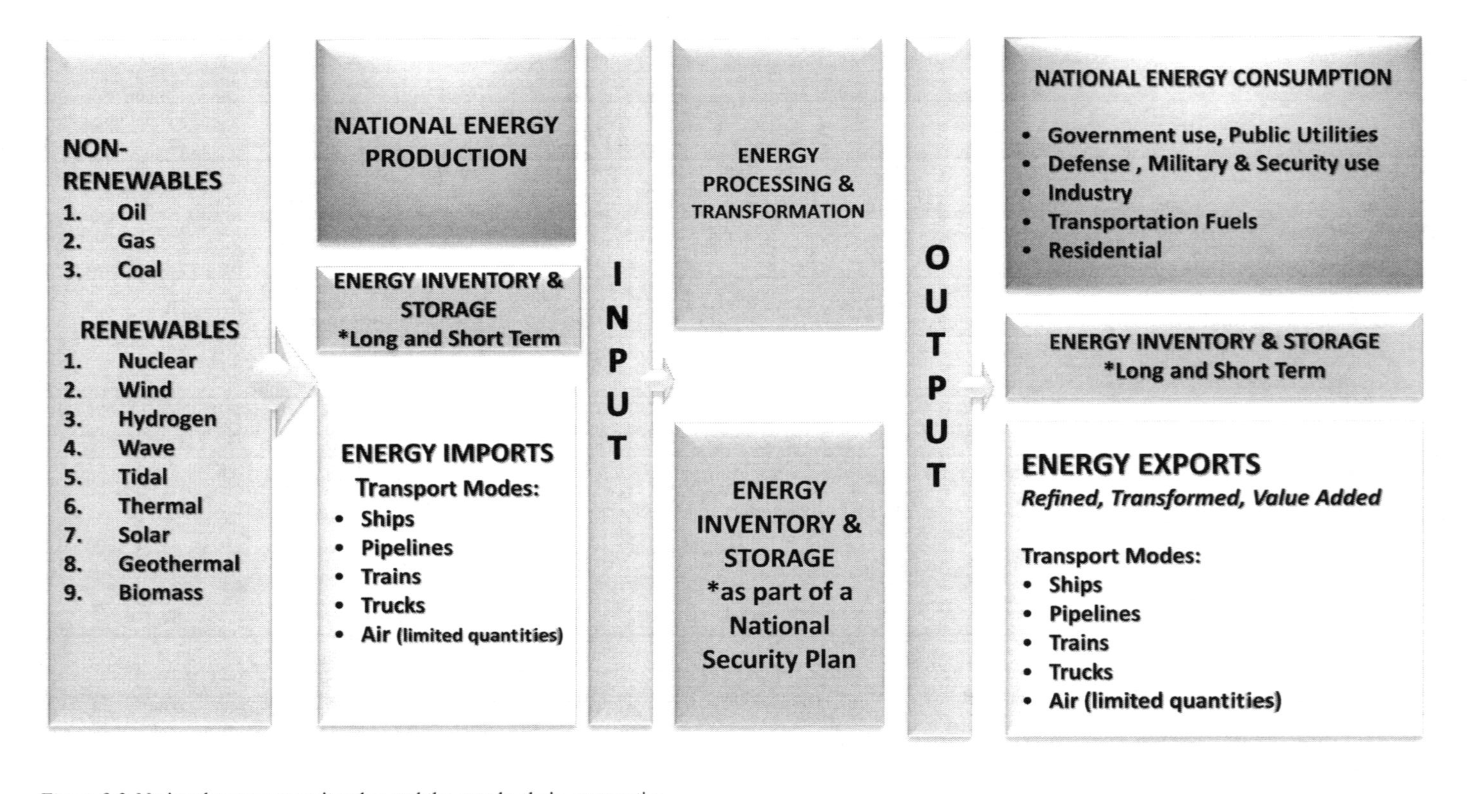

Figure 3.3 National energy security plan and the supply chain perspective.

Source: the Author.

c Even when a particular energy project seems financially lucrative, i.e., gross revenue at least equals and preferably exceeds costs, we still need to determine if the well or energy source is **technically recoverable**. This refers to the ability of science and technology to retrieve unconventional sources that, although potentially abundant and lucrative, are found in unconventional locations, e.g., subject to extreme weather, deep offshore, with challenging geological or oceanographic characteristics, etc.

d **Remaining oil and gas in place** pertains to the energy reserves that are less certain to retrieve, due to technological or financial reasons. As technologies and market conditions evolve, this amount also fluctuates.

Renewable Energy

Renewable energy is a growing part of the national energy mix, hence Figure 3.5 reflects the renewable energy resource categories with varying degrees of certainty. Renewables are found abundantly in the environment, yet our current challenges entail the capability to fully capture and store energy that is found naturally and to convert it to another form of energy, e.g., wind turbines convert kinetic energy to electricity, and solar panels convert thermal energy to electricity.

a **Captured renewable source energy** pertains to the retrievable amount of energy based on the technological ability to capture same.

b **Economically recoverable renewable energy** is the same as in the non-renewable section above. Yet, since the renewables are often state-funded, it must be determined how subsidies and tax exemptions will count in this estimation.

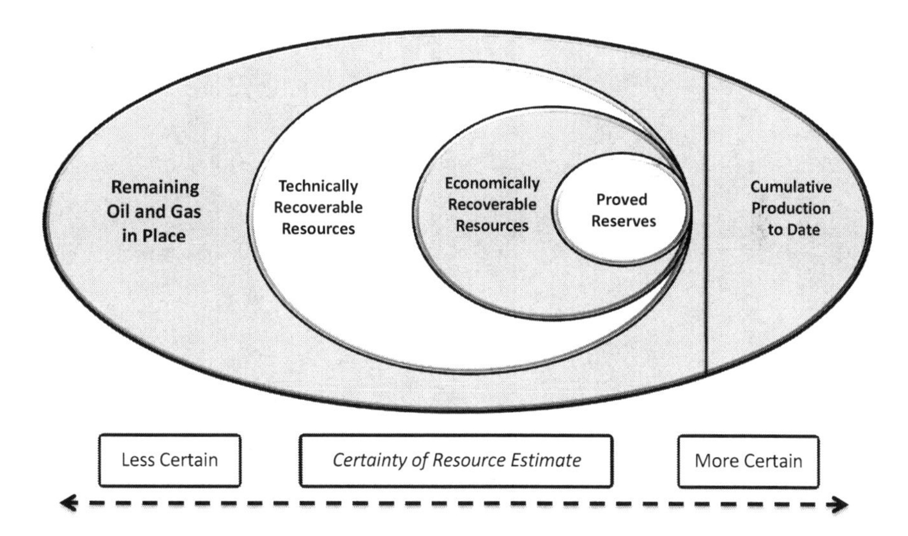

Figure 3.4 Oil and natural gas resource categories reflect varying degrees of certainty. Stylized representation of oil and natural gas resource categorizations (not to scale).

Source: EIA, 2014. *Oil and Natural Gas Resource Categories Reflect Varying Degrees of Certainty*. U.S. Energy Information Administration. Available at: www.eia.gov/todayinenergy/detail.php?id=17151, last accessed: October 9, 2016.

Note: Resource categories are not drawn to scale relative to the actual size of each resource category. The graphic shown in Figure 3.4 is applicable only to oil and natural gas.

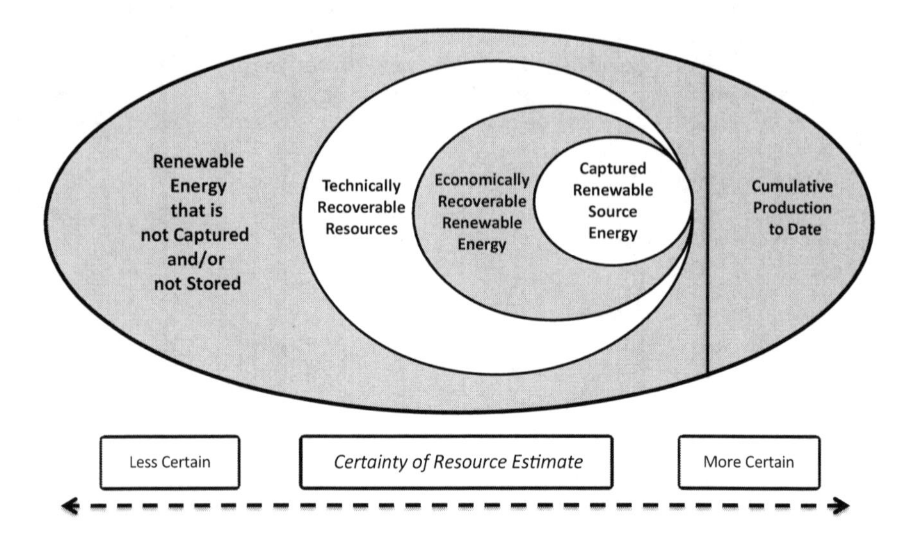

Figure 3.5 Renewable energy resource categories reflect varying degrees of certainty. Stylized representation of renewable resource categorizations (not to scale).

Source: the Author, modified renewables version, based on EIA, 2014. *Oil and Natural Gas Resource Categories Reflect Varying Degrees of Certainty*. U.S. Energy Information Administration. Available at: www.eia.gov/todayinenergy/detail.php?id=17151, last access: October 9, 2016.

Note: Resource categories are not drawn to scale relative to the actual size of each resource category.

c **Technically recoverable resources** are similar to the non-renewable section above. For example, solar energy is abundant, yet different solar panels have a distinct capability to attract, absorb and convert a portion of the energy available in nature.

d **Renewable energy that is not captured and/or not stored**. This division refers to the abundant, infinite or "lemniscate" amount of energy that is available in the universe, which Nicolas Tesla envisioned to someday capture and utilize. This division presents a challenge and an opportunity for future scientists to advance their technologies and gradually increase humankind's capacity to tame, capture and store larger scale renewable energy.

Energy Production and Consumption

Production and consumption statistics, or the Total Primary Energy Supply (TPES) and the World Final Energy Consumption indices are significant metrics that are utilized by economists, energy professionals and others in order to define a nation's geopolitical and socioeconomic power. The World Final Energy Consumption index pertains to the aggregate power consumed on an annual basis by the entire population. Statistical analysis of this data can help the energy professional understand the individual nation's strategy as to the allocation of energy in the domestic, industrial, public utility, commercial utility, storage or defense use segments.

Global energy production or Total Primary Energy Supply (TPES), on the other hand, reflects the wealth of a nation in terms of natural resources. It is the yard stick to measure annual growth indices, and hence to better understand the political, commercial, financial and social benefits in a country. It is the aggregate volume of all power sources used, both renewables and non-renewables, while it also verifies the technological capabilities of the energy companies in charge of the production.

Figure 3.6 covers the top petroleum and other liquids production (1980–2016). Through the prism of energy security, these indices reveal the security significance of a region, what is the energy production/consumption ratio, but also how much is at stake if the energy infrastructure of a nation is threatened.

Figure 3.6 provides data for six major oil-producing nations, namely the U.S.A., Saudi Arabia, Russia, China, Canada and Iraq. Their annual production rates classify them into two clusters; the cluster with the highest production comprises of the U.S., Saudi Arabia, and Russia, and the cluster with the second highest production is China, Canada, and Iraq. All these nations follow a linear annual growth, however each nation is subject to individual market cycles or production fluctuations. China and Canada's patterns are most linear with steady annual developments and minimal oscillations. The wider fluctuations of the other four nations reflect inconsistencies in annual production rates that influence the supply–demand equilibrium and impact the oil prices. Such shifts in supply could be a response to diminished demand, or a strategy to correct high oil prices, or even by-products of national economic and political turbulences. A very impressive reflection of this figure, when interpreted vertically, i.e., on a year-to-year basis, is that the global market share is truly volatile, with considerable variabilities in each nation's annual market share. Since this chart only reflects the upstream part of the supply chain, it would be safe to conclude that the global oil supply chains, trade agreements and trade routes are equally volatile on a year-to-year basis.

Figure 3.7 demonstrates the world energy consumption from 1990 to date, with subsequent projections through to the year 2040. Total world consumption of marketed energy expands from 549 quadrillion British thermal units (Btu) in 2012 to 629 quadrillion Btu in 2020 and to 815 quadrillion Btu in 2040 (EIA, 2017b).

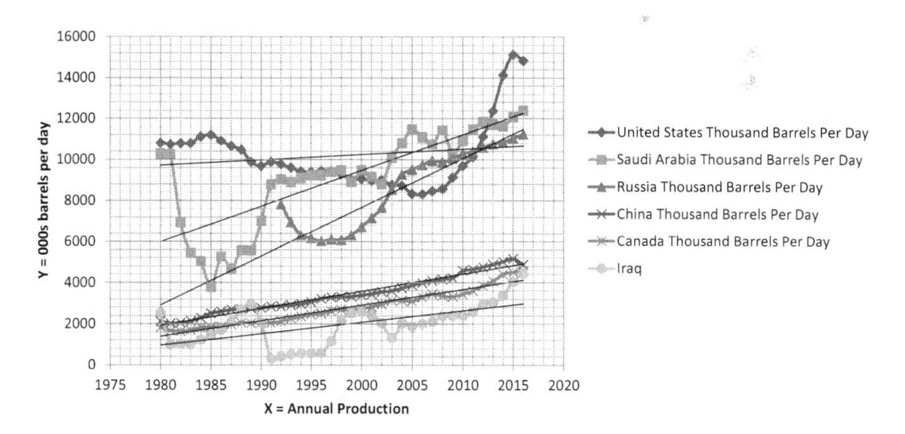

Figure 3.6 Top six petroleum and other liquids production (000s barrels a day) (1980–2016).

Source: the Author, based on data from:

EIA, 2016b. *U.S. and Other Top 5 Total Petroleum and Other Liquids Production*. U.S. Energy Information Administration. Available at: www.eia.gov/beta/international, last accessed: May 15, 2017.

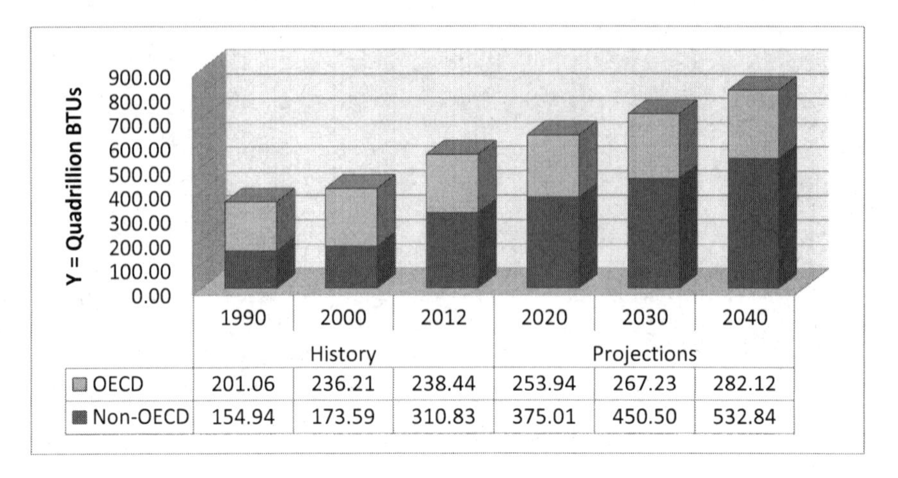

	1990	2000	2012	2020	2030	2040
	History			Projections		
OECD	201.06	236.21	238.44	253.94	267.23	282.12
Non-OECD	154.94	173.59	310.83	375.01	450.50	532.84

Figure: 3.7 Global energy consumption (quadrillion Btu). Years: 1990–2040.

Sources:

EIA, 2017a. *International Energy Outlook 2016*. Report Number: DOE/EIA-0484(2016). Release Date: May 11, 2016. U.S. Energy Information Administration. Available at: www.eia.gov/outlooks/ieo/world.cfm, last accessed: June 21, 2017.

EIA, 2017b. *International Energy Outlook 2016 With Projections to 2040*. U.S. Energy Information Administration. Available at: www.eia.gov/outlooks/ieo/pdf/0484(2016).pdf, last accessed: June 21, 2017.

The year-to-year fluctuations may be determined by geopolitical regulatory, technological and other factors, which in turn will impact the supply–demand equilibrium. However, when examining the big picture of the energy future, there are two observations that merit attention:

1 The years from 2015 to 2017 signify an abrupt shift of energy consumption from the OECD nations (comprising most European nations, America and Australia, South Korea and Japan)[1] to the Non-OECD nations (comprising the rest of the world, i.e., nations in Asia, Africa and Latin America, such as China, Russia, India, Brazil, Nigeria, Indonesia, Venezuela, etc.).
2 The aggregate world consumption shows a linear steady annual growth rate, which imposes opportunities for global growth. At the same time, it also reflects the increasing responsibility of nations to meet their energy security goals.

Electricity

Electric power entails the national grid network and the flow of electricity distributed to the domestic markets. It is considered as a secondary energy segment since it is not generated from a natural production state as we do with renewables and non-renewables. Instead, it is generated as a byproduct or derivative of the conversion of oil, gas, coal, nuclear, and renewable energy segments.

The electric power industry segment is crucial for a nation's energy security, as well as for its commercial and financial growth. For several decades now it has remained a profitable and rapidly developing energy sector with ample growth capabilities at a global level. As a general observation, the demand for electricity is rather steady, however what changes over time is the type of primary energy sources used by different nations to produce electricity.

While humankind can take pride in the 21st century's advancements, we need to reflect that, still, one out of seven people or 1.2 billion people among the global population has no access to electricity, either because they cannot afford it or because the national grid cannot reach areas that are remote, or deserted, or of particular geological and weather conditions (see Figure 3.8, as well as pages 109–112 and 136 and the energy matrix elements of Accessibility and Affordability).

Consequently, the way a nation utilizes its energy mix for the domestic electricity grid is a reflection of a country's energy policy, social policy, and long-term energy security agenda.

According to EIA (2017c), the global net electricity production is anticipated to grow by 69%, from 21.6 trillion kilowatthours (kWh) as of 2012 to 25.8 trillion kWh in 2020 and to 36.5 trillion kWh in 2040 respectively. The developing economies representing the non-OECD nations are forecasted to grow most in terms of global electricity production. Non-OECD electricity power growth exceeds 2.5% per annum from 2012 to 2040, due to population growth and increasing living standards (Figure 3.9).

This is reflected in a nation's domestic consumption due to the modern electric devices used, but also in the manufacturing sector as reflected in industrial activity, in addition to welfare, social and community services including government buildings, public utilities, education, the medical industry, department stores, etc.

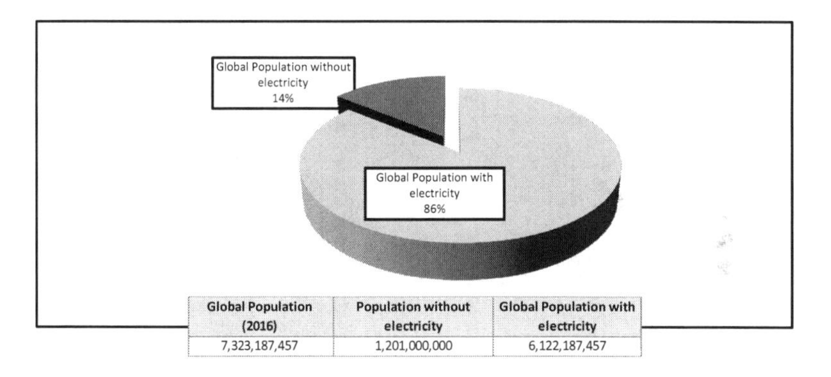

Global Population (2016)	Population without electricity	Global Population with electricity
7,323,187,457	1,201,000,000	6,122,187,457

FACT : One out of seven people globally has no access to electricity

Figure 3.8 Global population and the electric grid.

Source: the Author, based on data from CIA, 2017a. *CIA World Factbook, 2017*. Available at: www.cia.gov/library/publications/resources/the-world-factbook/geos/xx.html, last accessed: June 21, 2017.

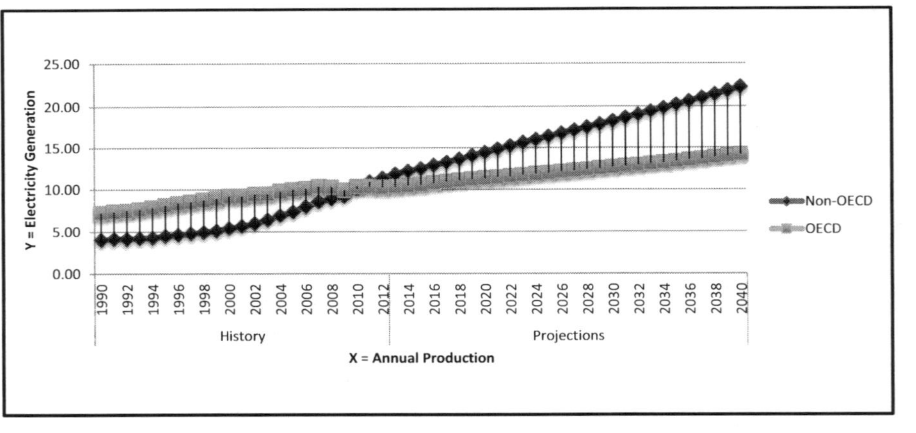

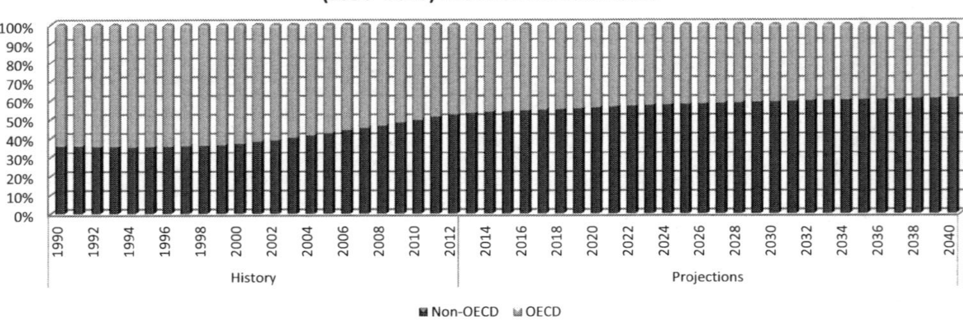

Figure 3.9 OECD and Non-OECD net electricity production (1990–2040) in trillion kWh.

On the other hand, the OECD nations combine a modest population growth and a fully developed but sometimes aging infrastructure. Hence, their electricity production is anticipated to grow modestly by approximately 1.2% per annum from 2012 to 2040. Among the different energy segments used for energy production, renewables represent the fastest-growing power supply for electrical power production, with average growth of 2.9% per annum from 2012 to 2040 (EIA, 2017b).

The Benefits of a National Energy Mix and Electricity Mix

Producing electricity from diverse sources is beneficial as it increases options and variety, and eliminates the risk deriving from

- depletion of resources or energy segment,
- price fluctuations,
- supply chain disruptions,
- future, permanent depletion of resources.

The key challenges entailed in the wide variety of the energy electricity mix pertain to the higher cost in energy conversion, and the overall operational cost of energy source diversification.

Figure 3.10 demonstrates how the top electricity-producing nations produce electricity from several energy sources, as part of their national energy security plan. And while the majority of power is produced from fossil fuels (mainly oil and coal), nations are still utilizing nuclear power, hydroelectric energy, solar, wind, biomass, etc., for contingency purposes. As a matter of fact,

- about 10% of total installed capacity of U.S., Canada and Russia, entails nuclear power;
- hydroelectric energy is popular in Brazil (69.30%), Canada (55.80%) Russia and China (over 20%). China in particular has committed to expanding its hydropower installations in the next years, as a backup plan for the future depletion of conventional energy sources (see Yangtze River and the Three Gorges Dam).
- Solar, wind, biomass and other renewable energy segments are popular in India (14.80%); China (13.70%); Brazil (10.50%); the U.S.; Canada and Australia by around 8%. (CIA, 2017a; EIA, 2017c).

A Geopolitical Landscape: New and Existing Global Players

> In a setting of changeable alliances and enmities, the lessons worth learning are nations' efforts for betterment, and their ingenuity in transforming economic and energy weaknesses into strengths.
>
> Maria Burns, National Academies Annual Meeting 2017

The Pillars of Energy Security

The notion of energy security is most important in our era, which is characterized by great geopolitical, economic, and technological developments. Governments, energy key players and policy makers need to address these radical challenges and opportunities.

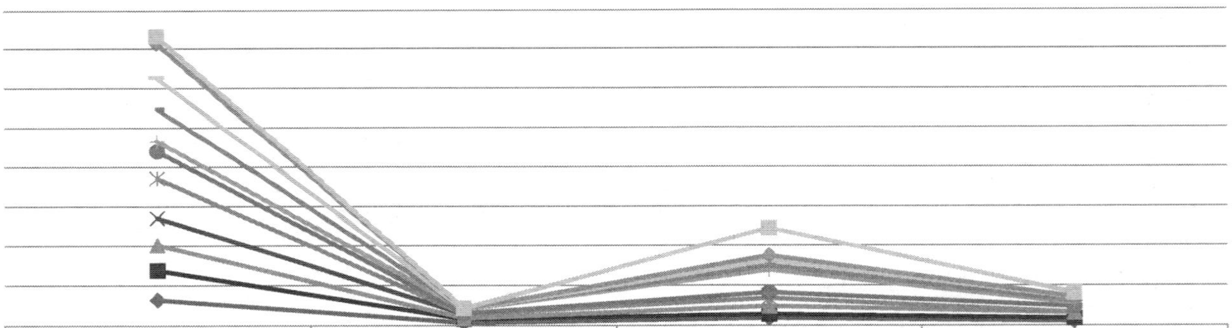

	Electricity – from fossil fuels:	Electricity – from nuclear fuels:	Electricity – from hydroelectric plants:	Electricity – from other renewable sources (solar, wind, etc.):
Brazil	18.70%	1.50%	69.30%	10.50%
Iran	85.60%	1.20%	12.40%	0.80%
Australia	78.50%	0.00%	12.70%	8.80%
Indonesia	83.20%	0.00%	11.00%	5.80%
Canada	25.70%	10.00%	55.80%	8.50%
India	69.30%	1.90%	14.00%	14.80%
S.Arabia	99.90%	0.00%	0.00%	0.10%
Russia	68.80%	10.10%	20.20%	0.90%
China	64.00%	2.00%	20.30%	13.70%
USA	73.50%	9.60%	9.50%	7.40%
World Mix	64.20%	6.80%	18.40%	10.60%

Figure 3.10 World electricity mix and top producing nations: % of total installed capacity (2012).

Source: the Author, based on:

1 CIA, 2017a. *CIA World Factbook, 2017*. Available at: www.cia.gov/library/publications/resources/the-world-factbook/geos/xx.html,
2 EIA, 2017a. *International Energy Outlook 2016*. Report Number: DOE/EIA-0484(2016). Release Date: May 11, 2016. U.S. Energy Information Administration. Available at: www.eia.gov/outlooks/ieo/world.cfm, last accessed: June 21, 2017.

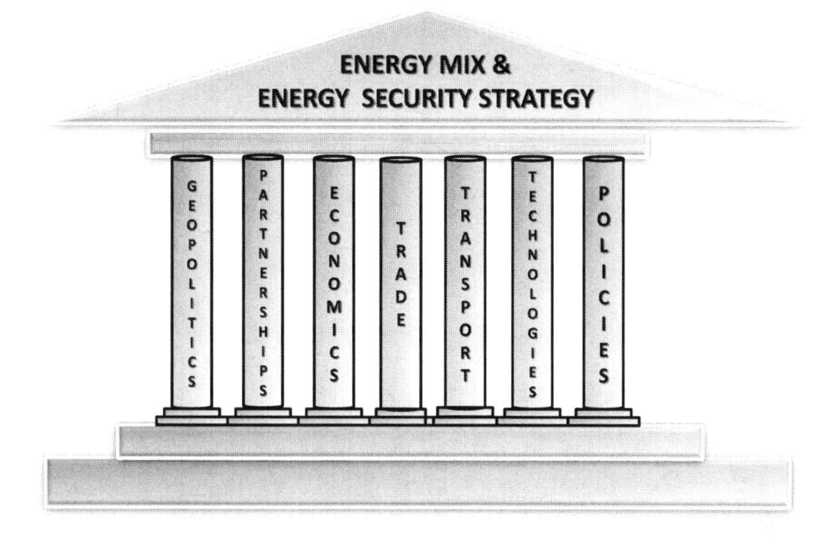

Figure 3.11 The pillars of energy security.

Source: the Author.

Figure 3.11 depicts the pillars shaping a nation's energy mix, which in turn determines their energy security strategy. These pillars include geopolitical, economic and policy factors, pertinent partnerships, trade and transport issues, as well as technological innovations. All of these factors are interrelated, as illustrated in the following analysis.

1 **Geopolitical developments**: The geopolitical security threats should be examined in correlation with the increasing energy consumption trends among blocks of countries such as the OPEC (Organization of the Petroleum Exporting Countries), OECD (Organization for Economic Co-operation and Development), BRICS (Brazil, Russia, India, China, and South Africa), VISTA (Vietnam, Indonesia, South Africa, Turkey and Argentina), and so on (Burns, 2014).

   Areas of concern should include indices such as GDP, external debt, currency fluctuations, unemployment, internal production/consumption, energy mix, etc.

   Geopolitical developments among new and existing global players include but are not limited to:

   a global partnerships and trade blocks;
   b challenges with national sovereignty, territory or energy-driven disputes, wars, social turmoil;
   c changes in political alliances, e.g., trade agreements or the formation of new trading blocks, shaping a new supply–demand equilibrium;
   d the role of state-owned banking systems, and the development of new banking, financial and investment institutions, creating a new market equilibrium;
   e the discovery of new energy sources, transforming a nation's role from major importer to a major exporter, e.g., America's shale gas;
   f the development of new technologies and processes, enabling unconventional energy production, or reducing the environmental impact;

g energy dependency due to i) wars, ii) supply chain disruptions, iii) the depletion of existing energy resources;
h safety, security or environmental regulations restricting the use of particular energy sources, and promoting alternative energy segments (Burns, 2016a; Burns 2016b).

2 **Partnerships**, including a) trade agreements among nations, deriving from geopolitical alliances, but also b) public/private partnerships. These partnerships will form joint strategies to facilitate trade, transport and security preparedness.
3 **Trade and transport** strategies eventually shape supply chain configurations including trans-continental haulage, pipeline systems, sea and land infrastructure, etc.
4 **Industrial developments** occurring in all energy segments, both non-renewables and renewables, including new technologies, new regulations, and public/private partnerships.
5 **Economic developments** deriving from energy-related, currency-related, or national economy-related factors. Energy economics entail a number of interrelated factors, including but not limited to: i) oil price fluctuations, ii) changes in energy supply and demand, iii) energy production and consumption, frequently driven by cost–benefit analysis.
6 **Policies** are driven by political agendas, trade agreements, industrial or commercial developments, but also by the necessity to mitigate security, safety and environmental challenges. Frequently, policies are inspired or empowered by novel technologies, e.g., green technologies enable policies that aim towards the reduction of greenhouse emissions.
7 **Technologies and process improvements** represent the catalyst that determines the outcome of cost-benefit studies and feasibility studies prior to the energy exploration/production stage. In our times, most conventional energy sources are depleted, and there is increasing pressure to retrieve unconventional sources. In this respect, technological innovations will determine a nation's energy mix by defining which energy sources are retrievable and at what cost.

And while the aforementioned pillars of security are the key tools to shape the energy strategy of nations and energy majors, it is the same pillars that affect energy security through inherent weaknesses or circumstantial vulnerabilities.

Geopolitical Alliances: Where Change is the Only Constant

The significance of energy security has been heightened during the 21st century and our modern era of geopolitical tensions. Energy-producing nations like Libya and Venezuela suffer from internal political tensions, whereas nations like Qatar and Iran may face instability and geopolitical isolation. Terrorism has impacted our society and the way we do business from Nigeria to the Philippines, and from Syria to Iraq and Afghanistan.

In seeing the bigger picture, energy security challenges and opportunities leave their own footprint and seal the destiny of nations and our planet as a whole. Geopolitical synergies become most impactful when countries take advantage of their group strengths and innovate, produce, and grow stronger together. And yet, change is the only constant: Today's allies may be tomorrow's enemies, and vice versa.

How Energy Bridges Politics and Economics

This section serves as a bridge between politics and energy economics. Facts and figures will enforce our arguments when positioning the political and economic influence of the major energy nations to be covered in this section.

Tables 3.1a and 3.1b provides economic, demographical and energy-related data for the top 20 energy nations, to be duly analyzed. The key terms used in these tables are explained herewith:

1 **Gross Domestic Product (GDP)**

Gross Domestic Product (GDP) is a primary economic index used to evaluate the size and annual growth of a national economy. It signifies the financial value of all commodities and services made in the physical borders of a country, over a fiscal year or particular point in time. Namely,

GDP = G + I + C + (Ex-Im)

G = Government expenditures on services and commodities;
I = Investments;
C = Consumption of the private and domestic sectors;
Ex-Im = the dollar value of net exports less net imports.

2 **GDP at purchasing power parity (PPP)**

Gross Domestic Product at purchasing power parity (PPP) or estimated gross domestic product based on purchasing power parity examines diverse costs of living and value levels, in comparison with the U.S. dollar. It estimates a nation's economy while incorporating inflation rates when estimating the price of domestic products and commodities. This index is useful to evaluate the domestic economy and purchasing power, whereas GDP is an index most useful for comparing the economies of different nations.

3 **Purchasing power parity (PPP)**

Purchasing power parity (PPP) is a financial index that compares the purchasing power of two or more countries, while estimating their exchange rate in equilibrium. The purpose of harmonizing the price level variations among nations is to attain an objective cross-country comparison.

When economists determine the PPP rates – which typically have a slow and predictable progress over a number of years – it is easier to focus on other market-changing events that cause fluctuations or shifts of power.

4 **Energy intensity**

Energy intensity pertains to the energy volume utilized as input to attain a specific service or activity. It is measured as energy per output unit. It also reflects the ratio among a nation's GDP and the Gross Inland Energy Consumption (GIEC), which is the aggregate amount of domestic consumption of the key energy sources: Oil, gas, nuclear, solid fuels and renewables.

5 **Balance of Trade (BoT)**

Balance of Trade (BoT) represents the difference in value between a nation's imports and exports within a fiscal year, or during a specific time. It is the most significant index within a nation's Balance of Payments (BoP).

A nation's trade is balanced in the very rare case that its imports and exports are almost of equal value, and thus reach equilibrium.

Typically nations have a Trade Surplus (TS), or a Trade Deficit (TD). Trade Surplus occurs when the value of a nation's exports surpass imports, whereas Trade Deficit occurs when the value of imports surpass exports, with the former (TS) being considered as most profitable for a nation. Countries with positive net exports enjoy a healthier economy, as their output grows to serve large global market segments, hence is not intended exclusively for their domestic buyers.

High demand for their products is not only a confirmation of quality and high value-for-money, but promotes job growth and the optimum utilization of the domestic factors of production. This leads to the growth of the GDP.

The Balance of Trade ratio estimation is very similar in the estimation of the Balance of Energy ratio: When a nation's balance of energy is positive, energy independence is achieved. Conversely, a negative balance of energy signifies energy dependence. A nation is importing or uses up energy from national reserves, which affects their future capacity to withstand increased demand or prolonged supply chain disruptions.

6 **World population and energy consumption**

It appears that the Balance of Energy is only one part of the equation. Economic and demographic indices are also worth observing. The population factor resembles the two sides of a coin: High population contributes to high workforce and high energy consumption, which are both positive aspects. However, high population imposes additional pressures upon governments to seek energy sources, and this pressure contributes to energy dependence. According to the United Nations, global population is estimated to reach 9.8 billion by 2050, with nine countries representing half of the global growth: China, India, Nigeria, Congo, Pakistan, Ethiopia, the U.S., Tanzania, Uganda and Indonesia. By 2100, global population will reach 11.2 billion (UN, 2017). At the same time, China's 13th five-year plan (2016–2020) reforms the one-child policy into a two-children policy, which will have a serious impact on the nation's population growth and subsequently its energy consumption and economy (USCC, 2017).

Energy Geopolitics: The Key Players

The world's greatest economies, namely the U.S., China, India, Russia, and the E.U., combine both growth and uncertainty: Their footprint is vast, hence the global market responds instantly to its fluctuations, for better or worse.

China is certainly a country of extremes: Its regime actively supports state-supported, mass industrial production and exports, while restricting private sector investments. In 2016 it was proclaimed the world's greatest economy, surpassing the U.S., and yet its currency and salaries (PPP) remain well below the global average, i.e., one US dollar = 6.62 yuan (RMB), and the Chinese PPP is $14,500 per year, compared with US PPP being $57,300 per year (CIA, 2017a). In terms of energy, it is positioned as the top energy-producing and consuming nation, with the greatest Balance of Energy deficit, reaching –461 in 2015 (Tables 3.1a and 3.1b). Hence, while the country produces and imports vast amounts of energy, its high population and industrial intensity feed an insatiable need for more power.

As China's fossil fuels are being used at a rapid pace, the nation sees alternative energy as a long-term solution to secure the nation's energy independence. China is the world's leader in renewables production, followed by India and the U.S.

China's 13th five-year plan (2016–2020) stipulates particular energy goals, namely a) a shift from fossil fuels to renewable energy, b) vast investment in hydropower with large-scale plants in the Yangtzi River, and c) greening, i.e., investment in environmental technologies and reduction of CO_2 emissions and greenhouse gases (USCC, 2017).

What China and India have in common is their high industrial activities and exports, with a moderate currency and a low PPP. Most important is the need to explore energy production opportunities, as their energy needs increase year after year.

Table 3.1a Top energy producing and consuming nations: Key indicators.

Rank	*Energy key players*	*Energy production 2015 (MTOE)*	*Energy consumption 2015 (MTOE)*	*Balance of Energy (2015)*	*Population (2016)*	*GDP (2016) (purchasing power parity)*	*PPP*	*Exchange rate per USD*	*Currency*
1	China	2640	3101	–461	1,373,541,278	$10.73 trillion	$14,500 (2016 est.)	6.626 (2016 est.) 6.2275 (2015 est.)	Renminbi yuan (RMB)
2	USA	2012	2196	–184	323,995,528	$18.56 trillion	$57,300 (2016 est.)	1	USD
3	Russia	1341	718	623	142,355,415	$3.751 trillion	$26,100 (2016 est.)	68.06 (2016 est.) 60.938 (2015 est.)	Russian rubles (RUB)
4	S. Arabia	650	215	435	28,160,273	$1.731 trillion	$54,100 (2016 est.)	3.75 (2016 est.) 3.75 (2015 est.)	Saudi riyals (SAR)
5	India	593	882	–289	1,266,883,598	$2.251 trillion	$6,700 (2016 est.)	68.3 (2016 est.) 64.152 (2015 est.) 64.152 (2014 est.)	Indian rupees (INR)
6	Canada	456	251	205	35,362,905	$1.674 trillion	$46,200 (2016 est.) $46,200 (2015 est.)	1.331 (2016 est.) 1.2788 (2015 est.)	Canadian dollars (CAD)

(continued)

Table 3.1a (continued)

Rank	*Energy key players*	*Energy production 2015 (MTOE)*	*Energy consumption 2015 (MTOE)*	*Balance of Energy (2015)*	*Population (2016)*	*GDP (2016) (purchasing power parity)*	*PPP*	*Exchange rate per USD*	*Currency*
7	Indonesia	401	227	174	258,316,051	$3.033 trillion (2016 est.) $2.888 trillion (2015 est.)	$11,700 (2016 est.) $11,300 (2015 est.)	13,483 (2016 est.) 13,389.4 (2015 est.)	Indonesian rupiah (IDR)
8	Australia	355	126	229	22,992,654	$1.189 trillion (2016 est.) $1.156 trillion (2015 est.)	$48,800 (2016 est.) $48,300 (2015 est.)	1.352 (2016 est.) 1.3291 (2015 est.)	Australian dollars (AUD)
9	Iran	289	244	45	82,801,633	$1.459 trillion (2016 est.) $1.397 trillion (2015 est.)	$18,100 (2016 est.) $17,600 (2015 est.)	30,462.1 (2016 est.) 29,011.5 (2015 est.)	Iranian rials (IRR)
10	Brazil	280	299	–19	205,823,665	$3.081 trillion (2016 est.) $3.192 trillion (2015 est.)	$14,800 (2016 est.) $15,400 (2015 est.)	3.39 (2016 est.) 3.3315 (2015 est.)	Reals (BRL)

Source: the Author based on:

1 CIA, 2017a. *CIA World Factbook, 2017*. Available at: www.cia.gov/library/publications/resources/the-world-factbook/geos/xx.html, last accessed: June 21, 2017.
2 EIA, 2017a. *International Energy Outlook 2016*. Report Number: DOE/EIA-0484(2016). Release Date: May 11, 2016. U.S. Energy Information Administration. Available at: www.eia.gov/outlooks/ieo/world.cfm, last accessed: June 21, 2017.

Table 3.1b Top energy producing and consuming nations, continued: Key indicators.

Rank	*Energy Key Players*	*Energy Production 2015 (MTOE)*	*Energy Consumption 2015 (MTOE)*	*Balance of Energy (2015)*	*Population (2016)*	*GDP (2016) (purchasing power parity)*	*PPP*	*Exchange Rate per USD*	*Currency*
11	Nigeria	259	134	125	186,053,386	$1.089 trillion (2016 est.) $1.108 trillion (2015 est.)	$5,900 (2016 est.)	305.00 (2016 est.) 192.73 (2015 est.)	Nairas (NGN)
12	UAE	215	81	134	5,927,482	$667.2 billion (2016 est.) $652.4 billion (2015 est.)	$67,700 (2016 est.) $68,100 (2015 est.)	3.673 (2016 est.) 3.673 (2015 est.)	Emirati dirhams (AED)
13	UK	118	215	–97	64,430,428	$2.788 trillion (2016 est.) $2.737 trillion (2015 est.)	$42,500 (2016 est.) $42,000 (2015 est.)	0.7391 (2016 est.) 0.6542 (2015 est.)	British pounds (GBP)
14	Norway	206	32	174	5,265,158	$364.7 billion (2016 est.) $361.7 billion (2015 est.)	$69,300 (2016 est.) $69,500 (2015 est.)	8.615 (2016 est.) 8.0646 (2015 est.)	Norwegian kroner (NOK)
15	Mexico	201	187	14	123,166,749	$2.307 trillion (2016 est.) $2.259 trillion (2015 est.)	$18,900 (2016 est.) $18,700 (2015 est.)	18.34 (2016 est.) 15.848 (2015 est.)	Mexican pesos (MXN)

(continued)

Table 3.1b (continued)

Rank	*Energy Key Players*	*Energy Production 2015 (MTOE)*	*Energy Consumption 2015 (MTOE)*	*Balance of Energy (2015)*	*Population (2016)*	*GDP (2016) (purchasing power parity)*	*PPP*	*Exchange Rate per USD*	*Currency*
16	Venezuela	186	59	127	30,912,302	$468.6 billion (2016 est.) $520.7 billion (2015 est.)	$15,100 (2016 est.) $17,000 (2015 est.)	56.57 (2016 est.) 13.72 (2015 est.)	Bolivars (VEB)
17	Kazakhstan	164	78	86	18,360,353	$468.8 billion (2016 est.) $464.2 billion (2015 est.)	$25,700 (2016 est.) $26,300 (2015 est.)	348.5 (2016 est.) 221.73 (2015 est.)	Tenge (KZT)
18	S. Africa	162	138	24	54,300,704	$739.1 billion (2016 est.) $735.4 billion (2015 est.)	$13,500 (2016 est.) $13,400 (2015 est.)	15.7 (2016 est.) 12.7581 (2015 est.)	Rand (ZAR)
19	Algeria	140	53	87	40,263,711	$609.4 billion (2016 est.) $588.4 billion (2015 est.)	$15,000 (2016 est.) $14,700 (2015 est.)	110.1 (2016 est.) 100.691 (2015 est.)	Algerian dinars (DZD)
20	Colombia	125	34	91	47,220,856	$688 billion (2016 est.) $674.5 billion (2015 est.)	$14,100 (2016 est.) $13,800 (2015 est.)	3,051.1 (2016 est.) 2,741.8 (2015 est.)	Colombian peso (CP)

Source: the Author, based on:

1 CIA, 2017a. *CIA World Factbook, 2017*. Available at: www.cia.gov/library/publications/resources/the-world-factbook/geos/xx.html, last accessed: June 21, 2017.

2 EIA, 2017a. *International Energy Outlook 2016*. Report Number: DOE/EIA-0484(2016). Release Date: May 11, 2016. U.S. Energy Information Administration. Available at: www.eia.gov/outlooks/ieo/world.cfm, last accessed: June 21, 2017.

India's energy policy is determined by its large population, leading economic and commercial activities. The country's scarcity of non-renewable activities compels the diversification of its energy mix, which comprises coal (4th largest global reserves), oil and gas, but also solar, biomass, wind and nuclear energy. The United States–India Peaceful Atomic Energy Cooperation Act is a part of India's commitment to invest in the development of its nuclear energy sector (see the case study in Chapter 6 of this book).

In 2017, an adjustment of the tariff cost of solar power generation positioned it significantly lower compared with other energy sources, whereas solar electricity became the standard price based on which all other energy sources (both renewables and non-renewables) are measured.

The U.S. is the second top nation with high production and consumption needs. Its population is about a quarter of China's, and yet the standards of living are higher, so that each American consumes eight times more power than the average Chinese, and gets paid four times more than the average Chinese. To make things more complex, the U.S. is an economy based on free trade, with the world's most dynamic and diverse private sector. Which means that, as U.S. corporations outsource and establish offshore legal entities in search of cheaper land and labor (and energy!), they actually feed the GDP of other nations, including China's. Which means that a nation's GDP reflects national wealth within a nation's physical boundaries, yet may not be the most reliable national wealth indicator as it does not include the private sector activities abroad (Burns, 2014).

What the U.S., Canada, Russia, Saudi Arabia and Australia have in common is the world's largest reserves of fossil fuels, relatively higher salaries, and their potential to retain a leading energy exporter's position for several decades ahead.

What Russia, Nigeria, Brazil, South Africa, Indonesia, and Iran have in common is that their energy production activities have boosted their economic influence upon energy-thirsty regions such as China, India, and the European Union, and are now influential members of global trade blocs. Russia and Iran in particular form a category on their own, as their energy reserves are so much needed that they have formed an intense alliance on nuclear power and geopolitical tactics in Syria, Afghanistan, etc. Needless to say, Russia and Iran are also military allies and share common strategies in critical political and war issues.

The Trade Bloc Power: Reshuffling the Energy Deck

In our globalized era, nations strongly recognize the "strength in unity" notion: Some nations are energy-rich, others enjoy cheap factors of production, others possess colossal physical capital, while others have technological and innovative capital. Tables 3.1a and 3.1b demonstrate how individual nations like China, the U.S., India and Brazil suffer from an energy deficit, yet enjoy immense profits from industrial and other global trade deals. Putting all this wealth together, what we have are trade blocs of immense global power that max out the impact of their individual factors of production.

Geopolitical alliances also known as trade blocs come to "reshuffle the deck" by rebalancing production/consumption ratios of energy and other major commodities and factors of production, and thus converting energy and other areas of deficit into surplus. (Also see Burns (2014) for a list of the major global trade agreements.)

The accumulation of such wealth would suggest that the sky is the limit, and that diplomatic tensions seem to be the only obstacle ahead. To be precise, conflict of interest could be an obstacle, bearing in mind that most of these nations have secured membership in more than one of these major trade blocs. Namely, Mexico is a member of NAFTA and MINTS; South Africa is a member of BRICS, VISTA and MINTS; both Indonesia and Turkey are members of VISTA and MINTS, and the list goes on. Another factor to determine the longevity and success of these partnerships entails the political environment of each member state, especially after each election term. Synergies always work best in an environment of similar political views and political stability, but also dissolve in dictatorial regimes or an unstable political environment.

Some of the world's most significant trade blocs are the following:

a NAFTA (North America Free Trade Agreement: USA, Canada, and Mexico)

The NAFTA agreement was implemented in 1994 with the purpose of promoting free trade while eliminating trade barriers and tariffs. The obvious benefits entailed the combination of each nation's competitive and comparative advantages. Capital and innovative technologies, energy reserves and low-cost labor and land were put to use to promote the growth of all three economies. Industrial activities, transportation, agriculture, labor and the environment are some key areas of collaboration. While controversial views exist as to the benefits and losses each nation encountered as a result of the NAFTA agreement, its continuation seems to be the only way ahead for the Americas, bearing in mind the intense competition and accumulation of power from the other trade blocs.

b BRICS (Brazil, Russia, India, China, and South Africa)

The BRICS trade block was formally established in 2010 and soon assumed a strong political influence, bearing in mind that all member states are also G-20 members. Diversity is key in this trade block:

- Its five members are located on three continents.
- China and India's populations combined exceed 2.5 billion people, i.e., one-third of the global population. On the contrary, South Africa's population is about 50 million people.
- Its members' financial and investment status is also characterized by diversity: India's economy seems to be the most preferred by foreign investors; China's economy has experienced moderate growth after ten years of growth frenzy. Brazil and Russia's economies are directly correlated with the oil price, hence suffered great losses in 2015 and 2016.
- In 2014, China formed the New Development Bank (also known as BRICS Bank), headquartered in Shanghai. While the bank's initial goal was to form a common development policy among members, the subsequent strategic decisions of the bank to fund electrification, trade and infrastructure initiatives appears to challenge the hegemony of the World Bank and the IMF institutions.

c VISTA (Vietnam, Indonesia, South Africa, Turkey, and Argentina

This is a trade bloc pertaining to emerging economies. Indonesia seems to be a significant energy provider, whereas China and South Korea have outsourced their manufacturing activities to Vietnam and Indonesia, hence passing on the baton of future industrial deals.

Argentina is a G-20 country member, with high oil and nuclear power activities, a well-developed transportation system and tremendous potential for tidal and hydroelectric energy due to the tidal La Plata river and seaport that was shaped by the convergence of the Parana and Uruguay rivers. Moreover, Argentina is among the 12 signatories to the Antarctic Treaty, and has territorial claims in Antarctica, the south pole region with ongoing evaluations of geothermal energy and future oil exploration.

The cases of South Africa and Turkey will be covered individually in a subsequent paragraph due to their memberships in more than one trade bloc, and their unique and multifaceted positioning.

d **MINTS (Mexico, Indonesia, Nigeria, Turkey, and South Africa)**

This bloc has gained increasing popularity due to the nations' steady economic growth, increasing attraction of foreign investment, natural resources in energy, minerals and manufacturing. This wealth is combined with cost-effective factors of production, i.e., labor and land. The nations' relatively high demographics also position them as emerging markets with buying power. Their leveraging status increases when other, more established, trade blocs such as BRICS face economic or sociopolitical challenges.

e **South Africa: A star is born**

The case of South Africa is pretty unique and merits attention: An active member in the BRICS and MINTS trade blocs, the country combines so much wealth and possibilities that it is considered as one of the future global game changers. South Africa has suffered recession and high unemployment rates, yet a flamboyant rebound is anticipated: South Africa combines diverse areas of wealth, reserves and factors of production, hence has the potential for tremendous growth and success for decades to come. Specifically, South Africa combines three significant energy sectors: Oil, coal and uranium, a vital mineral used in nuclear energy. Furthermore, it is the world's largest gold producer and has significant reserves of diamonds, platinum and palladium. Moreover, its unique position amidst four continents makes it a popular maritime transit and replenishment center: The world's major sea trade routes between America, Asia, Australia, the Middle East, and Europe pass through the nation's sea ports of Cape Town, Durban, Port Elizabeth, Richards Bay, etc. The wide spectrum of resources and activities boosts the country's foreign exchange and strengthen its currency, which is much stronger than China's and India's (see the section on Currency Wars, pages 146 and 369, and Currency Manipulation, pages 15, 24 and 109).

f **West Africa (Gulf of Guinea): The shift from LDC to Rapidly Developing Economies**

West Africa emerges as a prominent global energy player, with oil-rich Angola, Nigeria, Ghana and Equatorial Guinea (SW Africa), and uranium-rich Guinea (NW Africa) and Namibia being the major regional producers.[2] The initial years of exploration offered significant lessons learned for the energy conglomerates who invested in the region's upstream, and set up the foundations for a long-lasting infrastructure. Moving from less developed to rapidly developing economic status did not happen overnight, but it certainly was an experience that grew the political and tribal leaders, and populations out of their comfort zones. West Africa's energy reserves have initiated a new chapter in the region's culture of diversity, where tribal leaders needed to work with central government and multinational moguls, and swiftly train their people in response to the global demand. Efforts towards political stability could resolve many of the regional challenges such as bureaucracy, income inequality, poverty or corruption.

g **Turkey: Location is key**

The case of Turkey is also unique: The nation has gained most of its economic power due to its strategic geographic location amidst Asia and Europe, while surrounded by Russia, Iraq, Iran, Syria, and other Mediterranean and Middle Eastern countries. While its natural energy reserves are limited to lignite, the country has enabled the transportation of energy from Russia, Iraq, Iran, and Azerbaijan to Europe and Asia. Moreover, the country facilitates the landlocked Caspian sea's oil and gas trade, hence several Russian pipeline projects have involved Turkey (Burns, 2004). The country's first nuclear plant will be launched by 2020.

European Union: The Path to Energy Independence

The EU imports over 50% of all the energy it utilizes. Its import dependency is especially high for crude oil (over 80%) and natural gas (82%). The overall import bill exceeds €1 billion on a daily basis.

Europe's energy dependency makes the region vulnerable to supply–demand challenges, but also to geopolitical and supply-chain disruptions throughout the trade route from its suppliers to the European refineries. Here are some observations.

Reliance on Limited Energy Suppliers

Several European nations are heavily dependent upon a single supplier, e.g., several nations entirely depend on Russia for their natural gas supplies. Dependence on exclusive suppliers typically increases the risk of supply disruptions, regardless of the cause of hindrance, i.e., economic, political, weather, transportation, infrastructure or production delays.

Geopolitical Risks Leading to Supply Chain Disruptions: The 2009 Russia–Ukraine Gas Dispute

Russia's oil and gas trade flows frequently pass through Ukraine. Hence, any political dispute between these two countries will eventually result in delivery delays and energy shortages in Europe. In 2009 Gazprom, the Russian natural gas corporation, requested that Naftogaz, a Ukrainian gas corporation, settled prior financial obligations for gas supplies purchased. The prolonged dispute between Russia and Ukraine resulted in Ukraine delaying the natural gas cargo flows passing through Ukraine into Europe, thus leaving many countries of southeastern Europe with severe shortages.

Limited Oil and Gas Storage Capacities

The establishing of oil and gas storage capacities is another significant part of a nation's energy security strategy.

It is worth noting that oil and gas storage in most European nations is limited to a couple of days up to several weeks (maximum of 90 days). In 2002, Loyola de Palacio, EU Energy Commissioner, proposed the increase of petroleum inventory from 90 to 120 days (European Commission EUROPA.EU, 2017 and EUR-Lex, 2009).

As Europe enhances its long-term storage capacity, it will attain higher energy security levels, thus safeguarding its nations from power outage, diminished industrial activity, or even sociopolitical turmoil.

Cyprus: Europe's New Energy Partner

Cyprus is a prosperous energy-rich Mediterranean island whose small size[3] is disproportionate to its rich history, geopolitical turmoil, and impressive energy reserves.

An armed intervention from Turkey in 1974 occupied 36% of the country (northern part of the island), an act that was never accepted by the UN or any nation other than Turkey. The Republic of Cyprus became a Eurozone member, with its European Union status only applied to the 64% of the country (southern part of the island) that has the internationally recognized government (CIA, 2017b).

Over the past decade, substantial natural gas reserves were found in the southern part of the island in the Cyprus Exclusive Economic Zone (EEZ). According to CIA 2014 estimations, the proved reserves exceed 141.6 billion cubic meters (CIA, 2017b). According to Noble Energy's 2013 estimations, and as quoted by the U.S. Government, 4.5 trillion cubic feet (or 0.127 trillion cubic meters) were discovered specifically in Block 12 of the ROC Exclusive Economic Zone (EEZ) (DOC, 2016).

In 2016, pursuant to a few licensing rounds, the Cyprus government awarded four of the blocks to global conglomerates, namely ExxonMobil (USA), Noble Energy (USA), Eni (Italy/EU) and Total (France/EU).

The Cyprus case offers invaluable lessons for the energy economics and political science and logistics/transport disciplines. There is a win/win arrangement for all stakeholders, including Cyprus and the energy corporations, and the European Union markets, for the following reasons:

1. **Inter-regional trade facilitation**: Cyprus' EU membership facilitates importing/exporting procedures and eliminates delays.
2. **Energy security**: Contingency planning is vital for a nation's energy strategy. The Russia/Ukraine energy crisis case taught Europe that unforeseen supply/demand or transportation factors can have serious implications and halt vital national activities, even threatening national security.
3. **Supply chains** are more reliable when the supply and demand locations are closer. Considering that sea transport will be the prevailing haulage mode for Cyprus energy to be distributed in Europe, this translates into three days (to Italy) or up to ten days (to Western Europe, namely England and Germany).
4. **Cyprus is not landlocked**, as opposed to Russia or the Caspian regions that require high investment and years of infrastructure development for pipelines to haul their energy to the global markets.

   Maritime transportation infrastructure is readily available to move cargoes from Cyprus to its partners' refineries, with minimum delays or investment requirements.
5. Europe's need for increased **storage capacity** can be alleviated by the Cyprus offshore facilities, which can safeguard against any disruptions or emergency event that would require imminent energy supplies.
6. Cyprus enjoys a **strategic geographical position** that has been underutilized during the years of the island's occupation. It can now reach its full capacity once multinational energy companies establish a commercial basis on the island's EEZ.

In conclusion, the nation's energy wealth can leverage its geopolitical position and attract enduring alliances that will further empower the country.

A Case Study on Zimbabwe's Energy Independence

Zimbabwe is another energy-rich nation that has the capability to improve its status once it overcomes issues of political instability and global isolation. Back in 2010, the China–Africa Development Fund signed an agreement with a subsidiary of China National Nuclear Corporation, China's major nuclear power contractor, to develop a uranium plant partnership. There have also been claims of Zimbabwe's collaboration with Iran, Russia, and North Korea, yet these were never confirmed nor denied.

Over the past few years, Zimbabwe announced plans to nationalize its mining sector, including the uranium, diamonds, coal, and methane projects, after dissatisfaction with foreign governments harvesting the nation's wealth, with little profit left for the country and its people.

However, the nation's reform efforts proved futile with failure to pursue a production-driven growth model. The nation's infrastructure and development capabilities were not sufficient to pull through such projects, and the country's agro-based production was not enough to keep the economy healthy. The Zimbabwe dollar's devaluation did not make exports more competitive. Instead, diminished aggregate demand has driven production low, and hyperinflation has weakened domestic markets, which are cash-depleted.

The turning point for Zimbabwe will certainly not be easy for its people, as the nation's unemployment rate is 95%, i.e., among the highest globally, and more than half of the country's population has no electricity (i.e., 8.5 million out of 15.6 million people (CIA, 2017a).

A Shift from Globalization to Resource Nationalization and State-Owned Enterprises (SOEs)

> Resource nationalism has become a contagion impacting the mining and metals industries across the globe.
>
> Andy Miller, Ernst & Young,
> "Resource nationalism update". *Mining and Metals*, 2014

From the mid-1940s until the oil crisis of 1973, 85% of the global oil reserves were handled by "the Seven Sisters," i.e., seven energy conglomerates, five of which were based in the U.S. (Exxon, Mobil (now merged ExxonMobil), Chevron, Gulf, and Texaco), one in England (British Petroleum (BP)), and one in Denmark (Royal Dutch/Shell).

Four decades later, the globalization as we know it has been transformed in the name of resource nationalism. The roles are being reversed, as state-owned organizations, such as Saudi Aramco (Saudi Arabia), China National Petroleum Corporation (China), Gazprom and Rosneft (Russia), National Iranian Oil Company (Iran), Petrobras (Brazil), PDVSA, Petróleos de Venezuela, (Venezuela), and Petronas (Malaysia), claim to control 75% of the world's energy reserves.

Once these governments have established their own oil corporations, they place pressure during negotiations with the petroleum multinationals. By this new movement energy-producing countries seek to transition their resources' governmental and monetary leadership from international corporations to state-owned ones.

The new geopolitical reality gears towards "resource nationalism," which may be called "rediscovered statism" or "nationalistic mercantilism." State-owned enterprises (SOEs) are the legal entities that take ownership and control of the state's natural resources, from production to export. Nationalization of the energy industry empowers nations to take a leadership role, and to renegotiate contracts, concession agreements and export prices.

Russia, for example, has asked oil majors like BP and Shell to deliver majority stakes to Russia-owned Gazprom. At the same time, Bolivia's military has occupied the gas-rich regions, while international investors were given a six-month ultimatum to abide by their new nationalization rules or depart.

Regardless of its name, the rediscovered trend reflects the propensity of the energy-rich developing nations to profess sovereign control over their natural resources, in particular petroleum and gas, minerals and metals. The outcomes of nationalization pertain to political and commercial strategies that may exclude foreign interests, or retain the partnership with the risk of working in a hostile state.

With reference to the Energy Security Matrix (see page 109–112 and 119, this chapter), resource nationalism may grant national independence but it depends on the regime to grant energy affordability and accessibility for its people, and thus meet all objectives of a national energy strategy.

The South China Sea Oil and Gas Reserves Dispute

The South China Sea disputes entail both island territory and maritime claims concerning several sovereign nations within the South China Sea zone that is rich in fossil fuels. In fact, the Chinese call it "the Second Persian Gulf." South China Sea plaintiffs (i.e., Taiwan, Japan, the Philippines, Malaysia, Indonesia, Brunei, and Vietnam) are anticipating the judgment of an arbitration claim on the legitimacy of the Chinese government's actions. The decision will be determined by the United Nations' International Tribunal for the Law of the Sea. Meanwhile, China views that the possession strategy is allowed by international law.

The group of islands reclaimed by several South China Sea nations have substantial innate value. China's territory reclamation and man-made island constructing in the region is primarily due to offshore oil and gas resources.

The disputed area of the South China Sea Platform contains the Spratly Islands, the Reed Bank (or Recto Bank), and the Dangerous Ground. This is an oil-rich, carbon-rich region with the ideal geological conditions essential for hydrocarbon development, especially petroleum. State-owned oil major China National Offshore Oil Company (CNOOC), in charge of nearly all of China's offshore hydrocarbon projects, evaluates that the region contains approximately 125 billion barrels of oil and 500 trillion cubic feet of natural gas in undiscovered areas.

Due to the fact that a high proportion of the world's trade passes through the South China Sea, there are many non-claimant nations that want the South China Sea to remain as international waters, with several nations (e.g., the United States of America) performing "freedom of navigation" operations in close proximity to the Chinese islands.

In 2016 the Permanent Court of Arbitration determined that the region is part of the Philippine Exclusive Economic Zone (EEZ), however the region's disputes are ongoing as to the regional economic rights, keeping the Philippines from exploration activities.

An Economic Landscape

Arguably, the economic factor is an energy segment on its own, as money makes the world spin. Each year, the global energy industry invests at least 2 trillion dollars, hence contributes to the global GDP. To verify how the annual global energy production rates contribute to the global GDP, it is worth noting that for every $10 growth in the price of oil, an oil corporation gains about $4.5 billion in revenue each year. This direct investment amount is further enhanced by indirect investment from contractors and subcontractors supporting the global upstream, midstream and downstream supply chains, but also the recruitment, training and education of workforce.

Traditional economics praises the significance of stable trade agreements and predictable economic cycles, with hard currencies whose value will not fluctuate nor depreciate in the foreseeable future. And yet unpredictability is the main hard currency in our era, and economic forecasts are regarded with suspicion, or taken with a grain of salt. And there is a reason for it: Never before has the world seen so much aggregate volume of energy, coming from so many sources, yet with so many geopolitical and financial complications.

Timeline of Major Global Geopolitical Events, and Impact on Crude Oil Prices (1910–2017)

This section monitors the oil price fluctuations for the past century (Figure 3.12), in combination with significant socioeconomic events.

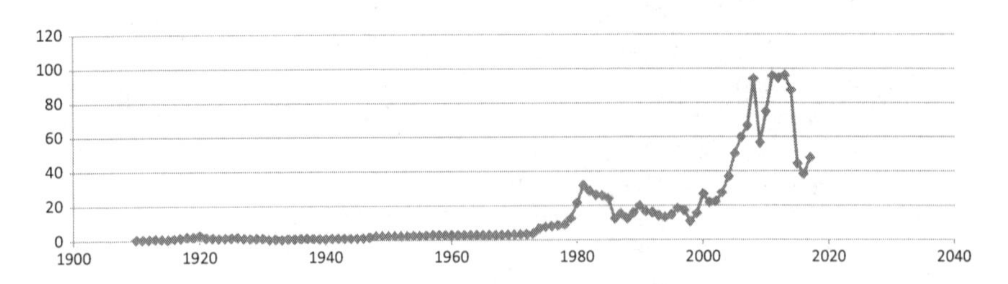

Figure 3.12 Crude oil price ($ per barrel).

Source: EIA, 2017c. *U.S. Crude Oil First Purchase Price*. U.S. Energy Information Administration. Available at: www.eia.gov/dnav/pet/hist/LeafHandler.ashx?n=pet&s=f000000__3&f=m, last accessed: June 21, 2017.

1919–1920: PEAK. The automobile industry is booming, generating a high demand for oil.

1929–1930: The Great Depression diminishes market demand for industrial products, thus eliminating oil demand. Prices of oil and commodities collapse.

1940–1945: WWII ignites an insatiable demand for oil, for military/defense purposes.

Military strategists declared that the entity (The Allies or Axis powers) with greater access to oil reserves would determine the outcome of the World War. As the Germans' access to Russian oil fails, so in 1945 the Allies win the war.

1946–1949: The aftermath of WWII creates a reconstruction period with high demand for oil for automotives and large-scale construction and manufacturing.

Oil shortage leads to peak demand and high prices.

1956–1957: Post-war reconstruction and ongoing growth raise high oil demand. Meanwhile, crisis in the Suez Canal eliminates oil supply, resulting in high oil prices. Oil buyers explore alternative producers beyond the Middle East region.

1973–1974: The Yom Kippur War between Israel and Egypt results in an Arab oil embargo against Israeli allies. Oil shortage leads once more to an oil price rise.

1978–1979: Iranian revolution and the change of regime causes contractual cancellations with western nations, prolonging the expansion stage of the market cycle.

1980: War between Iran and Iraq leads to oil shortage and an unprecedented oil price boom.

1980s: Alternative oil-producing markets increase production and global market share.

1985: Saudi Arabia increases exports. Oversupply leads to price collapse.

1988: The end of the Iran/Iraq war signified increased oil supply and price reductions.

1990: Gulf War. The invasion of Kuwait by Iraq disrupts Kuwait's oil production and the oil price peaks. During the same year, China's leader Deng Xiaoping plants the seeds of a "socialist market economy." The dragon awakens, for decades to come.

1997: The Asian crisis led to currency meltdown, private debt and diminished demand for oil. Russia's foreign exchange reserves are impacted therefrom.

Oversupply and diminished demand drop prices.

1998: OPEC attempts to adjust the price by strategically reducing production.

1997–1999: China keeps the yuan stable and joins foreign stock markets. The Asian and Russian economies also recover.

2000: The new century commences with positive market sentiment. Earlier underinvestment prevents suppliers from meeting production goals.

The global economy suffers from "unmaterialized growth" and subsequently modest profits in the energy sector.

2001: 9/11 terrorist attacks. America declares "Global War on Terrorism." The Afghanistan war in 2001 and the Iraqi war in 2003 lead to sociopolitical turmoil and market uncertainty.

2002–2003: The Venezuelan General Strike and Oil Lockout further destabilize the oil market.

2004–2007: Asia's impressive growth lifts the market to unprecedented heights while oil-producing nations use spare capacity in an effort to meet demand. The oil price approaches $100/barrel.

2008: Global economic meltdown driven by new market entrants and oversupply of oil and commodities. OPEC diminishes production to a two-year low.

2011: The Arab Spring introduced sociopolitical turmoil in Egypt, Libya, Syria, and other North African/Middle Eastern countries. Oil supply disruptions.

2015–2016: Innovative technologies enable unconventional drilling and shale exploration, causing oil oversupply. The U.S. is transformed from major importer to a major user, thus changing the market. OPEC protects status quo and retains production quotas.

Lessons Learned from the 2015–2016 Oil Price Collapse

Although the low oil prices were caused by oversupply, global energy demand also increased. As oil companies experienced diminished revenue, investment restrictions were imposed, and, as a consequence, there was inability to meet demand.

2016: Vienna Accord: OPEC agrees to reduce oil production.

2017: Market recovers. Vast gas reserves found in the U.S. and Russia. The market prepares for an economic rebound.

Economic Cycles

This section investigates the impact of economics in the energy industry and explains how these trigger new phases in economic cycles.

The energy industry, just like any other business sector, is subject to market fluctuations. An economic cycle or business cycle or market cycle describes the market's cyclical pattern from: 1) expansion, 2) peak or market boom, 3) contraction and 4) trough, or market collapse.

1 **Market expansion**

The expansion stage is characterized by increased demand for energy and a modest, or lower, energy supply or production rate. As energy producers attempt to meet the market demand, the production grows, banking and financial systems support the transactions to mitigate this demand, and the entire energy supply chain works tirelessly for the supply rate to meet with the demand anticipation. Producers that are capable recognize a market recovery.

2 **Market peak or market boom**

The peak stage reflects the optimum level of production and economic growth, generated by the buyers' insatiable need for the product. Again, this stage reflects high energy prices as the demand exceeds supply. Buyers' strategy may entail short-term consumption, long-term storage, or even speculative activities such as purchase and re-sale.

The peak is the critical point where investors must begin to sell assets, stocks, inventory, shares, etc., as the market cannot become any better. Experience and thorough study of the energy cycles are required for stakeholders such as nations and investors to recognize the actual peak of a market, and to take action just prior to the market contraction and trough. Also, preparedness is required as the transition time from peak to contraction occurs in a swift manner.

3 **Market contraction**

Contraction is the stage where supply has met demand, and gradually exceeds demands. The energy industry is capital-intensive and technology-intensive, and investment strategies take several months, even years, to materialize.

The contraction stage represents a market wake-up call that the once brilliant investment opportunity has now created a saturated market. Diversification of investment, or withdrawal from the particular market, or radical reduction of energy output is needed for the market to become profitable again.

4 **Market trough**

Trough is the ultimate stage of the cycle, where the market declines after a prolonged oversupply and limited market capability to absorb the excess inventory.

This is a significant stage that serves as a reality check for the global economy and individual companies.

The output has lost a great deal of its value as the supply has exceeded demand. This is a buyers' market, as energy importers can now leverage their position, negotiate with several sellers and buy "more with less."

This is a point where the prudent buyers benefit from economies of scale, can conduct long-term leasing contracts and purchase large volumes of energy for trade and inventory or storage purposes. Prudent companies have been preparing for this inevitable stage of the market cycle by saving resources and by strategically designing a contingency plan for how to pass this stage of the market cycle while suffering the least financial or commercial losses.

However, not all companies have the resources or the market leveraging status that will enable them to adequately prepare for the trough stage.

While the market sentiment is negative during a market trough, hasty decisions or radical market changes are observed. Market speculators may leave this particular energy market and enter other market segments. Banks may cease financial support towards the industry, thus leaving companies exposed to debt and insufficient funds, often forcing companies to exit the market or retreat from new business plans.

Several mergers and acquisitions (M&A) materialize during this stage, as companies pursue the "strength 'n' unity" doctrine.

Organizations contract their operations, meaning that they consolidate the supply chains, abandon production sites, or even sell their exploration rights and equipment to competitors.

This is a stage where pilferage or cargo theft is frequent, as abandoned production sites, storage units, equipment and other superstructures are common. Massive layoffs are also in store.

From a positive side, this is a moment of "catharsis," i.e., recycling, consolidation and renewal where lessons are learned. Companies tend to retrofit or sell or recycle (i.e., sell for scrap) assets that are no longer profitable in their current state. At the same time, professionals tend to upgrade their educational and training credentials in order to be competitive enough in the new market.

And while this clearance period may not be as promising or motivational as other stages of the market cycle, it provides an opportunity for companies and employees to strengthen their positioning by reorganizing a business strategy that is fit for the new market cycle.

During this critical stage, decisions made therefrom will plant the seeds that will determine the success of the subsequent market stages.

Prudent companies must be prepared for the next market cycle stage, and, as paradoxical as it may seem, companies must be ready for a transition from the trough to the market expansion stage. This is why, traditionally, this is a time to buy. Prudent speculators buy low-cost assets, anticipating the next stage of the market. Hence, companies that have been proactive and can survive this stage can anticipate great winnings in the other three stages of the cycle.

Energy Security and the Market Cycles

From an energy security perspective, it is the peak and the trough stages that present the highest risk.

a **Expansion and peak**: As a market expands, companies and professionals invest their time and resources into mitigating the new market needs. This rapid expansion affects not only individual companies but also the entire supply chain, while it tests the way business entities collaborate across the globe. Hence, market expansion finds corporations affluent in resources, but depleted of time to plan for the growth stage, and that lack experience of the new market conditions. Physical threats, cyber security threats, and personnel safety and environmental pollution incidents are likely during the market peak time, when companies have the resources but not the time required for situational awareness.

b **Contraction and trough**: The converse is also true. During a market contraction and trough stage, the industry is being deprived of its assets: Supply chains may collapse; energy companies may run on reduced budget and personnel; they may abandon their drilling or mining or harvesting sites, and vulnerabilities appear through the cracks of a resource-depleted industry. Most important, companies may be tempted to make shortcuts in regulatory compliance and operational performance.

Recognizing Business Cycles and their Stages

The optimum strategy for mitigating energy risks is to recognize the business cycles and their four principal stages. And while professional experience will help the keen observer recognize the patterns and distinct characteristics of a business cycle, there are two elements that make this recognition difficult:

1 Each business era differs in nature and bears its own unique characteristics based on tech advancements, supply–demand equilibrium, trade routes, geopolitics, and so on.
2 Although the expression "market cycle" is misleading, in reality there is never a single market cycle. Different regions, industries and trade routes may be subject to more than one market cycle.

This is why market cycles, that over a period of time vary in duration and intensity, are different and this is why the cycles cannot be recognized as mirror margins.

Fluctuations signify historical, technological, political, or socioeconomic triggers. Each peak and each trough signifies a point where the positive and negative factors finally evolve.

Classifying the Market Cycles

The following cycles are present in any and all industry segments, including the energy industry.

1 **Sir Isaac Newton's Pendulum Theory applications on investment**

In the 17th century, Sir Isaac Newton developed the "Pendulum Theory" of physics, motion, and gravity, and its applications in the stock market. As stock prices fluctuate from overvaluation to undervaluation, mirror-image patterns are observed among the stages of the cycle, i.e., the patterns and duration towards the peak are identical with the patterns and duration towards the trough.

While this theory does not determine the duration of a particular cycle, it is the first scholarly observation of an economic cycle with repetitive patterns from peak to trough and peak again.

2 **The cobweb wave, or agricultural model (1 year)**: Meeting the demand through seasonal fluctuations

This economic cycle reflects the seasonality of market fluctuations. Energy producers, just like farmers, must often invest, plan, and pre-decide the quality and quantity of energy to be produced before the supply/demand equilibrium reveals itself.

Hence producers may base their market plan on market rumors for a potential high demand, or preference for a particular energy product. If the rumors are accurate and several energy investors acted upon this demand, the result is oversupply, as many suppliers worked on mitigating a future market demand. On the other hand, if the rumors are inaccurate, the industry precipitates a market trough. As the supply exceeds demand, the industry benefits from a market peak only for a short time, whereas the trough is prolonged.

3 **Kitchin Cycles (3 years):** The way to a roaring economy leads to a financial meltdown

Joseph Kitchin observed the roaring economy of the 1920s, where mass industrial production and tremendous prosperity led to an economic growth of 42%, and in 1929 was followed by an unpreceded market meltdown. The waves reflect fluctuations in supply and demand equilibrium within about three years or 40 months (or two–five years according to different scholars), leading to an oversupply in commodities as several suppliers surge the market with a product to meet high demand. As a result, oversupply in production occurs, the prices are diminished, and eventually companies collapse or shrink their activities, resulting in mass employee lay-offs.

The cycle is characterized by high investment for excessive energy output and/or excessive inventory accumulated in storage units. The market recovers when producers are informed of the oversupply and take action towards inventory and production adjustment.

4 **Juglar Waves (7–11 years):** The cycles of fixed investments

In 1862 Clement Juglar observed the market cyclical patterns, where market growth, peak, and recession represent consecutive stages that are inevitable over a ten-year period. The market behavior in this cycle goes beyond inventory adjustment.

This theory may assimilate Sir Isaac Newton's Pendulum Theory, yet describe a ten-year cycle characterized by investment fluctuations presented in fixed capital.

5 **Kuznet's Waves (15–25 years):** The cycles of infrastructure and superstructure investments

Kuznet's Cycles are named after Dr. Simon Kuznet, who in the 1930s observed a 15–25-year cycle correlated with infrastructure and superstructure industrial investment. The investment swing may be initiated by an initial investment decision that enables or encourages mass industrial activities, such as energy exploration and distribution networks. This new investment decision may be initiated by numerous energy key players, hence the end of the assets' commercial life may occur concurrently in several companies, in the same region, supply chain or energy segment.

- The end of the cycle from a financial standpoint signifies the asset's return on investment or mortgage repayment. From an operational standpoint, it reflects the end of an asset's commercial life, re-sale or recycling.
- In the case of large structures such as offshore platforms, refineries, energy grids or energy processing plants, the end of a 25-year cycle could signify a major system upgrading, maintenance, retrofitting initiative, which hence signifies the beginning of a new 25-year cycle.

6 **Kondratiev's Waves (45–60 years):** The cycles of technological innovations

The Kondratiev Waves were observed in 1925 by N. D. Kondratiev and have a duration of about half a century (Figure 3.13). These cycles are triggered by technological advancements and/or banking investment cycles aimed at fuelling the market demand, which now peaks due to these investments.

These cycles are triggered by significant technological advancements, including but not limited to:

- the era of Industrial Revolution (1760–1820);
- the era of steam energy, steamships and railway systems (1800–1860);
- the era of steel and manufacturing (1850–1920);
- the era of fossil fuels, cars, Fordism and mass-marketing (1900–1960);
- the era of transatlantic flights and containerization (1960–2010);
- the era of the internet (1980–);
- the era of satellite-driven technologies: GPS, AIS, drones, dynamic positioning, etc. (1990–).

The aforementioned examples signify peak eras where the global market is literally driven by these innovations. The end of these eras does not signify a stage of saturation where these technologies are obsolete, but an era where new innovative systems are introduced and now lead the global economy, trade, and transport. This cycle also consists of peak and trough stages, which are spread over decades.

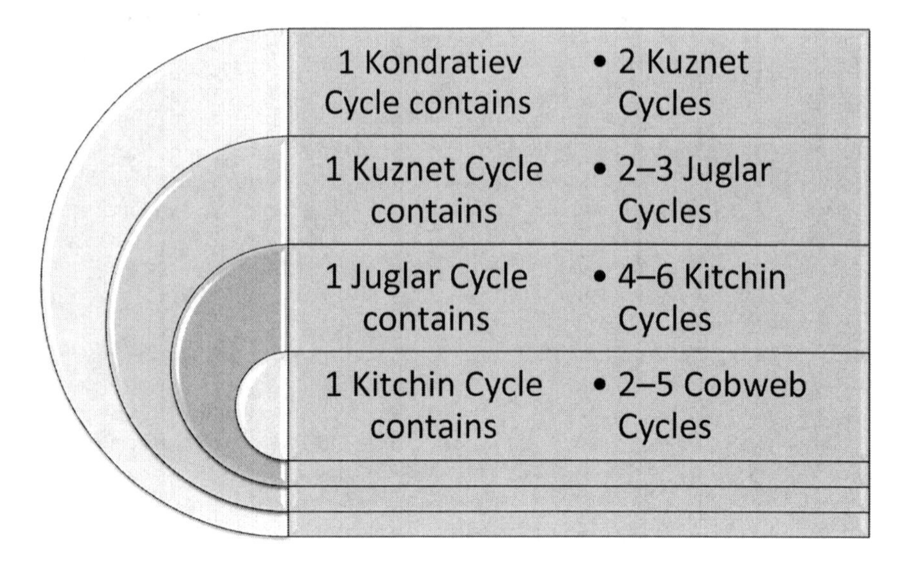

Figure 3.13 Classifying the market cycles.

Source: the Author.

Energy Dependence, Market Cycles and the Energy Mix

Based on the premise that an energy-dependent nation suffers greater financial losses during an energy crisis in a specific market segment, it is only reasonable to conclude that a nation with an efficient energy mix, i.e., with energy contingency planning, will overcome energy challenges faster and easier.

When a nation's energy mix and economic mix are diverse, they are more robust and can overcome market fluctuations. Once such a dependent economy collapses, its recovery is slower compared with more diversified economies, even when the energy market peaks. This is because of the supply chain, technological, and infrastructure damage, and because of the low morale that follows a massive economic crisis, unemployment and loss of income.

Energy Trade: Crude Oil and National Currencies

There is a strong association between crude oil and national currencies due to a number of factors, such as:

1 Commercial transactions, i.e., sale and purchase of energy contributes to the balance of trade of both the seller and the buyer.
2 Supply chain partners, e.g., refineries, manufacturers, warehousing, raw materials, service providers and any entities that may contribute to the energy input or output.
3 Market sentiment, i.e., speculators who observe the energy industry fluctuations and make commercial decisions, i.e., on investments, stocks and bonds, sale and purchase, accordingly.

Inelastic Supply and Demand

a **Inelastic supply** describes the scenario where any price fluctuations (upturn or downturn) will not lead to a matching increase or reduction in supply.

Oversupply causes price decline. As shown in Figure 2.3, oversupply causes low prices. However, diminished investment may cause delays in production and inability to properly maintain or purchase new technologies. Energy disruptions are caused by supply-chain inefficiencies.

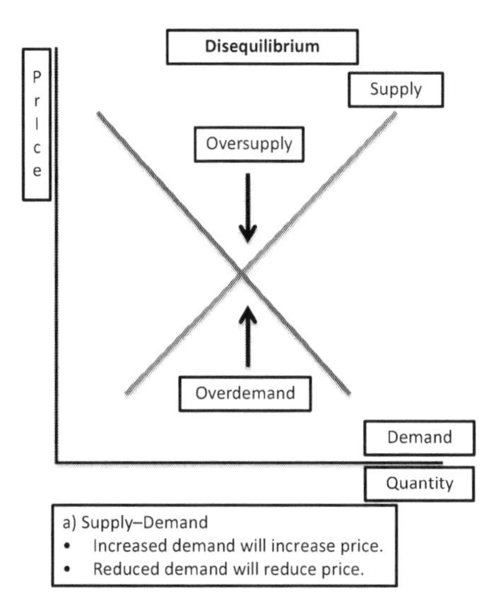

Figure 3.14a Market disequilibrium.

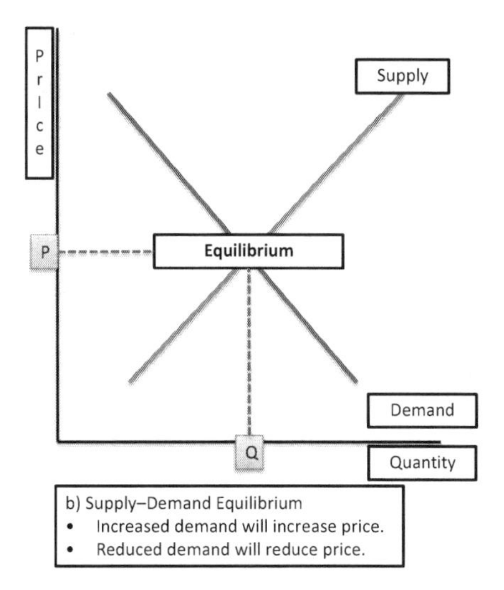

Figure 3.14b Market equilibrium.

Source: the Author.

b **Inelastic demand** describes the scenario where any demand fluctuations (increase or reduction) will not lead to a matching price adjustment. This favors the supply side of the market because a price upsurge leads to the overall income levels, regardless of reduced quantity purchased, or the suppliers' inability to fully meet market demand.

A Case Study on Oversupply

As new energy players undertake mortgages and other financing arrangements as part of their corporate expansion, oversupply occurs in the global markets, i.e., the national energy mix arrangements are now more enhanced in term of choices and contingency plans, yet the market surges with a high volume of energy.

Eventually, oversupply causes a market crisis: Jobs are lost, volumes of energy remain unused, and even national economies collapse. When the latter occurs, the repercussions are grave, as it is much harder for a nation to recover from a market collapse.

Following a currency depreciation, a nation's population will experience a diminished purchase power, diminished revenue, loss of jobs, sociopolitical or socioeconomic turmoil, etc.

Moreover, it is tougher for a bankrupt nation to recover and be ready for production, even when the market recovers, grows, and reaches a peak.

Global Energy Markets and Currency Wars

Currency wars pertain to the artificial currency depreciation as exporting nations attempt to boost exports at an attractive rate of exchange. The "currency war" strategy gained popularity in the late 20th century, and is in contrast to prior national strategies to maintain a strong currency value. A strong national currency typically benefits importing activities as it offers a strong hard currency and increased buying power for the consumers. Conversely, a weak currency benefits major exporting economies. Although nations weaken their local consumers' buying power, global buyers prefer to import from countries with weak currencies.

National currencies are influenced by the fluctuations in demand of their major exporting goods. Hence a nation with energy-exporting activities will inevitably see its currency fluctuations closely aligned with the energy market fluctuations.

Figure 3.15 Market disequilibrium where oversupply causes price decline.

Source: the Author.

It is worth noting that a nation does not need to be a major energy player for its currency to be impacted by the oil price fluctuations. Currencies of small and medium countries are equally impacted, as long as a high portion of their GDP, e.g., at least 40%, derives from a particular energy or industrial segment. This observation is useful for market forecasting and speculation purposes, as long as the researcher identifies specific energy or other industrial segments associated with particular currencies of the buyers' or the sellers' side.

Hence the term "petrocurrency" arises for nations heavily impacted by the petroleum industry, i.e., either as exporters or importers. Petrocurrency is the dollar payment to a petroleum exporting country, recognizing that the country's rate of exchange will be closely aligned with the energy market fluctuations. This "pegging" arrangement is profitable for the exporter when the oil prices are high and detrimental when the oil prices are low.

Exporting Economies: Benefits and Risks

A major exporting economy enjoys several benefits, and two main risks:

1. When a nation's production relies mainly on foreign buyers, the domestic economy is vulnerable, as it can be easily manipulated by foreign supply–demand power games.
2. When a nation's exports increase, there is a positive correlation with its currency.

   As its currency increases in value, it will eventually become more expensive, hence unprofitable and no longer an attractive option to foreign buyers. Consequently, the domestic factors of production (labor, land, capital mechanisms) become more expensive. Hence, the country gradually loses its *comparative advantage* as other countries now have a lower opportunity cost.

Historically, major global economies lost their opportunity cost and consequently their economic influence as their currency grew stronger. Nations with strong currencies and high-cost factors of production, in particular labor and land, tend to lose the buyers' bid out of a competing economy with cheap labor, cheap land, and weaker currencies.

The energy industry is no exception to this rule: This is a capital-intensive industry, where a project's return on investment will determine if a particular project is profitable enough for the exploration to be materialized. At times of energy oversupply, the projects that are cancelled or discontinued first are the high-cost low-output ones.

Energy Security and the Significance of Economic Diversity

When economies are involved in diverse industrial or service-providing activities their economy is healthier, as the less an economy relies on a single industry the more diversified, the stronger, and more independent it is. Also, the less they will be impacted by a commodity's price fluctuations or the trough of a particular industry.

In conclusion, this chapter bridged geopolitics with economics and the energy markets, and provided numerous examples of national energy strategies towards energy security in line with national goals for sustainability, independence, efficiency, affordability, and accessibility.

Notes

1 The Convention on the Organisation for Economic Co-operation and Development (OECD) was established after World War II in order to promote reconstruction and mutual growth (OECD, 2017).

2 Reference is made to two different countries: Oil is extracted from Equatorial Guinea, which is located at the southern region of West Africa, between Cameroon and Gabon. Uranium is mined from Guinea, which is located in the northern region of West Africa, between Senegal and Guinea-Bissau.
3 Facts: The area of Cyprus is 3,572 mi², i.e., ten times smaller than Ireland (32,595 mi²), or 72 times smaller than Texas (268,597 mi²), or a million times smaller than the US (3.8 million mi²), or Canada (also 3.8 million mi²) or China (3.705 million mi²).

References

Burns, M. (2004). "The Development of New Oil Supplies in the Caspian and Black Sea Regions: What are the Key Obstacles that Hold Back the Oil Exports?" Post-Graduate Research Thesis, London Metropolitan University, UK.

Burns, M. (2014). *Port Management and Operations*. Boca Raton: CRC Press, Taylor & Francis.

Burns, M. (2016a). *Logistics and Transportation Security. A Strategic, Tactical and Operational Guide to Resilience*. Boca Raton: CRC Press, Taylor & Francis.

Burns, M. (2016b). "Global Projections for Shipping. Pathways to Sustainability". NAMEPA Conference, ExxonMobil Campus, Houston, Texas, February 5.

Burns, M. (2017). National Academies Annual Meeting 2017, TRB Committee on Critical Transportation Infrastructure Protection. Agenda available at: https://sites.google.com/site/trbcommitteeabe40/meetings, last accessed: June *21*, 2017.

CIA (2017a). *CIA World Factbook, 2017*. Available at: www.cia.gov/library/publications/resources/the-world-factbook/geos/xx.html, last accessed: June 21, 2017.

CIA (2017b). *CIA World Factbook, 2017: Cyprus*. Available at: www.cia.gov/library/publications/resources/the-world-factbook/geos/cy.html, last accessed: June 21, 2017.

Department of Commerce (DOC) (2016). "Cyprus – Oil and Gas Exploration and Exploitation", *Cyprus Country Commercial Guide*. US Department of Commerce, International Trade Administration. Available at: www.export.gov/article?id=Cyprus-Oil-and-Gas-Exploration-and-Exploitation, last accessed: June 21, 2017.

Energy Information Administration (EIA) (2014). *Oil and Natural Gas Resource Categories Reflect Varying Degrees of Certainty*. U.S. Energy Information Administration. Available at: www.eia.gov/todayinenergy/detail.php?id=17151, last accessed: October 9, 2016.

Energy Information Administration (EIA) (2015). *Forecasts. World Incremental Liquid Supplies 2011–2016*. U.S. Energy Information Administration. Available at: www.eia.gov/forecasts/steo/outlook.cfm March 2015, last accessed: May 12, 2017.

Energy Information Administration (EIA) (2016a). *U.S. Crude Oil First Purchase Price (1850–2010)*. U.S. Energy Information Administration. Available at: www.eia.gov/dnav/pet/hist/LeafHandler.ashx?n=PET&s=F000000__3&f=A, last accessed: June 21, 2017.

Energy Information Administration (EIA) (2016b). *U.S. and Other Top 5 Total Petroleum and Other Liquids Production*. U.S. Energy Information Administration. Available at: www.eia.gov/beta/international, last accessed: May 15, 2017.

Energy Information Administration (EIA) (2017a). *International Energy Outlook 2016*. Report Number: DOE/EIA-0484(2016), release date: May 11, 2016. U.S. Energy Information Administration. Available at: www.eia.gov/outlooks/ieo/world.cfm, last accessed: June 21, 2017.

Energy Information Administration (EIA) (2017b). *International Energy Outlook 2016 with Projections to 2040*. U.S. Energy Information Administration. Available at: www.eia.gov/outlooks/ieo/pdf/0484(2016).pdf, last accessed: June 21, 2017.

Energy Information Administration (EIA) (2017c). *International Energy Outlook 2016, Chapter 5: Electricity*. Available at: www.eia.gov/outlooks/ieo/electricity.cfm, last accessed: June 21, 2017.

Energy Information Administration (EIA) (2017d). *U.S. Crude Oil First Purchase Price, (1974–2016)*. Energy Information Administration. Available at: www.eia.gov/dnav/pet/hist/LeafHandler.ashx?n=pet&s=f000000__3&f=m, last accessed: June 21, 2017.

EUR-Lex (2009). Minimum Stocks of Crude Oil and/or Petroleum Products Directive (2009/119/EC), Council Directive 2009/119/EC of 14 September 2009 Imposing an Obligation on Member States to Maintain Minimum Stocks of Crude Oil and/or Petroleum Products. Publications Office of the EU. Available at: http://eur-lex.europa.eu/legal-content/EN/ALL, last access: June 21, 2017.

European Commission EUROPA.EU (2017). *EU Oil Stocks. Minimum Stocks of Crude Oil and/or Petroleum Products*. Brussels: European Commission, Energy. Available at: https://ec.europa.eu/energy/en/topics/imports-and-secure-supplies/eu-oil-stocks, last accessed: June 21, 2017.

Organisation for Economic Co-operation and Development (OECD) (2017). *OECD History*. Paris: Organisation for Economic Co-operation and Development. Available at: www.oecd.org/about/history, last accessed: June 4, 2017.

UN (2017). *The World Population Prospects: The 2017 Revision*. New York: UN Department of Economic and Social Affairs. Available at: www.un.org/development/desa/en/news/population/world-population-prospects-2017.html, last accessed: June 25, 2017.

U.S.–China Economic and Security Review Commission (USCC) (2017). *The 13th Five-Year Plan. K. Koleski, Staff Research Report*. February 14. US-China Economic and Security Review Commission. US Government Publishing Office. Available at: www.uscc.gov/sites/default/files/Research/The%2013th%20Five-Year%20Plan.pdf, last accessed: June 6, 2017.

Part II

Non-Renewable Energy Security

Energy security is about sustainable production, accessibility, and the use of reliable logistics systems and technologies. Chapters 4–6 will cover these areas, and the operational, regulatory technological factors that contribute to a nation's energy security.

Renewable and Non-Renewable Energy

There are two main categories of energy resources:

1. **Renewable energy sources**, like solar, wind, geothermal, hydroelectric, tidal energy, and so on, that are replenished naturally and comparatively faster than non-renewable energy sources.
2. **Non-renewable energy sources**, whose natural substitution is slow compared with their consumption. Non-renewables represent resources with non-sustainable utilization, as their formation takes millions of years.

The non-renewables category comprises:

a. **Fossil fuels**:
    - petroleum or crude oil (liquid) and natural gas (gaseous) (Chapter 4);
    - coal (solid) (Chapter 5).

b. **Nuclear energy or atomic energy** is also considered a non-renewable source as it uses superfuels based on minerals, i.e., chemical elements such as uranium, thorium, and plutonium, etc. (Chapter 6).

Fossil Fuels

Fossil fuels are composed predominantly of hydrocarbons, i.e., organic molecules that contain carbon and hydrogen atoms. For this reason the time period that fossil fuels formed (about 360–300 million years ago) is called the Carboniferous period. This heterogeneous mixture requires special geological conditions such as high pressure and temperatures, resulting in metamorphism, i.e., the change of chemical composition. Crude oil is refined to generate liquid petroleum energy products such as gasoline, diesel fuel, and heating oil. These are hydrocarbon deposits that were naturally formed by the decomposition of organic materials such as prehistoric organisms, plants, and animals that lived millions of years ago.

4 Oil and Gas Security

Among the two man-made objects visible with a naked eye from the moon, the one is the Great Wall of China, and the others are the world's tallest offshore platforms.

(rephrased from NOIA, 2012)

Oil and Gas: An Overview

This chapter is dedicated to oil and gas, the prevailing energy segment for the past century or so.

Oil and gas security can be better understood when the different products and processes are classified, and their individual risks analyzed. However, one particularity of the petroleum products is the fact that they are very diverse, with numerous classifications.

Energy security professionals will need to better understand the diversity and common threads, and identify security vulnerabilities, strengths and recommendations, by comparing and contrasting elements in this complex supply chain.

Elements to be considered include the different technologies used:

- the chemical composition, showing the safety measures needed during processing, storage, transport, and use;
- supply chain characteristics that are specific to the geographical region, the geological and well-specific particularities, government leasing issues, and so on.

Most important, this chapter explains the security and safety challenges of different classification segments, as follows:

a Upstream, midstream, downstream.
b Supply chain particularities in land and offshore exploration.
c Conventional vs unconventional energy, in terms of geographical remoteness, temperatures, deep offshore conditions.
d innovative technologies.
e Crude oil and gas vs shale oil and gas (Chapter 4), vs biofuel, synthetic fuel, syngas (Chapter 10).

In order to better encompass the wide spectrum of security, each stage of the supply chain is explained in this chapter (as seen in Figure 4.1) and individual risks discussed, and the Benefits, Risks, and Recommendations section provides conclusions on the oil and gas security risks, with useful observations. Energy conglomerates and global organizations are also featured in this chapter, to better demonstrate best practices and the impact of large-scale synergies.

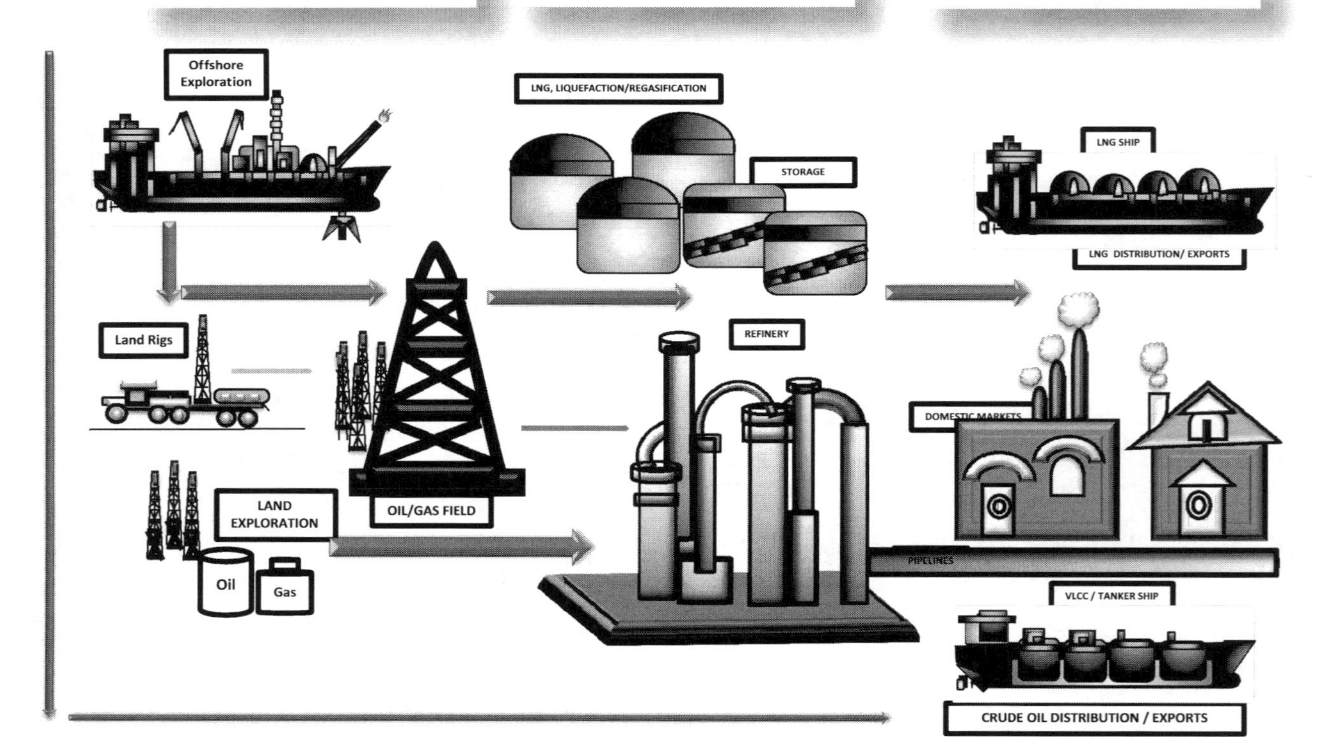

Figure 4.1 Oil and gas supply chain.

Source: the Author.

Petroleum

Where oil is first found is in the minds of men.

Wallace Pratt (1885–1981)

Petroleum, also referred to as crude oil, is a heavy, combustible, blackish to yellowish blend of primarily liquid, but also solid and gaseous hydrocarbons that are formed naturally underneath the earth's surface in rock strata. The primary form of organic matter used to produce petroleum derives from photosynthetic microorganisms referred to as plankton, namely phytoplankton (aquatic plants), and zooplankton (aquatic animals).

The chemical composition of crude oil varies depending on the geographical location. Nevertheless, its main constituents are include:

- carbon (about 90%);
- hydrogen (about 10%);
- traces of nitrogen, sulphur and oxygen;
- traces of metals.

This intricate mixture of hydrocarbons can also be categorized as follows:

- alkanes (i.e., saturated hydrocarbons such as methane, ethane, propane, etc.);
- naphthenes (i.e., cyclic aliphatic hydrocarbons such as cyclohexane);
- aromatic hydrocarbons (i.e., arenes).

API Classification: Light, Medium, Heavy Oil

According to the American Petroleum Institute (API) crude oil is also categorized as light, medium, or heavy in accordance with its assessed API specific gravity or density.

a Light and very light oils and distillates have an API gravity greater than 31.1° (i.e., lower than 870 kg/m3).

The "Very light oils and distillates" classification includes: Jet fuel, kerosene, gasoline, light and heavy naphtha, etc. These are more volatile, hence any environmental or toxicity risks are acute, i.e., requesting instant mitigation, but with limited long-term effects.

The "Light oils and distillates" category includes fuel oils grade 1 and 2, diesel fuel oils, marine diesel oils, and light crude marine gas oils (EIA, 2012a).

b Medium petroleum has an API gravity between 22.3 and 31.1° (i.e., 870 to 920 kg/m3).

This is a category encompassing many commonly used crude oils and intermediate refined products. Due to their low volatility, any incident such as spillage has long-term effects on the environment and human health, while emergency response and cleaning operations become more complicated.

c Heavy oils are crude hydrocarbons that possess API gravity cover values between 10° and 20°. These include crude oils of grades 3, 4, and 5, as well as heavy marine fuel oils. The terminology "extra heavy oil" and "bitumen" pertain to petroleum with specific gravity less than water density, or under 10°API. Due to the condensed nature of these products, risks pertaining to environmental and health hazards are increased.

Crude oil is relatively stable, which enables its long-term storage. On the other hand, refined products may lose their fuel efficiency if not used within a few months after refinery process, and special chemical additives are used to prolong the commercial life of refined products.

Crude oil is extracted and refined to generate fuels such as petrol, naphtha, kerosene/jet fuel, diesel oil, gasoline, fuel oil, extra heavy oil/oil sands bitumen, etc. Typically, a 42-gallon barrel of refined crude oil includes 47% gasoline, 23% diesel, 10% jet fuel, 4% liquefied petroleum, 3% asphalt, and 18% various other distillates. It is worth noting that, due to the refinery processing gain, the 42-gallon barrel of crude oil will yield about 45 gallons of products (EIA, 2017a). This is demonstrated in Figure 4.2.

Gasoline or Petrol

Gasoline or petrol is a volatile flammable liquid blend of hydrocarbons, mostly methane (CH_4) and benzene (C_6H_6), derived from crude oil and other petroleum fluids. In the U.S., approximately 19 gallons of gasoline are generated from each 42-gallon barrel of refined crude oil. It is widely used as fuel for engines of internal combustion (e.g., in transportation), with the main grades being regular, mid-grade and premium. It is also used as a solvent to remove grease, waxes, oils, etc. (EIA, 2017a). Global and national environmental regulations mandate the use of additives in *finished* gasoline fuels sold in gas stations and selected industries, in order to comply with fuel emission and air pollution standards. For example, the U.S. law stipulates that at least 10% of ethanol should be used as an additive.

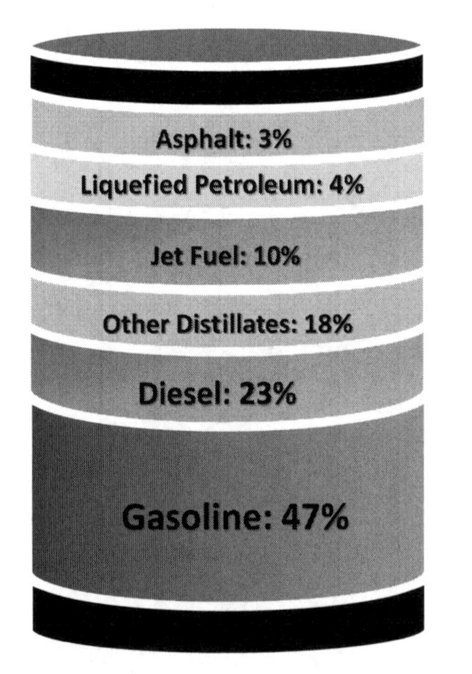

Figure 4.2 Percentage of petroleum products made from a 42-gallon barrel of refined crude oil.

Source: the Author, based on *Petroleum Supply Monthly*. U.S. Energy Information Administration, February 2017, preliminary data for 2016.

Diesel Fuel

Diesel fuel is a combustible petroleum product consisting of over 70% of paraffins and other hydrocarbons, and about 20–30% of aromatic hydrocarbons such as benzene, naphthalenes, and so on. It is used in diesel engines, mainly in sea and land transportation.

Liquefied Petroleum (LP) or Liquefied Petroleum Gas (LPG)

Liquefied petroleum (LP) or liquefied petroleum gas (LPG) pertains to a flammable mix of hydrocarbons such as propane, butane, and other gaseous compounds. These are used as fuel in sea and land transportation. Together with liquefied natural gas (LNG), they enjoy increasing popularity in the maritime industry, with over 1,000 ships globally using LNG or LPG fuel.

Kerosene and Jet Fuel

Kerosene or paraffin is a mixture of flammable liquid hydrocarbons produced by refining oil or bituminous shales. It is used to produce jet fuel, or other fuel types, or solvents. Its advantages include cost efficiency, chemical stability, and effectiveness as a lubricant. It is safe during storage and transportation when handled properly. When mixed with diesel, it expands the low temperature tolerance during the winter.

To reduce air pollutants, in particular sulfur content, different kerosene products, i.e., 1-K or 2-K, are processed. Highly flammable and explosive, safety measures should apply as ignition may occur in high atmospheric temperatures. Since its vapors are heavier than air, vapors can penetrate confined areas such as basements or tanks (NOAA, 2017; EIA, 2016d).

Jet fuel or aviation-turbine fuel (ATF) is a blend of various hydrocarbons widely used in diesel engines (i.e., internal combustion engines, compression ignition engines) or jet engines (gas turbine engines), in both civil and military aviation. An increase in demand has been experienced for the past 30 years or so.

Jet A and Jet A-1 are based on unleaded kerosene, and have 8 to 16 carbon atoms per molecule. Their main difference is the freeze points.

Jet B is a blend of kerosene and naphtha-type jet fuel, with 5 to 15 carbon atoms per molecule.

Extra-Heavy Crudes and Bitumen

Extra-heavy crudes and bitumen belong to the conventional energy classification, and thus play a significant role in supplying the global markets with sustainable energy. Their overall production reaches 4,700 billion barrels (Bb), among which approximately 1,000 billion barrels – the counterpart of the global oil reserves for conventional oil – are technologically, financially, and lawfully feasible to extract. Hence these are defined as proven, recoverable reserves.

Natural Gas

> Natural gas obviously brings with it a number of quality-of-life environmental benefits because it is a relatively clean-burning fuel. It has a CO_2 footprint, but it has no particulates. It has none of the other emissions elements that are of concern to public health that other forms of power-generation fuels do have: coal, fuel oil, others.
>
> Rex Tillerson, former United States Secretary of State
> and former CEO of ExxonMobil

Natural gas is a combustible mixture rich in hydrocarbons (i.e., hydrogen and carbon) primarily consisting of methane (about 80%) and lesser portions of propane, ethane, nitrogen, and butane. Also:

- liquefied natural gas (LNG) is used in maritime transportation;
- compressed natural gas (CNG) is used in vehicles.

It is typically developed by two natural processes:

a **Biogenic**, from methanogenic species found and extracted in landfills.
b **Thermogenic**, i.e., generated when anaerobic organisms are buried for thousands of years and exposed to high pressures and temperatures. Typically present near petroleum deposits.

The natural gas reserves are found in porous sedimentary rocks in the vicinity of other fossil fuels such as oil and coal. Its molecules are captured within solid rock formations, i.e., in cracks or structural sockets. Natural gas derives from:

a **"Dry gas" or "conventional non-associated gas"** derives from wells that are purely gaseous, i.e., not related to liquid petroleum reserves.
b **"Wet gas" or "conventional associated gas"** derives from wells that also contain crude oil. It is either found as a gas cover over the petroleum well, or dissolved in petroleum, hence, by its sheer pressure, both oil and gas are pushed out to the well exterior (see also Figure 4.5).
c **"Tight sand gas"** formations pertain to unconventional natural gas sources that are found in sandstone or limestone. Among its benefits are its abundance in vast reserves globally, and the lower exploration costs compared with shale gases.

Natural gas is a combustible clean-burning fuel with low emissions. Some of the risks pertain to its chemical composition. Due to its volatility, there is a risk of combustion and explosion. In its natural form it is colorless and odorless, hence leakages are not identified in a timely manner. Suppliers and exporters find its low density a challenge in storage and hauling. For this reason, liquefaction, i.e., its conversion into liquefied natural gas (LNG) is a solution to increase safety in storage and transportation.

The use of natural gas is becoming increasingly popular. It expands to industrial and domestic applications, including electricity generation, cooking, air conditioning, and heating systems; fertilizers; pharmaceuticals, plastics, fabrics, and many everyday items.

Natural gas plays an increasingly important role in the energy mix. Environmentally friendly, it releases 60–90% lower emissions compared with other fossil fuels and smog-releasing pollutants. It meets the environmental threshold regulations on low sulfur, carbon dioxide (CO_2) and nitrogen oxide (NO). It is relatively safe to transport and store. In case of leakage, natural gas is dispersed in the atmosphere.

Natural Gas Liquids (NGLs)

Natural gas liquids are hydrocarbons, i.e., generated by hydrogen and carbons, hence belonging in the same family as oil and gas. Natural gas is about 80% methane (CH_4), 10% ethane (C_2H_6), and may also contain propane (C_3H_8), butane (C_4H_{10}), and pentane (C_5H_{12}).

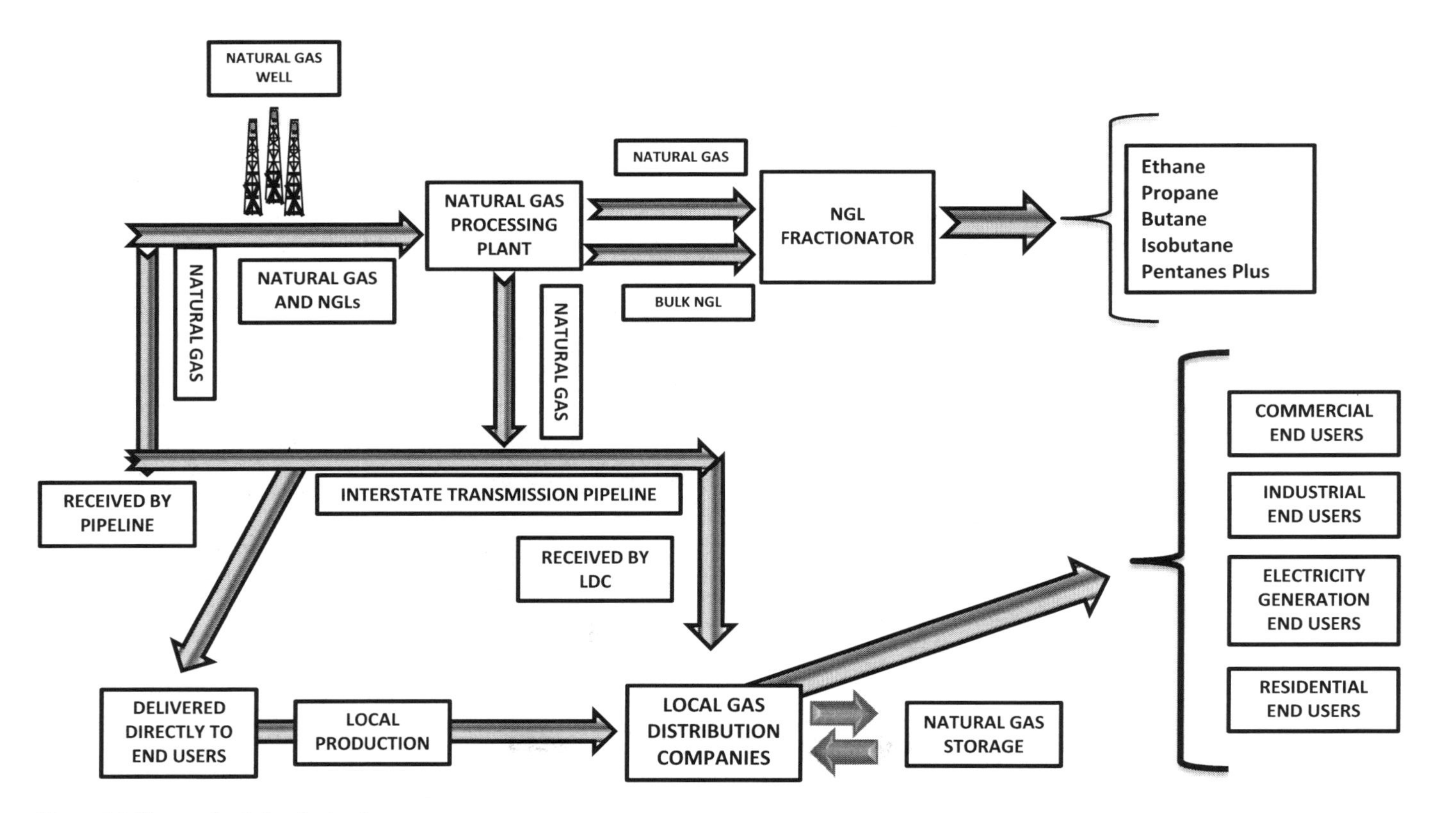

Figure 4.3 The supply chain of natural gas.

Source: the Author, based on EPA, 2015. *GHGRP 2015 Supplier of Natural Gas and NGL Supply Chain*. U.S. Environmental Protection Agency. Available at: www.epa.gov/sites/production/files/styles/large/public/2016-10/ghgrp2015suppliersgaswellchart.png.

Table 4.1 Natural gas composition.

Paraffinic		*Non-hydrocarbons*	*Aromatic*
Natural gas (dry gas)	Natural gas liquids(NGLs)	Water (H_2O)	Benzene (C_6H_6)
Methane (CH_4)	**a Ethane (C_2H_6)**	Carbon dioxide (CO_2)	
	b LPGs	Hydrogen sulfide (H_2S)	
	Propane (C_3H_8)	Helium (He)	
	Butane (C_4H_{10})		
	c Heavier fractions		
	Pentane (C_5H_{12})		
	Hexane (C_6H_{14})		
	Condensates & heavier fractions, etc.		

Source: the Author.

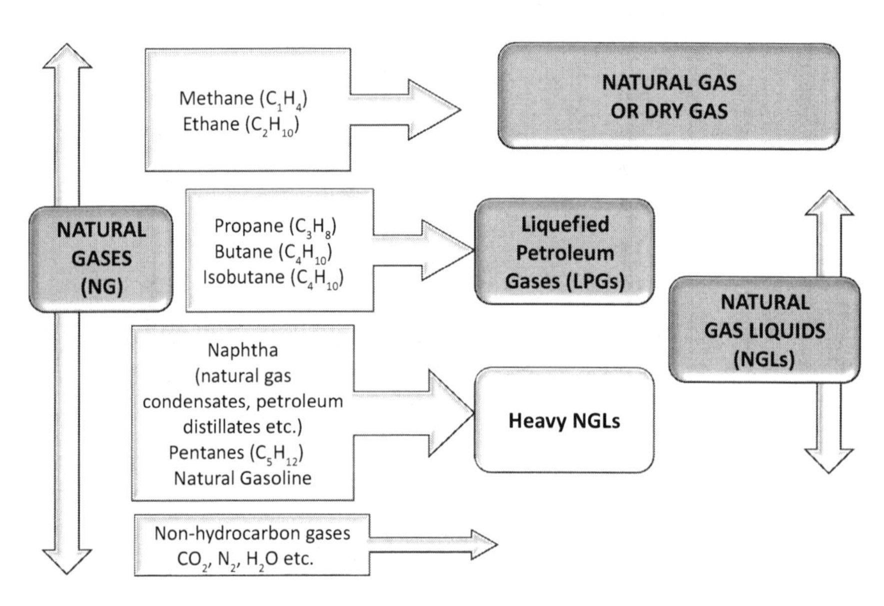

Figure 4.4 Comparison between dry gas, LPGs, and NGLs.

Source: the Author.

Before natural gas is distributed downstream, most of the above liquids are extracted for other commercial products, whereas water, sulfur, and other impurities are also removed. Despite their analogous chemical composition, their use is diverse:

- **Methane (CH_4)** is a colorless odorless gas and the fundamental compound in natural gas. Its chemical composition of a single carbon atom means it has a lower boiling point, hence it ensures safe storage in tanks at higher pressure, and it is relatively tolerant of low and high temperatures. Its initial distillation form is gaseous, which entails risks of explosion. The process of liquefaction eliminates this risk, thus increasing safety and also reducing its volume.
- **Ethane (C_2H_6)** is a two-carbon alkaline and a colorless and odorless gas, produced during the petroleum-refining process. It produces ethylene and plastic products.

- **Propane and Butane (C_4H_{10})** and "autogas" (a blend of propane and butane) serve as popular heating fuels. Their applications and use include fuel and biofuel for engine combustion, indoor and portable heating systems, etc. (NIH, 2017).
- **Gasoline and pentanes plus** are used as combustion engine fuels but also for fossil fuel extraction, e.g., oil sands (EIA, 2012b).

Liquefied Petroleum Gas (LPG)

Liquefied petroleum gas entails hydrocarbon gases generated from fossil-fuel-refining processes, i.e., either from the refining of crude oil or from the natural gas production process. It was first produced in 1910 by Dr. Walter O. Snelling of the U.S. Bureau of Mines, and, by 1912, its products became widely consumed (EIA, 2016a). Liquid petroleum gases include ethane, propane, and butane, which are found both in natural gas and crude oil. They are refined and maintained in a liquid form, hence are known as liquid petroleum gases (EIA, 2016b).

Propane (C_3H_8) and Butane (C_4H_{10}) are flammable colorless hydrocarbons in the family of LPGs. They are gaseous by-products of petroleum refining and natural gas treatment that can be compressed and liquefied for storage and transportation convenience. As seen in their chemical formulas, propane contains three atoms of carbon whereas butane contains four, meaning that propane has a lower boiling point and is safely stored in tanks at higher pressure, and it can better tolerate low and high temperatures.

These are some of the benefits of LPGs:

- lower environmental impact, especially no ground or water impact, due to limited content of sulfur or other pollutants; air pollution is possible;
- high calorific value (HCV), i.e., pertains to high amount of heat released per unit (weight or volume) during combustion;
- to eliminate safety risks, it is liquefied at low pressure levels;
- safe and easy storage and transportation in LPG cylinders (i.e., propane and butane bottles) further facilitates the logistics processes;
- sustainable, reliable combustion performance (DOE, 2017).

The use of LPG is becoming increasingly popular for industrial use (electricity production, manufacturing, agriculture), domestic use (hot water, heating and cooking), and transportation, i.e., on sea and land.

Table 4.2 LPG and NG comparison.

Characteristics	*LPG*	*Natural gas*
Chemical description	C_3H_8	CH_4
Chemical composition	Made up mostly of propane (C_3H_8) and butane (C_4H_{10}) but also methane (CH_4), butane, isobutene (i-butene), and a trace mixture of pentane and other chemicals.	Made up mostly of methane (CH_4), but also propane, butane, and ethane.
a Physical description:	a Flammable hydrocarbon gas.	a Explosive & flammable hydrocarbon gas.
b Odor and color in natural state:	b Odorless and colorless in natural state. Industrial odor caused by ethyl mercaptan.	b Odorless and colorless in natural state.
c Pressurization:	c Pressurization causes its liquefaction.	c Pressurization causes its liquefaction.

(continued)

Table 4.2 (continued)

Characteristics	*LPG*	*Natural gas*
Origin	Petroleum distillation and gas processing.	Unassociated petroleum and gas reservoirs.
Molecular weight	44–56	17–21
Superior heating power (Kcal/m3)	1.50–2.0	0.58–0.72
Energy/calorific values	93.2 MJ/m3	38.7 MJ/m3
Optimum burning requirements: Oxygen to gas ratio	25:01:00	10:01
Density to oxygen	1.5219:1 * heavier than oxygen	0.5537:1 *lighter than oxygen
Combustion	Ratio of 25:1	Ratio of 10:1
Principal uses	Industrial, commercial (fuel), and residential.	Industrial, commercial (fuel), and residential. * Automotive and thermoelectric generation (fuel); industrial (fuel); petrochemical and metallurgical.
Storage pressure	15 ATM	200 ATM
Storage and transportation	• Tanks, cylinders, pipelines. • Liquefaction (compression) increases safety and reduces space.	• Tanks, cylinders, pipelines. • Liquefaction (compression) increases safety and reduces space.

Source: the Author.

Shale Oil and Gas

The chemical composition of shale fossil fuels does not differ from the composition of conventional carbonates, i.e., oil and gas. It is the geological morphology of the reservoir rocks that varies. Shale oil and gas are developed in non-permeable sedimentary solid formations in pores and cracks, at a depth of at least 3,300–16,000 feet (1,000–5,000 meters). As unconventional technology evolves, operations are feasible at an increased depth.

Risks

WATER USE IMPACT

Shale gas exploration and the fracking process necessitates the extensive use of water. Energy companies typically conduct studies to ensure the water allocated does not deprive the community of other uses, such as agricultural utilities. Also, environmental studies can verify that the ecosystem is not negatively impacted.

The fracking process entails high volumes of waste water, which may need treatment prior to recycling or discarding due to the chemicals and impurities that may be present in the water. The water treatment and purification process may be costly, both due to the high volume of water and also due to the intricate process.

Incidents related to leaks and spillage of the fracking process liquids may pollute the ecosystem. Health hazards could occur if toxic chemicals find their way into public drinking water systems.

A Case Study on Shale Gas Partnerships: How George Mitchell used Federal Maps to Strike Unconventional Shale Gas Reserves

This case study presents a public–private synergy, where an entrepreneur used government mapping in combination with innovative extraction techniques in order to explore shale gas reserves that, until then, were considered unconventional.

George Phydias Mitchell (1919–2013) was an American entrepreneur, philanthropist, and real estate magnate, recognized as the "Father of shale gas exploration." He initiated the hydraulic fracking process and enabled the large-scale commercial production of shale gas in the Barnett shale field in Texas. This is the biggest onshore gas field in America, expanded over an area of 5,000 square miles (or 13,000 km^2) and over 18 counties (RRC. TX, 2017).

In the 1980s a public/private partnership between the U.S. Department of Energy and Mitchell Energy (a small independent company) developed a multi-level directional drilling project in the Devonian Shale. The government provided the geological mapping of tight shale gas wells and Mitchell invested – for several years – in an ambitious R&D project that freed the mammoth reserves confined in underground shale stones.

The commercialization of such projects could only be materialized by the 1990s, when the demand for gas products boosted prices and ensured the commercial feasibility.

Mitchell's R&D experiments involved:

1 the implementation of hydraulic fracturing, i.e., the inoculation of high-pressure solutions to frack shale rocks; and
2 the introduction of horizontal and directional drilling techniques, which entail "drilling at an angle."

Mitchell's contribution in global energy security is immeasurable: Nations abundant in shale oil and gas, such as the ones shown in Table 4.4, can now un-tap their reserves and become energy independent, with significant financial growth and jobs growth potential.

The combination of the above techniques resulted in excellent well yield results, that are summarized in Table 4.3.

To measure the impact of this innovation, it is worth observing Table 4.4 and the major shale-oil/gas-producing countries. Many of these leveraged the global need for energy security and obtained global economic prominence by exporting shale oil and gas.

Table 4.3 The benefits of hydraulic fracturing, horizontal and directional drilling.

Innovations	*Aggregate benefits*
Rock wells that were previously depleted or unrewarding could now be drilled and produce high volumes of oil and gas.	• Well yield increase; production optimization.
Directional drilling could reach underground targets in populated regions and unconventional locations.	• Elimination of surface geological impact.
A single drilling pad could now be used to retrieve shale fuel from several wells simultaneously.	• Decrease of time and overall upstream costs.
Pressure relief wells and the implementation of directional drilling could remedy high-pressure wells or out of control wells.	• Simplification of the drilling permit process.
Water injection significantly eliminated the use of chemicals previously used.	• Increased safety by minimum surface operations, and minimum surface installation of utility lines.
	• Elimination of environmental impact.

Source: the Author.

Table 4.4 Top ten shale oil and shale gas producers of technically recoverable shale oil resources.

Rank	*Country*	*Shale oil (billion barrels)*	*Rank*	*Country*	*Shale gas (trillion cubic feet)*
1	Russia	75	1	China	1,115
2	U.S.A.	58	2	Argentina	802
3	China	32	3	Algeria	707
4	Argentina	27	4	U.S.A.	665 (1,161)
5	Libya	26	5	Canada	573
6	Australia	18	6	Mexico	545
7	Venezuela	13	7	Australia	437
8	Mexico	13	8	South Africa	390
9	Pakistan	9	9	Russia	285
10	Canada	9	10	Brazil	245

Source: EIA, 2014. *Shale Oil and Shale Gas Resources are Globally Abundant.*

U.S. Energy Information Administration, U.S. Department of Energy. Available at: www.eia.gov/todayinenergy/detail.php?id=14431, last accessed: June 12, 2017.

Oil and Gas: A Historic Overview

Petroleum exploration has existed for over a century, and has experienced a remarkable growth in terms of technological innovations used. As a result, oil is drilled from unconventional sources and production growth rates have the potential to increase exponentially. At the same time, it is the oversupply that drives down petroleum prices and this spurs energy companies to find innovative ways to get more petroleum for less money.

In the early 19th century, natural gas was mined from coal and used in street lampposts and for illumination fuel. Shortly after, scientific and technological advancements made oil and gas the most popular fuel globally, and for decades to come.

In the 20th century the oil and natural gas market grew rapidly, and the industry tried to mitigate the known risks of flammability (oil) and explosivity (gas). For example:

a the use of methanethiol (also known as mercaptan), a benign yet sharp-smelling gas, to timely alert natural gas users of leakage;
b robust pipeline systems ensured safe transportation.

1848: The world's first oil well at Baku, Azerbaijan. Major Aleveev drilled the world's first oil well by the use of a simple cord-tool drilling method inspired by ancient Chinese practices.

1859: Titusville, Pennsylvania (USA). "Colonel" Edwin L. Drake conceived the design of a handmade rig, created with the help of Mr. Smith, a blacksmith. Drake drilled down 70 feet and emerged coated with oil. At the time, petroleum was mainly used to make kerosene for lighting functions, as whale oil was relatively scarce and hence expensive. The imminent boom in automobile manufacturing generated a large-scale demand for oil and sparked production from 150 million barrels developed worldwide in 1900 to over one billion barrels in 1925. By 1859 Colonel Edwin Drake became famous for drilling the first oil well.

1862: oil production had exceeded three million barrels. The oil boom era commenced. Proactive entrepreneurs who bought vast plots of land in Pennsylvania made a fortune. The petroleum industry supply chain is expanded, and the first refineries were built.

1901: The Texas oil boom. Spindletop is to Texas what Titusville was to Pennsylvania. Patillo Higgins was the "Prophet of Spindletop:" his practical geological knowledge convinced him of the vast amounts of oil in the "Big Hill" region, Spindletop, at a time when public opinion rejected this possibility. Higgins partnered with Anthony Lucas and used fishtail pieces, water-based drilling soil, and a steam-driven rotary drill rig. This is how the Texas oil thrived.

Oil and Gas Transport: The First Tanker Ships

The extensive production of oil triggered the need to construct an efficient logistics network, hence new trade and transport methods were developed. Initially, oil was stored in wooden barrels that were carried on barges, break-bulk boats and trains. However, the use of barrels was costly (50% of production costs), unsafe (leakage), and impractical (transportation time was prolonged, with several manual labor hours). This is when tanker ships were constructed.

1878: Tanker ship *Tollefsen* of Norway was the first tanker constructed globally, and, shortly after, Swede Ludwig Nobel constructed *Zoroaster*, the world's first successful oil tanker. Nobel suggested an innovative technique whereby, rather than filling drums with the drilled oil and subsequently loading them onboard ships, a pipeline could instantly load petroleum into the ship's tanks. This is how the concept of tanker ships was widely applied. The *Zoroaster* was designed to trade in the landlocked Caspian Sea, through Baku, Azerbaijan, and Astrakhan, Russia.

1891: Gasoline engines. The Daimler Motor Company commenced the production of gasoline engines in the U.S. for tram cars, carriages, quadricycles, fire engines, and ships.

Fluid Catalytic Cracking (FCC)

Distillation or cracking is one of the most significant petroleum conversion techniques employed in refineries, as it converts the hydrocarbon fractions of petroleum crude oils from a high-boiling high-molecular weight status into more beneficial products such as gasoline, olefinic gases, and several other useful products.

Initially, the thermal cracking process was implemented, which was soon totally substituted by catalytic cracking as the latter generates more valuable products, including gasoline fuel with an increased octane rating, more olefinic gases, and other byproducts.

In **1920** Standard Oil of New Jersey considered the concept of fluidized catalytic cracking.

In **1922** Eugene Jules Houdry, a mechanical engineer, and E. Prudhomme, a pharmacist, developed a small laboratory in France, experimenting on the catalytic conversion of lignite coal to gasoline. By **1931** Houdry was invited to the U.S. to work in the Vacuum Oil Company. Over the next decade Vacuum Oil underwent mergers and acquisitions, yet Houdry developed his innovative process and was publicly acclaimed by 1938, shortly before World War II.

By 1942 the Fluidized Catalytic Cracker (FCC) technique was commercialized. The technique was pivotal in developing aviation jet fuel during the war, and evolved into producing propylene and other refined products.

In 1929 controlled directional drilling techniques were introduced, involving the directing and adjusting of a wellbore alongside a fixed trajectory, to at least one subterranean well at specified horizontal displacements (HD) and true vertical depths (TVD) from the source point. Directional drilling proved to be more beneficial than vertical drilling, hence becoming the prevailing drilling method as greater portions of the production development can be extracted.

In 1949 the hydraulic fracturing process was first implemented in the Oklahoma and Texas oil wells, and has since been used to extract tight gas from reservoir rocks of low permeability, i.e., shales, carbonates, and coal seams. The hydraulic fracturing method involves the injection of a fluid mixture of water and chemicals into the rocks.

Offshore Drilling

Semi-Submersible Rigs, or Mobile Offshore Drilling Units (MODUs)

1949: The Breton Rig 20 was the first submersible mobile offshore drilling platform, and was developed by John Hayward. Its submersible pontoon barge design could only drill at a depth of 20 feet of water. By **1954**, Kerr-Mc Gee's Kermac 44 drilling platform could drill at depth of 40 feet, and was stable enough during extreme weather.

The **1950s** was a time when different pioneering global rig designs were launched. And although their dimensions, capabilities, performance, and sophistication has greatly evolved since then, it is worth noting that the types and preliminary functions remain the same: Floating units, submersibles, semi-submersibles, barges, drilling tender rigs, etc.. The first barges and platforms were naturally ship-shaped as they were either converted tankers or greatly influenced by naval architecture designs.

1954: Jack-up drilling rig. After WWII Leon Delong designed docks built with pilings attached to the pier hull once a pneumatic cylinder was used to jack them down into the seabed. The whole pier emerged out of the water to the desired level. Delong Corporation expanded in Cape Cod and Venezuela.

Pipelines, Onshore and Offshore

1850s: The first onshore pipelines appeared in two regions:

1 Baku, in modern-day Azerbaijan, by the Branobel company; and
2 Pennsylvania, USA, by the Oil Transport Association.

The first pipelines were made of iron and carried oil. Modern pipelines are made of steel or plastic, and carry both oil and natural gas liquids.

1954: The first commercial offshore pipeline was constructed. This was a ten-mile-long structure designed by Brown & Root at the Gulf of Mexico, laid at a depth of 20 to 32 feet. In 1958, Brown & Root also designed the first purpose-built pipe-lay vessel.

In the 21st century, oil and gas and their byproducts remain the most widely used energy segments globally. Technological innovations and improvements support the mitigation of growing demand: Deep offshore technology, hydraulic fracturing (i.e., hydrofracking), horizontal drilling and directional drilling technologies develop, enabling unconventional production.

All these technologies and methods lay out the foundations of modern energy upstream, midstream and downstream processes, and established oil and gas as the prevailing energy source of our era.

Conventional vs Unconventional Oil and Gas

Conventional or unconventional oil and gas have dual meanings. They may refer to:

a The type of energy sources that are difficult to drill due to low permeability, low porosity, and so on. Unconventional energy includes: Coal seam gas, tight gas or oil deposits, shale gas/oil, hydrates, coal bed methane, oil sands, and so on.

b The methods or technologies used to extract conventional energy reserves, including projects in very deep offshore locations, projects in extreme weather environments like the Arctic, or floating LNG equipment (EIA, 2016c). In this case, the term "unconventional" may also refer to:

- the downstream (exploration) and midstream (production) technologies;
- economic parameters, including unconventional partnerships, formation of consortiums, trade agreements, or financing terms;
- the supply chain particularities, such as the level, consistency, and project duration.

Figure 4.5 demonstrates the difference between conventional and unconventional drilling, but also the difference between associated and non-associated reservoir, as discussed on pages 115 and 123–124, and also pages 9, 20, 29, 31–35 and 115, and the formation of shale oil and gas reserves, as discussed on pages 162–164.

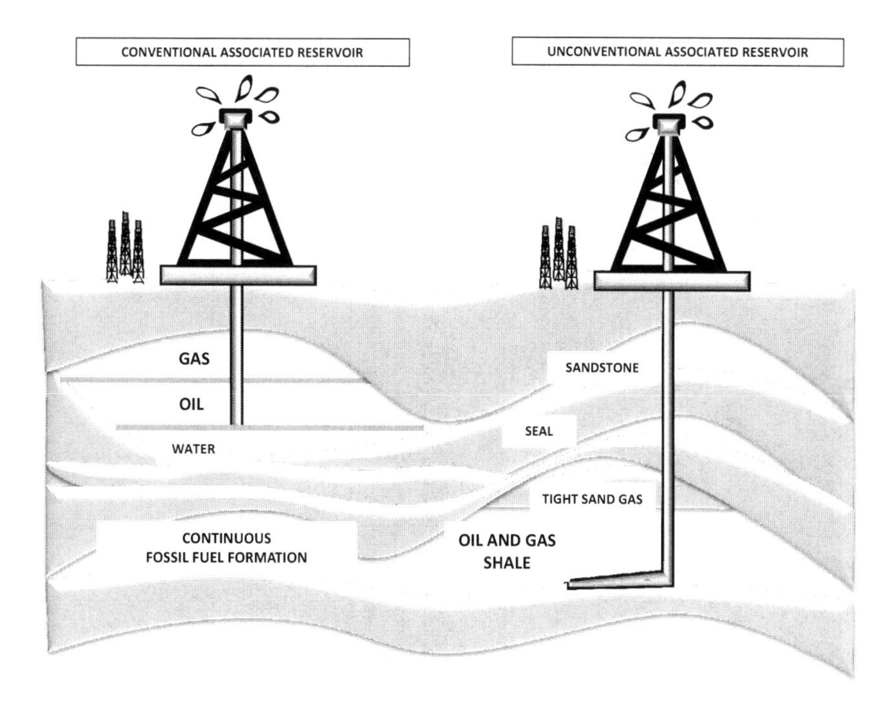

Figure 4.5 Conventional vs unconventional oil and gas exploration.

Source: the Author.

Unconventional energy production increases the risk factors, both in terms of likelihood of risk (e.g., extreme weather is more likely in deep offshore environments), but also in terms of impact of risk, e.g., a large-scale multi-drilling project may impose higher risks compared with a simple, smaller-scale exploration. From a security perspective, in cases where unconventional drilling is located in remote regions, any supply chain disruptions or challenges become more impactful.

Technology can help energy producers identify if specific energy wells are conventional or unconventional, as stratigraphic and lithographic data is interpreted to develop reservoir characterization prototypes. Namely, a project may be considered as conventional for the standards of an advanced economy and unconventional for the standards of a less-developed economy.

Similarly, the designation of unconventional changes over a certain period of time as unconventional innovative technologies eventually become obsolete.

The geological formation of the energy wells also varies: There is a distinction between conventional and unconventional oil that will be examined in several sections of this chapter, from an economic, commercial, technological, and operational perspective.

Conventional Drilling (Vertical)

- Conventional drilling or conventional production typically refers to vertical drilling and entails crude oil, natural gas and their condensates. Conventional oil and gas wells are composed of permeable sandstone while unconventional wells pertain to non-porous, reduced permeability rocks that make the extraction process more challenging and costly.
- Conventional resources can be extracted, after the drilling operations, just by the natural pressure of the wells and pumping or compression operations. Eventually, when the well production is reduced, a conventional well can employ an artificial lift or water and gas injections to sustain production.
- After the depletion of maturing fields, the natural pressure of the wells may be too low to produce significant quantities of oil and gas.
- Once production reaches a position in which the earnings are not adequate to cover expenses (i.e., surpass or at least reach the break-even stage), the drilling operations will be stopped and the well will be abandoned. Different techniques may be used to boost production, mainly water and gas injection or depletion compression, but these oil and gas fields will still be conventional resources.
- If techniques to extract oil are used beyond the conventional methods, the project is classified at unconventional.

Unconventional Drilling (Horizontal and Directional Drilling, or Fracking)

- The unconventional wells are covered with solid, non-porous rock. This rock confines the gas and prevents it from seeping through the surface layer. The extraction is carried out by drilling bore holes within the cap rock. Consequently, traditional equipment cannot penetrate the sturdy and challenging geologic structures. Unconventional reservoirs require unconventional technologies to drill beyond the conventional deposits.
- Horizontal or directional drilling are the most common techniques to retrieve unconventional wells, making it possible for oil and gas to be streaming from tight sands, which typically would not be retrieved by conventional technologies and processes.

These methods play an important role in a country's energy mix, since unconventional deposits help a nation un-tap its existing energy reserves, and present an alternative supply source when conventional fields are depleted.

- Unconventional oil and natural gas, especially shale gas, have been referred to as the future of gas supply in the world, especially in North America, with substantial financial opportunities.

The classification of a well as conventional or unconventional changes over time, and is subject to technological innovations, such as:

a the ability for precision mapping and geological surveys,
b the ability to drill under extreme weather or challenging geological conditions,
c the ability to increase efficiency and eliminate cost in deep offshore operations.

From an energy security perspective, if the "high peak oil" theory is correct, unconventional drilling techniques will prolong the availability of non-renewable energy, thus giving science additional time to experiment and fully develop new platforms of sustainable, secure and environmentally friendly energy.

Classifying the Unconventional Hydrocarbon Products

Unconventional hydrocarbon products include extra-heavy oil, oil shales, oil sands (natural bitumen), kerogen oil, gases and liquids deriving from chemical processing of natural gas, i.e., gas to liquid (GTL), coal to liquids (CTL) and additives.

Tight oil and gas reserves pertain to unconventional, solid, non-absorbent shale and sedimentary geologic structures. The oil in intact shale is motionless since petroleum molecules are greater than shale rock pores. In the past such reserves were regarded as unrecoverable due to the pricey and intricate drilling operations required. However, tight oil extraction has been enabled due to fracking and horizontal drilling (multi-stage drilling) breakthroughs. These techniques capture tight oil and gas that is tightly enclosed in extremely condensed wells of sedimentary non-permeable stones, for example limestone or sandstone, at depths exceeding 3,200 feet (or 1,000 meters). Since innovative technologies increasingly enable the extraction of unconventional shale hydrocarbons, the corporate decision to extract is mainly based on the cost/benefit analysis, and is less due to technological factors.

The benefits of shale extraction are not purely commercial and economic. A nation's energy security and independence may rely on these energy reserves. The U.S. is a nation that, due to the extraction of shale gas reserves, has been shifted from energy dependence to energy independence, and from being a major energy importer to now being a major energy exporter.

As confirmed by the U.S. Energy Information Administration, fracking now makes up over 50% of all U.S. oil production compared with 2000 when it comprised less than 2% of U.S. oil output. Hence, in 15 years, it added about 4.3 million barrels per day of unconventional shale oil to the U.S. production (Hirsch et al., 2016).

Commercial and Economic Development

Oil has grown to be the most influential commodity utilized by nations all around the world.

The energy markets categorize crude oil by means of the geographical location of its origin and through its relative weight or viscosity, into "light," "intermediate" or "heavy."

Table 4.5a Oil and gas key players: Key indicators.

Rank	Oil and gas key players	Crude oil production	Crude oil exports	Crude oil imports	Crude oil proved reserves	Natural gas production	Natural gas consumption	Natural gas exports	Natural gas imports	Natural gas proved reserves
1	China	3.983 million bbl/day (2016 est.)	58,650 bbl/day (2016 est.)	7.599 million bbl/day (2016 est.)	25 billion bbl (2016 est.)	150 billion cu m (2016 est.)	224 billion cu m (2016 est.)	3.918 billion cu m (2014 est.)	75.1 billion cu m (2016 est.)	6 billion cu m (Dec. 2016)
2	USA	9.415 million bbl/day (2015 est.)	1.162 million bbl/day (2015 est.)	8.567 million bbl/day (2015 est.)	36.52 billion bbl (2016 est.)	766.2 billion cu m (2015 est.)	773.2 billion cu m (2015 est.)	42.87 billion cu m (2014 est.)	76.96 billion cu m (2015 est.)	10.44 trillion cu m (2015 est * new discovered in 2017)
3	Russia	10.83 million bbl/day (2016 est.)	5.116 million bbl/day (2016 est.)	15,110 bbl/day (2016 est.)	80 billion bbl (2016 est.)	635.5 billion cu m (2015 est.)	453.3 billion cu m (2014 est.)	222.9 billion cu m (2016 est.)	8.9 billion cu m (2016 est.)	47.8 trillion cu m (2016 est.)
4	S. Arabia	10.05 million bbl/day (2015 est.)	7.416 million bbl/day (2013 est.)	0 bbl/day (2013 est.)	269 billion bbl (2016 est.)	102.4 billion cu m (2014 est.)	102.4 billion cu m (2014 est.)	0 cu m (2014 est.)	0 cu m (2014 est.)	8.489 trillion cu m
5	India	761,000 bbl/day (2015)	0 bbl/day (2013)	3.785 million bbl/day (2013)	5.675 billion bbl (2016 est.)	30.4 billion cu m (2014 est.)	52.1 billion cu m (2014 est.)	0 cu m (2014 est.)	21.7 billion cu m (2014 est.)	1.489 trillion cu m (2016 est.)
6	Canada	3.677 million bbl/day (2015 est.)	3.21 million bbl/day (2015 est.)	581,300 bbl/day (2015 est.)	171 billion bbl (2016 est.)	151.5 billion cu m (2014 est.)	116.5 billion cu m (2014 est.)	77.96 billion cu m (2014 est.)	21.89 billion cu m (2014 est.)	1.996 trillion cu m (2016 est.)
7	Indonesia	831,000 bbl/day (2016 est.)	310,100 bbl/day (2013 est.)	463,000 bbl/day (2013 est.)	3.7 billion bbl (2016 est.)	75 billion cu m (2015 est.)	39.7 billion cu m (2015 est.)	31.78 billion cu m (2014 est.)	1.8 billion cu m (2013 est.)	2.83 trillion cu m (2016 est.)
8	Australia	322,300 bbl/day (2015 est.)	248,400 bbl/day (2015 est.)	332,800 bbl/day (2015 est.)	1.2 billion bbl (2016 est.)	62.64 billion cu m (2014 est.)	38.51 billion cu m (2014 est.)	31.61 billion cu m (2014 est.)	6.938 billion cu m (2014 est.)	860.8 billion cu m (2016 est.)
9	Iran	3.3 million bbl/day (2015 est.)	1.042 million bbl/day (2013 est.)	87,440 bbl/day (2013 est.)	157.8 billion bbl (2016 est.)	174.5 billion cu m (2014 est.)	170.2 billion cu m (2014 est.)	9.86 billion cu m (2014 est.)	6.886 billion cu m (2014 est.)	34.02 trillion cu m (2016 est.)
10	Brazil	2.532 million bbl/day (2015 est.)	397,100 bbl/day (2013 est.)	394,400 bbl/day (2013 est.)	16 billion bbl (2016 est.)	20.35 billion cu m (2014 est.)	20.35 billion cu m (2014 est.)	100 million cu m (2014 est.)	17.32 billion cu m (2014 est.)	471.1 billion cu m (2016 est.)

Source: the Author, based on:

1 CIA, 2017. *CIA World Factbook*. Available at: www.cia.gov/library/publications/resources/the-world-factbook/geos/in.html, last accessed: June 18, 2017; and
2 EIA, 2016a. *Energy Explained*. U.S. Energy Information Administration. Available at: www.eia.gov/energyexplained/index.cfm?page, last accessed: June 3, 2017.

Table 4.5b Oil and gas key players, continued: Key indicators.

Rank	*Top energy-intensive nations*	*Crude oil production*	*Crude oil exports*	*Crude oil imports*	*Crude oil proved reserves*	*Natural gas production*	*Natural gas consumption*	*Natural gas exports*	*Natural gas imports*	*Natural gas proved reserves*
11	Nigeria	2.317 million bbl/day (2015 est.)	2.231 million bbl/day (2013 est.)	0 bbl/day (2013 est.)	37 billion bbl (2016 est.)	43.84 billion cu m (2014 est.)	18.84 billion cu m (2014 est.)	25 billion cu m (2014 est.)	0 cu m (2013 est.)	5.111 trillion cu m (2016 est.)
12	UAE	2.82 million bbl/day (2015 est.)	2.637 million bbl/day (2013 est.)	0 bbl/day (2013 est.)	98 billion bbl (2016 est.)	54.24 billion cu m (2014 est.)	66.32 billion cu m (2014 est.)	8.066 billion cu m (2014 est.)	20.14 billion cu m (2014 est.)	6.091 trillion cu m (2016 est.)
13	UK	893,300 bbl/day (2015 est.)	699,700 bbl/day (2015 est.)	1.047 million bbl/day (2015 est.)	2.8 billion bbl (2016 est.)	38.58 billion cu m (2014 est.)	70.45 billion cu m (2014 est.)	10.55 billion cu m (2014 est.)	42.83 billion cu m (2014 est.)	205.4 billion cu m (2016 est.)
14	Norway	1.61 million bbl/day (2015 est.)	1.255 million bbl/day (2015 est.)	22,400 bbl/day (2015 est.)	5.1 billion bbl (2010 est.)	108.8 billion cu m (2014 est.)	5.87 billion cu m (2014 est.)	114.4 billion cu m (2015 est.)	0 cu m (2014 est.)	1.922 trillion cu m (2016 est.)
15	Mexico	2.154 million bbl/day (2016 est.)	1.193 million bbl/day (2016 est.)	11,110 bbl/day (2015 est.)	9.7 billion bbl (2016 est.)	44.37 billion cu m (2014 est.)	72.77 billion cu m (2014 est.)	52 million cu m (2014 est.)	28.84 billion cu m (2014 est.)	432.9 billion cu m (2016 est.)
16	Venezuela	2.5 million bbl/day (2015 est.)	1.548 million bbl/day (2013 est.)	0 bbl/day (2013 est.)	300 billion bbl (2016 est.)	21.88 billion cu m (2014 est.)	23.72 billion cu m (2014 est.)	0 cu m (2014 est.)	1.839 billion cu m (2014 est.)	5.617 trillion cu m (2016 est.)
17	Kazakhstan	1.621 million bbl/day (2016 est.)	1.292 million bbl/day (2016 est.)	145,800 bbl/day (2014 est.)	30 billion bbl (2016 est.)	21.38 billion cu m (2016 est.)	13.1 billion cu m (2016 est.)	13.7 billion cu m (2016 est.)	2.2 billion cu m (2014 est.)	2.407 trillion cu m (2016 est.)
18	S. Africa	3,000 bbl/day (2015 est.)	0 bbl/day (2013 est.)	466,100 bbl/day (2013 est.)	15 million bbl (2016 est.)	950 million cu m (2014 est.)	4.75 billion cu m (2014 est.)	0 cu m (2013 est.)	3.8 billion cu m (2014 est.)	15.01 billion cu m (2012 est.)
19	Algeria	1.37 million bbl/day (2015 est.)	1.146 million bbl/day (2013 est.)	2,920 bbl/day (2013 est.)	12 billion bbl (2016 est.)	83.29 billion cu m (2014 est.)	37.5 billion cu m (2014 est.)	40.8 billion cu m (2014 est.)	0 cu m (2013 est.)	4.504 trillion cu m (2016 est.)
20	Colombia	1.019 million bbl/day (2015 est.)	859,000 bbl/day (2016 est.)	0 bbl/day (2016 est.)	2.3 billion bbl (2016 est.)	11.86 billion cu m (2015 est.)	10.9 billion cu m (2015 est.)	1.102 billion cu m (2015 est.)	0 cu m (2013 est.)	134.7 billion cu m (2016 est.)

Source: the Author, based on:

1 CIA, 2017. *CIA World Factbook*. Available at: www.cia.gov/library/publications/resources/the-world-factbook/geos/in.html, last accessed: June 18, 2017; and
2 EIA, 2016a. *Energy Explained*. U.S. Energy Information Administration. Available at: www.eia.gov/energyexplained/index.cfm?page, last accessed: June 3, 2017.

The relative content of sulfur in natural petroleum deposits also leads to categorizing oil as "sweet," for oil that consists of comparatively low sulfur levels, or as "sour," for oil that consists of considerable amounts of sulfur.

Considering the oil wells situated all around the globe, the role of energy security has become pivotal in safeguarding the petroleum that is being gathered, refined, stored, and distributed.

The majority of regions abundant in oil and gas entail security risks due to sociopolitical turmoil, sabotage, terrorism, or warfare. This heightens the necessity for security, and at the same time verifies the significance of energy. Several oil-producing nations maintain strategic petroleum supplies as a buffer to be used in case of energy shortage, political turmoil, or economic crisis.

The world's major oil and gas key players are classified in Tables 4.5a and 4.5b, in terms of producing and consuming, importing and exporting rates, but also volumes of reserves.

The top crude oil producers are Russia, Saudi Arabia, the United States, Iraq, and China. The top natural gas producers include the United States, Russia, Iran, Qatar, and Canada. These are among the world's greatest economies, hence the impact of energy upon their economic growth is apparent.

The nations with the largest oil reserves include Venezuela, Saudi Arabia, Canada, Iran, and Iraq, and the world's largest natural gas reserves are in Russia, Iran, Qatar, the U.S., and Saudi Arabia. The long-term energy security designation gives strategic significance to these nations, in terms of energy independence and the ability to invest in long-term industrial infrastructures with socioeconomic benefits. China and India are the superpowers that consume and import much more fuel than they produce, thus making them vulnerable to external energy factors such as fluctuations in oil and gas production, prices, and geopolitical and other events that may cause supply chain disruptions. Since oil and gas trade agreements necessitate close diplomatic bonds and strategically formed trade agreements, it is interesting to observe China and India's energy supply from major producers such as Russia, Saudi Arabia, Iran, and West Africa.

Tables 4.6a and 4.6b demonstrate the global statistics for crude oil (2013–2016 est.) in million bbl/day, i.e. show the aggregate global statistics in terms of production, consumption, imports, exports and reserves. The refined production-consumption balance is negative, i.e. minus 5.71, whereas the imports/exports ratios for oil and gas are also negative. This might indicate two things: First, the supply is restrained, in order to retain the oil price at satisfactory levels. Second, there is a negative balance in refined petrol products production/consumption equilibrium, indicating either a past surplus that is currently consumed at a subsequent fiscal year. Finally, the tight equilibrium between production and consumption does not indicate contingencies in storage in case of future supply chain disruptions.

Table 4.6a Global statistics for crude oil (2013–2016 est.) in million bbl/day.

Crude oil – production:	81.8 million bbl/day (2015 est.)	
Crude oil – exports:	46.6 million bbl/day (2013 est.)	Balance = –0.78
Crude oil – imports:	47.38 million bbl/day (2013 est.)	
Refined petroleum products – production:	87.69 million bbl/day (2013 est.)	Balance = –5.71
Refined petroleum products – consumption:	92.79 million bbl/day (2014 est.)	
Refined petroleum products – exports:	27.74 million bbl/day (2013 est.)	Balance 1.6
Refined petroleum products – imports:	26.14 million bbl/day (2013 est.)	
Crude oil – proved reserves:	1.665 trillion bbl (1 January 2016 est.)	

Table 4.6b Global statistics for natural gas (2013–2016 est.) in million cubic meters.

Natural gas – production:	3.544 trillion cu m (2014 est.)	Balance = 0.016
Natural gas – consumption:	3.56 trillion cu m (2014 est.)	
Natural gas – exports:	1.157 trillion cu m (2013 est.)	Balance = –0.315
Natural gas – imports:	1.472 trillion cu m (2013 est.)	
Natural gas – proved reserves:	193.9 trillion cu m (1 January 2016 est.)	

Source: the Author, based on:

1 CIA, 2017. *CIA World Factbook*. Available at: www.cia.gov/library/publications/resources/the-world-factbook/geos/in.html, last accessed: June 18, 2017; and
2 EIA, 2016a. *Energy Explained*. U.S. Energy Information Administration. Available at: www.eia.gov/energyexplained/index.cfm?page, last accessed: June 3, 2017.

An Overview of the Oil and Gas Supply Chain: Upstream, Midstream, and Downstream

The production of oil and gas follows similar supply chain patterns, not only because they are both fossil fuels but also because, frequently, a single well contains both oil and gas. The three key stages are as follows (Figure 4.6):

1 Upstream

This includes the exploration and production stages; seismic surveys and mapping, and oil and gas extraction and drilling. The government is in charge of oil and gas leasing, with the leasing contracts being issued for three key types of land: a) onshore public land; b) offshore public land (seabed); and c) tribal land (e.g., for nations with indigenous populations, such as the Native Americans in North America, the Inuits in Canada, the Maoris in New Zealand, and so on).

Investment decisions will be determined by the financial, commercial and technological feasibility studies. Preliminary investment decisions may be altered and budget allocation may become more detailed as the following studies are conducted and discussed:

Financial and commercial strategy:

- feasibility study
- cost/benefit analysis
- return on investment study.

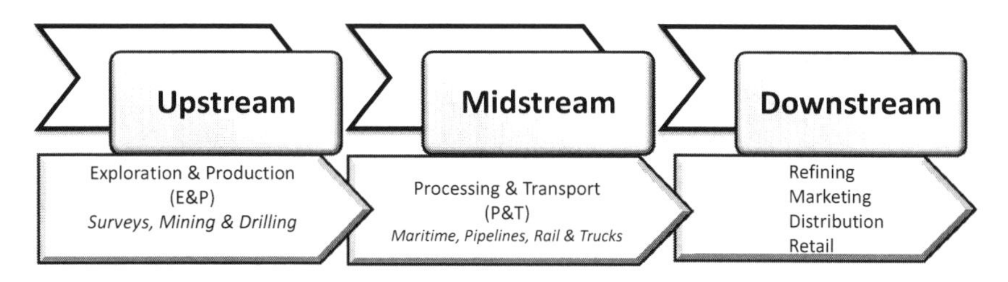

Figure 4.6 The three energy market segments: Upstream, midstream, and downstream.
Source: the Author.

Risk management and regulatory compliance strategy:

- risk management (see Chapter 1 of this book), including contingency plans and risk mitigation and recovery plans
- HSQE plan (see Chapter 2 of this book).

Project management strategy:

- resource management
- time management
- operational management
- benchmarking and performance appraisal studies.

The leasing process may vary according to the nature of the agreement and the national policy. In general, the key contract types entail: Public domain leases, right-of-way leases, renewal/exchange leases, etc.

An environmental impact assessment is conducted by the government and/or the energy consortium, to ensure the project will comply with environmental rules, and will retain its environmental footprint.

2 Midstream

Midstream encompasses the processing and transport process (P&T), i.e., transportation from the well to oil refineries or natural gas processing plants, and storage. The particularity of the midstream stage pertains to the wide geographical scope of operations. In this stage, transportation connects the dots between crude, unprocessed energy, and value-added activities.

3 Downstream

Downstream is a stage where the refined/processed commodities move closer to their potential markets and are streamlined to their final consumers, be it industrial, public utilities or domestic. Some energy products remain in storage, others are locally distributed, whereas a portion of the energy is exported and further distributed overseas.

Strategic Energy Reserves and Long-Term Storage

This is a separate category beyond the typical commercial supply chain of fuel. Finally, a segment of the crude unprocessed energy is stored as part of government or corporate strategic planning.

Crude energy products are ideal for long-term storage for two reasons:

1. Their chemical composition does not change or deteriorate as much over time, thus allowing better quality in the future refinery output.
2. The future economic and market cycles will dictate the demand for particular refined products. Therefore, decisions on processing and refining should be taken at a later stage for optimum utilization of the available products.

Security risks are present in every stage of the aforementioned supply chain, i.e., from initial planning, to exploration, transportation, processing, distribution, but also storage. This is a high-risk high-profit industry, where both intentional and unintentional threats may impact a project.

Public–private partnerships are frequently involved in the above stages as part of national or corporate strategic planning.

Technologies, Operations, and Processes in the Oil and Gas Upstream, Midstream, and Downstream Stages

The energy industry is technology-intensive and capital-intensive throughout its three principal market segments, i.e., 1) upstream, 2) midstream, and 3) downstream.

This section describes in a concise manner the complex operational and technological protocols, but also the challenges that modern energy professionals need to overcome, including safety, security, and environmental risks. And yet, upon reflecting on the industry's evolution, it appears that the industry has the capacity to master such risks and challenges through the use of revolutionized processes and technologies.

Upstream

The upstream sector, also called the "exploration and production" (E&P) sector for crude oil and natural gas, involves four principal phases involved in the upstream industry: 1) exploration, 2) well development, 3) production and 4) well abandonment (EPA, 2000).

1 Exploration (Pre-Drilling Preparation Process)

The exploration group 1) selects specific regions, 2) examines deeds and acreage, and 3) negotiates leasing contracts. Hydrocarbon exploration is a high-risk investment and risk assessment is paramount for successful exploration portfolio management.

Exploration and production professionals conduct geophysical surveys to determine the location, characteristics, and volume of reserves. Some of the tools they use include:

a **Geological mapping** is a specialized chart created to illustrate geological characteristics of the oil and gas wells. In digital geologic mapping systems, different features are depicted by different colors and signs to display the location, depth, dimensions, physical, and chemical characteristics, etc.

b **The seismic survey process (used in offshore oil and gas exploration)**. Sound waves assist scientists map the seabed and geological formations underneath.

- Surveyors unleash pressurized air into the water to produce brief sound waves that echo off subsurface geological levels and are recorded by cables with sound sensors that are being pulled behind the vessel.
- Data evaluation is used to create detailed comprehensive maps of the oil and gas wells.
- Seismic surveys are typically performed at a reduced speed of roughly four–five knots, or about five miles per hour (mph), and the acoustic sound source is usually triggered at 10–15-second time periods.
- **Passive seismic methods** employ the low frequency, low resolution content of the subsurface seismic sources. While considered environmentally friendly, they

are frequently disregarded as inefficient and may not be accepted in a dynamic seismic strategy.

- **Reflection seismic methods** generate reflective seismic waves to conduct seismic imaging. To improve the seismic imaging previously offered by 2D space, the industry renders in 3D, and recently created 4D seismic mapping, which translates sounds into more precise image coordinates and therefore offers design optimization.

  To identify the position and dimensions of oil and gas wells, exploration companies use reflection seismology (conduct seismic surveys or seismic reflection), to generate precise mapping of different wells and identify the geological morphology under the seabed and under land. Similar to the ultrasound techniques used in medicine, geophysicists record the sound waves echoed from the energy wells as they return to the surface.

  Due diligence is required for the licensing/permits procedure to ensure that seismic surveys are appropriately monitored and controlled with increased precision, eliminating environmental damage.

c **Data mining for estimating reserves and production volume**

- Four-dimensional quantitative and qualitative data analysis. Data mining in the energy industry entails the employment of a data-driven approach throughout the entire supply chain, from upstream to midstream and downstream. Oil and gas extraction and the upstream operations in general use data mining techniques in order to quantify both the overall volume of reserves, but also the individual oil wells or pockets, the air chambers etc.
- Four-dimensional seismic engineering is employed to keep track of alterations brought on by development or even the impact of fluid and gas treatment to enhance restoration. Until recently, the industry's groundwork has been concentrated on *qualitative* elements, such as:

  - distinguishing among depleted and preserved undrained reservoir locations to identify residual oil and gas compartments;
  - mapping problematic wells and diagnosing root cause and remedy;
  - verifying if defective areas enable liquid to circulate through them or not.

- A shift from qualitative data into big data and quantitative analysis. One of the principles in statistics purports that the smaller the sample population analyzed, the greater the sampling error. Hence, in order to enforce the statistical data accuracy and production-related forecasts, technological research projects increasingly shift their attention on to *big data and quantitative data* collection and analysis. The findings of such analyses greatly strengthen the forecasting accuracy on reservoir characteristics and reaction during the exploration and development stages. The findings also reveal details on geological configurations, but also the timestamp evaluation, i.e., petro-physical elements of porosity, permeability, density, water saturation, and lithology.

d **Four-dimensional (4D) seismic monitoring technologies**

- The energy industry's reservoir monitoring capabilities are transitioning from qualitative 3D seismic monitoring into quantitative time-lapse 4D seismic monitoring, where a particular oil well can be examined during two or more points in time, through precision contrasting of 3D seismic images acquired months or even years apart. The time-lapse, or 4D, seismic method entails obtaining, evaluating, and presenting repetitive seismic surveys on the developing hydrocarbon field. A standard ultimate processing technique entails a time-lapse difference dataset (i.e., the seismic information from Survey 1 is deducted from the information from Survey 2). The variation should be near to zero, apart from cases where reservoir alterations have taken place.

Hence, while the 3D technologies manage to pinpoint depleted sections of a reservoir, the new 4D technology capabilities manage to generate saturation maps.

Time-lapse or 4D seismic technologies consider time as the fourth dimension: The systems contrast the outcomes of 3D seismic surveys recurring at extended time intervals, i.e., commence prior to a field starts off upstream activities, throughout midstream development phases. Significant disparities observed amongst the survey outcomes are caused by fluid modifications and/or alterations in reservoir pressures.

e **Other upstream exploration tools**

Contemporary petroleum geologists use satellite systems, seismology, geodesy, geophysical surveys and a plethora of innovative methods and technologies to study reserves found deep offshore, or in surface area rocks and soil. They employ gravimeters in oil wells and other subterranean geological sites. These are very sensitive types of accelerometers used to identify the slightest alterations in acceleration due to the Earth's gravitational field.

Similarly, magnetometers are used to determine even minor modifications in the Earth's magnetic field resulting from oil flows.

While the size, weight and sensitivity of these devices continuously improves, new technologies were discovered to fine tune the soil and water exploration processes.

f **Fuel sniffers**

Fuel sniffers are devices that can smell molecular footprints of captured hydrocarbons evaporated in the atmosphere at concentrations of only ten parts per trillion. Sniffers are placed onboard a low-flying aircraft (about 1,000 feet) flying around suspected oil reserves. The device sniffs the atmosphere to contrast that data with air samples gathered in other locations, while taking into consideration the soil and climate particularities.

New technologies offer improved alternatives, relatively cheaper, accessing deeper offshore, remote areas or locations with extreme weather.

2 Exploratory Drilling

Exploratory drilling is necessary to validate the presence of sufficient hydrocarbon deposits and that the well(s) can generate adequate oil or gas to ensure it is financially feasible to pursue.

The team evaluates the operational and financial feasibility at different stages of the project until the well's depletion. Nevertheless, this is a critical stage that will validate or reject the decision to proceed with the next stage.

This phase introduces a logistics plan where flows of equipment and tools need to enter the exploration area. The drilling area is prepared, i.e., leveled, with any obstructions removed.

A drill pad is constructed to host the wellheads for several horizontally drilled wells. Its major advantage for operators is that they can drill several wells much faster, as opposed to drilling one well per site.

Mud pits are used, i.e., several steel tanks used to hold water, cycle the drilling mud and store sand and sediments. The drilling rig is installed, with its generators, pumping systems, and other devices.

3 Well Development

Spudding is the starting point of drilling one or several new wells and reservoir pockets. The drilling bit permeates the surface employing a drilling platform suitable for drilling to the authorized depth.

4 Production

The oil and gas production process entails the extracting of hydrocarbons, typically from more than one well. As a next stage, the blend of liquid substances (i.e., oil, gas, water fluids, mud, and solids) are isolated. The components of non-commercial value are treated for disposal, whereas the commercially profitable oil and gas products are retained, stored, and transported for the midstream part of the process.

5 Vertical vs Horizontal Drilling

Typically, conventional oil and gas can be extracted by using vertical drilling, on terrestrial or aquatic wells, at the superficial shallow continental shelf, whereas a well is drilled all the way down to the soil or seabed. Unconventional oil and gas extraction requires horizontal and directional drilling.

Hydro-fracturing, or hydraulic fracking or jetting, is a well-development procedure associated with shale oil and gas. It is a preferred method implemented to optimize the oil and gas wells' productivity, and entails the injection of high pressure water (typically 5–7,000 pounds per square inch) into a soil understructure through the well to extract energy from the wellhead to the pipelines.

The method employs numerous horizontal water jets from inside the well filter in order that high-speed water flows release particulate matter and drilling mud deposits. The released matter enters the well filter so that it is later pumped out of the well. Optimum efficiency is attained on larger diameter wells, i.e., of at least four inches. Jetting is especially effective in forming highly stratified unconsolidated sediments and natural formation wells. Different technologies and operational diligence will eliminate the entry of potential impurities in the aquifer and prevent issues with subsoil drainage.

Extracting is easily achieved, merely by the innate well stress combined with average compression or pumping. Another benefit of unconventional techniques is their use in mature or abandoned oil and gas wells. When the innate well's pressure is extremely low due to depletion, high-tech drilling methods prolong the well's lifespan and enhance profitability of the project (return on investment).

6 Site and Well Abandonment

Decisions on site and well abandonment must take place when the proven oil or gas reserves have been depleted and the well reaches a point where its operating expenses are not covered and the field has reached its economic limit. The abandonment procedure will be dictated by the current regulatory framework in the specific country, state, and industry segment.

The four stages of site abandonment are:

1. Abandonment planning from an engineering and safety management perspective.
2. Discontinuation of oil or gas production and the well's abandonment.

3 Partial or entire offshore structure removal.
4 Equipment abandonment, removal or recycling.

Site abandonment requires decisions about leaving, removing, and/or recycling installations and infrastructures, such as the topside and the substructures of the offshore equipment.

The operator takes away piping and pumping systems, taps the well with concrete, and detaches crucial structures and components such as the wellhead, the tanks, and the gravel pad.

The topside of the offshore platform covers the steel tubular jacket, i.e., the hull area above the water line that covers the oil production plant, the drilling rig, and the crew accommodation space.

The substructure includes the supporting structural aspects of ground or offshore oil and gas platforms that secure the drilling rig constituents, and may offer space for storing under the main deck.

Midstream

In the energy industry, the midstream segment is viewed as a low risk and relatively uncomplicated process that connects upstream with downstream, or supply with demand. Its success depends on regulatory compliance, roughly focused in occupational health, safety, security, and the environment.

Transportation is a significant aspect of the midstream operations, and comprises of pipelines, sea, land, and railway haulage. This stage also handles the storage and marketing aspects of oil and gas.

Downstream

The downstream segment encompasses petroleum refineries, petrochemical plants, and finally retail and distribution channels. This stage is closer to the buyers' markets, and covers several countries and regions, subject to trade agreements and the demand for specific energy products. Some of the high-demand non-renewable energy products include fuel oil, diesel oil, natural gas, gasoline, jet fuel, heating oil, propane asphalt, lubricants, synthetic rubber, plastics, fertilizers, pesticides, pharmaceuticals, etc.

Technologies and the Refinery Processes

A petroleum refinery is a technology-intensive well-synchronized industrial process plant that is specially designed to generate natural and chemical modifications in crude oil to transform it into high-demand commodities such as diesel oil, jet oil, lubricating oil, fuel oil or liquefied petroleum gas.

The petroleum-refining procedure aims to convert crude oil into value-added products, subject to the market demand. Crude oil is comprised of numerous hydrocarbon compounds that are segregated during the refining process. Refining consists of three principal levels, i.e.: 1) separation, 2) conversion, and 3) treatment.

Global Oil Refineries

There are about 700 refineries around the world, most of which operate 24/7, 365 days a year. However, refineries vary, as some facilities use state-of-the-art technologies while others use older, even obsolete technologies.

Also, the size of the facilities, and the demand of the markets they serve, may determine:

a refinery complexity, i.e., how simple or complex are the processes they follow;
b refinery capacity, i.e., optimum daily refinery potential, product separation and treatment, i.e., how many different grades they produce;
c regulatory compliance, especially the possession of technologies to eliminate environmental emissions, such as low sulfur, NOx, and so on.

Refineries' geographical position depends to a great extent on factors such as:

a The location of supply chains: Refineries are strategically situated in the vicinity of the producers, the exporters, the importers, or significant transportation, storage or distribution centers.
b Occasionally, the location is selected based on low-taxation low-cost property costs, or low operational costs.
c Locations with high industrial activities or high population density, the vicinity of oil transportation routes, economic potential for refinery construction, etc. The Asia Pacific region hosts a large number of refineries, followed by Europe, Eurasia, and North America.

The chemical composition of crude oil entails a mixture of hydrocarbon substances and compounds, or particles of oxygen, sulfur, nitrogen, hydrogen, HO_2 (water) or sodium chloride (salt), etc. Once the oil is transported to the refinery, a large number of these non-hydrocarbon components are taken away and the oil is separated into different compounds, to produce different grades of marketable, high-demand petroleum and/or petrochemical products. Crude oil and natural gas are of minimal use in their unprocessed condition, yet through the refining process value is added as the generated products (i.e., fuels, lubricating oils, waxes, asphalt, petrochemicals, etc.) are streamlined into the markets.

The first refining process is basic distillation. Due to the fact that crude oil is made up of a mixture of hydrocarbons, this first and fundamental refining process has the purpose of separating the crude oil into its "fractions," the extensive groups of its element hydrocarbons. Crude oil is warmed up and placed into a still, i.e., a distillation line, where diverse products boil off and may be recovered at specific temperatures.

The lighter products, such as liquid petroleum gases (LPG), naphtha, and "straight run" gasoline, are recovered at the lowest temperatures. Middle distillates, such as jet fuel, kerosene, and distillates such as home heating oil and diesel fuel, will derive at the next stage. Eventually, the heaviest products such as residual fuel oil (RFO) or heavy fuel oil (HFO), will be retrieved, typically at temperatures over 1,000°F (538°C). The simplest refineries, called topping or hydro-skimming plants, will stop at this stage. Their principal purpose is to distil crude oil into a modest range of products. More advanced refineries will re-distil the heavier components into lighter products in order to maximize the production of the most appealing products (EIA, 2016c; EIA, 2016d).

The Refining Process

The purpose of the refining process is to create high-demand high-value products by following the process of:

a separation – remove the non-hydrocarbon materials;
b conversion – distinguish the hydrocarbons into various grades; and
c treatment – break the oil down into its various constituents, while transforming it into high-demand and useful products.

Figure 4.7 reflects a more detailed refining process, in six significant stages.

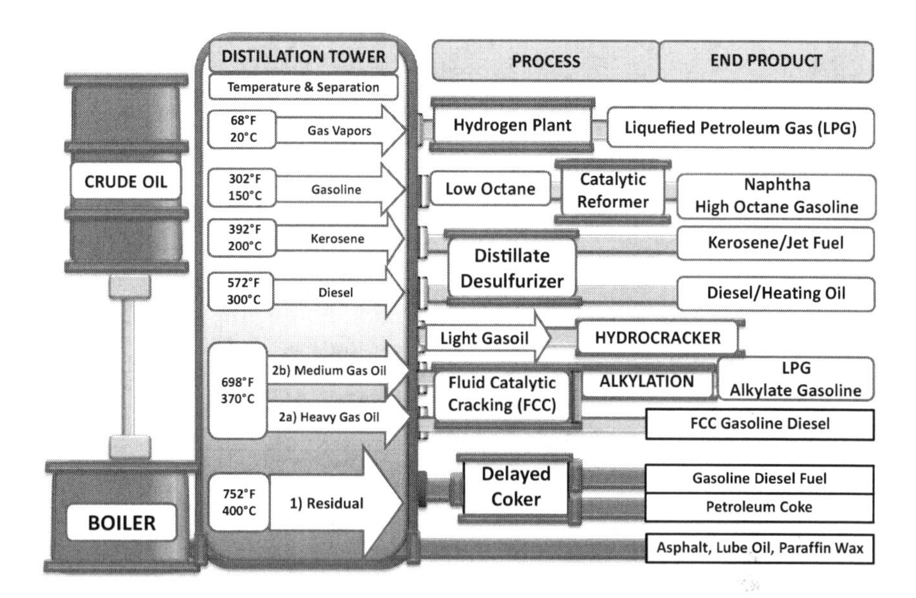

Figure 4.7 The crude oil refining process and end products.

Sources: the Author, based on data from:

1 DOT, 2011. *The Fuel Refinery Process*. U.S. Department of Transportation, Pipeline & Hazardous Materials Safety Administration, Pipeline Safety Stakeholder Communications. Available at: https://primis.phmsa.dot.gov/comm/factsheets/fsrefinery.htm, last accessed: February 8, 2017.

2 OSHA, 2017. *Petroleum Refining Process*. Section IV: Chapter 2. U.S. Department of Labor, Occupational Safety and Health Administration. Available at: www.osha.gov/dts/osta/otm/otm_iv/otm_iv_2.html, last accessed: February 8, 2017.

1 Separation

a. In the first stage, the goal is to separate the crude oil molecules in accordance with their molecular weight, at a normal atmospheric pressure, with the process of simple distillation. Hence, during the "topping" or "hydro-skimming" process, input flows into the distillation tower. The boiler is used to increase the temperature to approximately 752°F or 400°C.

b. Vapors cause the different petroleum molecules to be separated according to their distinguished densities and boiling points. Hence, the residuals, which have the heaviest molecules, do not vaporize but stay at the bottom of the distillation tower. The lighter molecules separate themselves from the residuals as they vaporize, and their molecules get liquefied depending on their boiling points.

c. The vacuum distillation is the next step after the simple distillation, in which the different products now vaporize as the temperature drops.

d. As a result, the "petroleum cut" occurs when high viscosity hydrocarbons remain at the bottom, while gaseous products evaporate.

e. At 302°F or 150°C, gas vapors make it to the top, whereas the liquids get separated and flow into different trays found at different height levels of the distillation tower.

f. A second distillation of the heavy residuals separates molecules of medium density to create fuel and diesel oil (EIA, 2016d; DOT, 2011; OSHA, 2017).

2a “Catalytic Reforming” (CR) process

Reforming is a chemical method designed to transform heavy naphtha (i.e., the low octane naphtha that has been distilled from crude) into reformates, i.e., higher octane products of finished gasoline. The distinguished specifications of the fuel blend will determine the quantities generated for each grade.

2b Isomerization

Just as in reforming, isomerization is used to modify the molecular formation, i.e., from normal pentane to higher octane – branched molecules, or from linear normal butane into iso-butane.

3 Treating

This process uses catalysts, electrolysis and hydrogen to chemically remove contaminants such as sulfur, nitrogen oxides, salts, nickel, and so on. For example, hydrodesulfurization (HDS) is a catalytic chemical process in which a hydro-treater unit in a refinery is used to remove sulfur, a pollutant, from refined petroleum products and natural gas. Given that the environmental regulations globally for low sulfur are becoming stricter each year, refiners conduct “hydrotreating” to separate sulfur from the petroleum products.

Coking, residue catalytic cracking (RCC) and de-asphalting are deep conversion methods, intended to transform the heaviest molecules, such as residues, into lighter, cleaner, higher-value products.

4 Fluid catalytic cracking (FCC) or simply cracking

In this crucial conversion practice catalysts or other methods are used to convert the distilled crude (which is now in a grade somewhere lighter than heavy fuel such as diesel oil), and it is further converted into more valuable products such as gasoline, olefinic gases, and other products.

5 Coking

Coking is a refinery unit procedure that generates petroleum coke, a coal-like substance, while making use of the low-value low-demand residues (bottoms) from the atmospheric or vacuum distillation column by converting them into higher-value commodities. Coking is a rather profitable option, as petroleum coke exports exceed 19% of America’s petroleum product exports, with most of the product sold to China and other Asian nations.

Petroleum coke is being used in electricity, manufacturing, and the production of electrodes in the aluminum and steel industrial sectors (EIA, 2013).

Energy Controlling Systems

Energy infrastructure encompasses advanced controlling systems such as:

1. **Distributed Control Systems (DCS)** are modern instrumentation and control platforms. These are high-tech automated command systems for power plants or refineries, with a central management control function and independent remote controls allocated in the entire network. Some of the benefits of DCS include:

   a. real-time data sharing both to operators and selected supply chain partners such as vendors and consumers;
   b. synced surveillance systems and alarms;
   c. historic database recording and sharing;

d ERP operations and batch management. In an energy production setting, output is typically completed in batches. Furthermore, several resources taking part in the operations are handled independently in batch sets with distinctive requirements. Batch management is therefore employed to separate different production lots. This assists in monitoring the batches throughout their life-cycle, and also in documenting the distinctive features of each product, such as chemical qualities. This system may be implemented to different supply chain stages and individual upstream, mainstream, and downstream processes.

2 **Programmable Logic Controllers (PLCs)** are robust computer systems that monitor and control production inputs and outputs, and generate logic-based judgments for computerized procedures or equipment. They browse data and alerts from diverse receptors and input units including security alarms or switches, and execute accordingly. The PLCs are ideal for the energy industry as they can operate in harsh weather and extreme temperature conditions, sandstorms or humidity.

3 **Advanced Metering Infrastructure (AMIs)** consist of advanced automated software and hardware that incorporate process data evaluation by means of readily accessible remote controls. These products facilitate monitoring of comprehensive real-time data and regular data accumulation and sharing to diverse supply chain entities.

The AMIs typically comprise of on-site data collection and analysis, communication channels among the buyers and sellers pertaining to energy, i.e., electric power, or other energy utility and computer data collection and sharing systems.

The primary advantages of AMI use may be typically classified as:

- program functionality advantages;
- customer support benefits;
- economic advantages.

4 **Automated Inspection and Maintenance Systems (AIMS)** enable all standard procedures to be combined in an efficient single-window protocol, as opposed to manually working on several dozens of inspection checklists. As RFID chips are installed in every single piece of machinery on the drilling rig, automated report and audit data is logged in and available for analysis and further corrective action to be taken. Pursuant to the inspection, all data, inspection findings and corrective action is available online, both on the energy production site and remotely in the company premises.

Oil and Gas Benefits, Risks, and Recommendations

Approximately 80% of the world's energy derives from oil and gas. The upstream, midstream, and downstream processes are rather intricate, as this is a high-tech industry with a high number of unconventional projects, both onshore and offshore. The industry has invested in advanced research, technological innovations, process improvement methods, and highly effective training programs to eliminate risks and ensure sustainable safety, security, and environmental protection.

We live in an era of proactive risk management, and for this reason most industrial products we use are classified as "Dangerous Goods," (DG) or contain DG ingredients. Construction materials, domestic cleaning detergents, cosmetics, processed foods and medicines are products of a DG or "HazMat" industrial process. Similarly, most energy products, both renewables and non-renewables, are classified as DGs or HazMat at some point of their transformation from one energy type to another. This special designation was developed by the United Nations and ratified by governments and the global energy industry in order to better protect our planet and its inhabitants.

Oil and gas products require special storage, handling and transportation protocols. Yet, the safety, security, and environmental statistics demonstrate that the inherent risks do not make the oil and gas industry an unsafe industry. Quite the contrary, for over a century energy companies have validated their dedication to safety and security and environmental standards by complying with the national and international HSQE regulations, and investing in technologies and processes that can mitigate the threats.

Preventive risk management is the key, hence the following risks are intended to be used as risk management checklist points, with the purpose of proactively enhancing national and corporate standards of safety, security, and the environment.

A list of the benefits, risks, and recommendations for upstream, midstream, and downstream are found herewith.

1 **Seismic surveys: Air-gun blasting**
The purpose of seismic surveys is to reduce financial risks of overinvestment or misinvestment, i.e., by drilling in an area of reduced reserves. The actual seismic survey process entails air-gun blasting that is used to identify and map oil and gas reserves.

2 **Project-related risks** entail:

   a poor subsurface seismic imaging (low resolution of the velocity profile or poor spatial sampling);
   b insufficient oil and gas reserves, leading to project cancellation; or
   c unconventional oil and gas reserves, requiring expensive or advanced exploration technologies. Lack of advanced technologies or diminished return on investment may lead to project postponement.

3 **Safety risks**: The process imposes modest operational and safety risks for most sites. The risk factor is amplified in unconventional environments such as:

   a isolated regions, such as deep ocean or desert areas;
   b extreme weather conditions such as ice, hurricanes, deep ocean waves, etc.

4 **Environmental risks** entail the damage to marine life (i.e., mammals like whales and dolphins as well as smaller fish and zooplankton). Particular regulations entail marine estuaries and the mapping of regions where seismic surveys and energy exploration is prohibited.

5 **Resource management risks** entail the availability of factors of production (land, capital, labor) and other resources (such as technologies, software, processes, and special capabilities), in the following areas:

   a cost and budget allocation of resources;
   b physical availability of resources; timely delivery, i.e., just in time, with limited disruptions, delays or deviations.
   c supply chain security, i.e., protection of the entire supply chain from all types of threats, both physical and cyber.

6 **Permits, consents, and license blocs for onshore and offshore agreements**
All exploration projects are speculative by nature, and consequently there is no guarantee that the exploration will be physically viable, or commercially successful

or financially profitable. In fact, the trade prices are volatile, subject to the market supply/demand equilibrium, as stipulated in the next section on commercial and financial risks.

7 **Commercial and financial risks**

Price fluctuations are characterized by highly cyclical patterns, based on two factors: Crude oil prices and the natural gas supply–demand equilibrium. Namely:

- fluctuations are "pegged" with crude oil prices;
- demand for fuel peaks during the winter, hence prices peak as well.

The commercial, financial, operational risks of:

- producing the right products for the right markets;
- selling the energy commodities at profitable contract terms, with long-term financial and commercial benefits;
- observing the market cycles, and anticipating the next stage of the global market; adjusting production to meet future changes in demand;
- overall, adopting a proactive stance, and expanding, retaining or revising plans for future production.

8 **Human factor risks**

Lack of familiarization, training or situational awareness; fatigue or poor judgment are some of the risks related to the human factor. These can be prevented or eliminated through:

- management committed to promoting a risk-free company culture;
- the development of processes that ensure incident prevention, training, and efficient mitigation;
- careful HR and recruitment, personnel familiarization, training and safety/security drills;
- crew recycling, i.e., the recruitment of permanent personnel that are trained in a company-specific, product-specific, energy-plant specific manner;
- the development of leadership and teamwork, and a no-blame culture where incidents and near-misses serve as lessons learned.

9 **Geological risk**

The risk of failure, i.e., not finding oil. The well's geological morphology may determine the anticipated complexity of a project.

Safety risks as well as cost and completion timelines pertinent to geological and geophysical particularities.

10 **Oil well management risks (monitor and control)**

Risk factors pertain to the optimum management, i.e., the monitoring and controlling of wells, with focus on the well production and the drilling intensity ratio. Oil and gas wells combine the risk of high pressure (blowout) and explosion (gas) and fire (oil). The main hazards are related to:

- abnormal well pressure, reservoir tightness, over-pressured zones;
- formation fluid (oil or gas) discharge into the atmosphere;
- formation fluid entry into the wellbore, thus halting well production and building pressure;
- reservoirs susceptible to damage, reservoirs with low water saturation, etc.

11 Well abandonment (temporary or permanent)

a Temporary well abandonment may occur for a number of reasons, such as:

- safety and environmental threats, e.g., extreme weather, or structural or geophysical issues related to the well;
- security threats, e.g., geopolitical turmoil, diplomatic tensions, war or terrorism; commercial challenges, e.g., financial meltdown, oil price collapse, global oversupply, diminished demand, etc.

b Permanent well abandonment may occur in any of the following cases:

- when the causes of temporary well abandonment have caused lasting damage or risks, with no solution in the foreseeable future;
- when the well is depleted;
- when the well is unconventional and technological remedies are either not available or too costly.

12 Drilling relocation risks

Superstructure is hereby defined as the drilling rigs for onshore or offshore platforms comprising the drilling, production, storage, and processing units. These solid structures have a commercial lifespan of approximately 100 years. Relocation risks pertain to the decision to permanently abandon the current project and to initiate a new project. Hence, the risks entailed in this new endeavor comprise all the upstream, midstream, and downstream risks stipulated in this section.

a The physical risks of abandoning the previous well have been examined in the "well abandonment" section.
b The financial risks should also encompass the cost of relocating when a well is depleted, the high exploration cost associated with unconventional wells, and any market factors that would make the project financially unviable.

13 Midstream and downstream risks

These hazards pertain to commercial and financial decisions but also safety, security, and environmental issues, such as:

a the safe and environmentally friendly transportation, storage, and refinery processing of oil and gas;
b commercial decisions related to refinery processing, such as

- deciding on the geographical location of a refinery, e.g., an optimum location near the production region or the buyers' region,
- deciding on the refined byproducts, subject to the market demand;

c security risks at a physical, cyber, and supply chain level, i.e., prevention from intentional threats such as terrorist attack, hacking, sabotage, geopolitical conflicts, etc.

In addition to natural disasters, weather risks, and intentional damage, two more risks should be added: i) geopolitical risks in upstream or midstream locations, and/or any point in transit. For example, elections, military coup, invasion, social turmoil, etc. ii) The commercial and financial risk of moving the cargoes to a specific location. Local governance, policies, protectionism, markets, etc., are areas deserving a proactive approach.

14 **Fracking and refracking**

Innovative fracking technologies have generated double the production outcomes in unconventional oilfields. The re-fracking or re-fracturing techniques are the solution to tight oil, by employing the most novel methods of fracking in order to drill in wells that have previously been fracked with less innovative technologies. This technique will prolong the tight oil production peak and retain a booming market (EIA, 2017c).

Energy Transportation Security

From a security perspective, the element of risk is inherent in transportation. The oil and gas industry, and the non-renewable energy segments in general, involve lengthier transportation routes, several transit areas, and more "predators" attempting to sabotage the system and steal energy products and supplies.

Monitoring and controlling a stationary asset entails a fair amount of hazards, yet when this asset is in constant motion, the risk increases both in terms of likelihood and in terms of impact.

The energy haulage segment consists mainly of sea and land transportation. The aviation industry plays a secondary role due to the high cost and volume restriction for hazardous materials (dangerous goods).

It is worth noting that energy transportation using sea, land and air transport is not restricted to the energy products' haulage, but also to upstream and midstream equipment and any resources required throughout their supply chain.

Maritime Transportation

The maritime industry has a long and meaningful history to share with the energy industry. The first offshore platforms, designed a century ago, were converted tanker ships. Oil and gas storage units resemble ship tanks in terms of construction processes, materials used, architecture and structural engineering. Land pipeline systems resemble tanker ship piping systems and the notorious "spaghetti configuration," referring to the intricate multiple-miles-long and yet highly efficient pipeline system in both industries.

Sea transport encompasses at least 80% of global trade, one third of which pertains to petroleum products. There are about 100,000 ships in the world: 60,000 larger ocean-going vessels and 40,000 smaller coastal vessels. One third of these pertain to the carriage of petroleum products.

Crude oil cargoes typically navigate on larger crude oil carriers, whereas refined products navigate in smaller product tanker ships.

Liquefied natural gas carriers (LNG ships) and liquefied petroleum gas carriers (LPG ships) carry gaseous products that have been liquefied in order to reduce their risk of explosion. By reducing their temperature, pressure, and volume security is enhanced.

While the risk of warfare, sabotage, and terrorism exists, the key security risk pertaining to maritime transportation entails cargo theft and sea piracy, both motivated by financial incentives. Cargo theft takes place at port areas, i.e., load port, discharging port, and any transit areas.

Sea piracy occurs in most continents globally, with Southeast Asia and West Africa regions being the most piracy-prone due to the high volume of energy and other products moved.

Just imagine a mega tanker (Ultra Large Crude Carrier, ULCC), a ship whose length exceeds the height of the Empire State Building, carrying 550,000 metric tons of crude oil. Tankers are rather expensive assets, carrying a quite expensive cargo. Modern sea pirates actually target high-value assets and commodities in search of ransom. Oftentimes they hijack the vessel, kidnap crew members, and seize the ship for re-sale or conversion. (See Burns (2014) *Port Management and Operations*, for more details on maritime operations and security.)

Land Transportation

This segment entails energy products hauled via truck or rail transport. Due to volume restrictions, the cost per unit is higher in trucks and on rails when compared with sea transport.

Rail transportation has the flexibility of increasing the number of wagons, whereas they are restricted in terms of the route (i.e., there is a predetermined route on rail).

Trains can enhance both the levels of security and propulsion by adding more locomotive engines and distributed power units.

Truck transportation has the flexibility of unrestricted driving, yet faces limitations related to the cargo volume/weight.

Both rail and truck transportation segments entail shorter journeys and smaller cargo volumes compared with ocean-going ships, i.e., either continental or domestic, with the longest journeys conducted within Eurasia, i.e., Europe – Russia – Far East Asia.

Crude oil, gas and their products are carried in container tanks or bottles. Any impact on the vehicle, e.g. a car collision, increases the risk of leakage. Ignition, e.g., triggered by the use of a metallic object, may cause fire (especially for petroleum) or explosion (especially for gas).

Petroleum leakage can cause environmental damage.

Poor road infrastructure can amplify the risk of collision or ignition or leakage. Also, human error, fatigue, equipment breakdown, extreme weather, or careless drivers in vehicles near the energy vehicle may impose risks.

Pipeline Systems

Pipeline transportation entails the haulage of liquid, gaseous, or chemically stable products via a piping system. Products with chemical stability can be hauled via pipelines, i.e., materials that remain unaltered and are not prone to decomposition or chemical change, and neither internally nor externally influenced by air, light, extreme temperatures, pressure, etc.

The global pipeline system expands throughout 120 nations and 2,175,000 miles (3,500,000 km). Since the infrastructure is both costly and requires high tech, it is worth noting that 75% of the world's pipeline system is accumulated in three nations: 65% in the U.S., 8% in Russia and 3% in Canada (CIA, 2017).

The commodities hauled by pipelines include but are not limited to petrochemical and gaseous products as well as chemicals (such as propylene, ethane, ethylene, etc.), and are used by chemical plants and manufacturing companies to make computers, furniture, telephone devices, food containers, prescription drugs, spectacles, bike tires, life vests, fertilizers, etc.).

Pipeline Classification

Pipelines are classified in terms of the commodities they carry. Within the liquid petroleum and gas pipeline network there are crude oil lines, refined product lines, natural gas lines, highly volatile liquids (HVL) lines, and carbon dioxide lines (CO_2).

There are two broad categories that distinguish between the carriage of a) crude oil and its refined products, and b) natural gas and its processed commodities. Namely:

1 **Liquid petroleum or oil pipelines or hazardous liquid pipelines** are involved in the transportation of crude oil and refined products (including jet fuel, diesel, and gasoline), as well as ethane, butane, propane, etc.
2 **Natural gas pipelines** pertain to the transportation of natural gas, which mainly consists of methane, and smaller ratios of ethane, propane, butane, pentane, and so on.

Pipelines are also subdivided according to their commercial or utility purpose, i.e.:

a **Production lines** are the first pipelines located close to the oil or gas well. Their task is to move these relatively unprocessed and unrefined energy segments from the production areas to the refineries and chemical plants.
b **Gathering lines** are very small pipelines, usually from two to eight inches in diameter, in the areas of the country where crude oil is found deep within the earth. The larger cross-country crude oil transmission pipelines or trunk lines bring crude oil from producing areas to refineries. There are approximately 72,000 miles of crude oil system lines (usually 8 to 24 inches in diameter) in the United States that connect regional markets.
c **Transmission lines** carry refined or processed products throughout the nation at high pressure: Gasoline, jet fuel, home heating oil, diesel fuel, but also natural gas and other commodities.

  These refined product pipelines vary in size from relatively small (8- to 12-inch diameter lines), to much larger ones that go up to 42 inches in diameter. These pipelines deliver oil and gas products to large fuel terminals with storage tanks that are then loaded into tanker trucks for the final few miles, i.e., the delivery to gas stations or distribution lines.
d **Distribution lines** offer the final delivery stage of transport to the final consumer, be it industrial, utility, or residential. These function at lower pressures for safety purposes, but also due to the relatively smaller volume of energy they move.

Pipeline Security, Safety and other Risks

When focusing on the causal factor of a potential risk, physical risks can be distinguished into intentional (security) and unintentional (safety) related:

a Security risks where intentional threat occurs, e.g., as an act of terrorism, vandalism, industrial competition, diplomatic or social hostility, warfare, and so on.
b Safety risks where unintentional damage or threat is likely. Weather or climate conditions, human error or negligence, corrosion, are among the most common risks (Burns, 2015).

Since the majority of pipeline systems are located underground there is low risk. Their 24/7 operating schedule makes them a highly reliable and efficient transport mode.

Pipelines move products from upstream to mainstream and downstream, i.e., liquid and gaseous products from the production sites to refineries and chemical plants for value-added processing, then to storage units, and finally products are distributed to the markets and retail outlets.

Pipelines entail a fairly safe, secure and reliable means of sustainable, just-in-time transportation for fuels. Since the majority of pipelines is subterranean, they provide optimum utilization of land, but, at the same time, their maintenance and repairs may be challenging. This is where technology comes in to rectify matters.

PIPELINE INTERVENTION GADGETS (PIGS) USED IN ALL OIL AND GAS PIPING SYSTEMS

Although pipelines are considered as one transportation segment in the oil and gas industry, in reality the industry is heavily reliant upon pipelines in every stage of its supply chain.

Pipeline Intervention Gadgets (PIGs), commonly known as "pipeline pigging systems," pertain to inspection and maintenance equipment. They are used in "pigging and blowing" operations for all pipeline systems, both transcontinental and shorter pipes, e.g., connecting refineries with port terminals and tanker ships, or refineries with rail and trucking transport systems.

The PIGs enter the pipeline from a launching station, conduct the inspection or unblocking operations without slowing or interrupting production flow, and exit at the receiving station.

Here is a list of the tasks conducted by these devices:

- i they identify locations where the flow is slowing;
- ii they rectify these transmittal disruptions by cleaning/blasting any fuel residues or debris found in the pipelines;
- iii they optimize performance without slowing down or disrupting production.

The use of PIGs commenced in the oil and gas industry initially in piping systems of large diameters. However, their high effectiveness and efficiency made them quite popular devices in piping systems of diverse industries, and in pipes of most diameters.

Case Study: U.S. Energy Infrastructure

America's energy infrastructure comprises an impressive network of 2.6 million miles of interstate and nationwide pipelines, over 640,000 miles of energy transmission cables, 414 natural gas storage centers, 330 seaports managing crude oil and products, and over 140,000 miles of railroad for non-renewable energy transportation. The U.S. pipeline system covers over 190,000 miles (CIA, 2017; Burns, 2014; Burns, 2016).

While the existing system is exemplary according to global standards, there is an annual reoccurring cost for maintenance, and also investment in state-of-the-art technologies for both infrastructure and superstructure. New technologies, emerging security threats, cyber security, and extreme weather conditions necessitate investment for the renewal of the existing infrastructure. In 2013 the U.S. evolved into a leading global oil and gas producer, exceeding Russia and Saudi Arabia. The nation's oil output attained a 24-year peak, growing over the consolidated global growth. Improving the U.S. infrastructure could create a projected $1.14 trillion in capital funds, enhancing jobs growth and energy independence up to 2025, generating 1.15 million jobs each year, among which 830,771 jobs are associated with pipeline investment (FERC OEIS, 2016).

***

In concluding, the previous sections of this chapter describe the processes, hazards, and the industry's efforts to monitor and control its supply chain and mitigate all inherent risks.

Planned maintenance systems (PMS) are required to mitigate the oil and gas risks, be it spillage, corrosion, or risks of fire and explosion, while safeguarding the overall structural integrity. Extreme temperatures or fluctuations may cause material contraction and expansion, hence attention is needed. The implementation of social responsibility strategies and due diligence protocols, with a robust site-specific training and familiarization system, will ensure that the company safeguards its employees, contractors, communities, and the ecosystem.

The next section is a case study on ExxonMobil; a world-renowned energy conglomerate that adheres to the highest industry standards. Its exemplary performance and achievements make this a most suitable case study for energy professionals and scholars wishing to learn more about strategic thinking and best practices.

Case Study: ExxonMobil

ExxonMobil, the world's greatest publicly-owned oil and gas conglomerate, employs innovative state-of-the-art technologies to fulfill the ever-growing global energy demand. This is a highly innovative energy corporation, descending from John D. Rockefeller's Standard Oil Company. During the last 125 years ExxonMobil has developed from a regional kerosene trader to the greatest publicly-traded energy conglomerate globally.

ExxonMobil maintains an industry-leading supply of assets and sources among the world's greatest incorporated refiners, traders of oil products, and chemical suppliers. The corporation's dedication to the highest moral principles, compliance and ethics is mirrored in its safety and environmental regulations and procedures (ExxonMobil, 2017a).

Historic Overview

For the past 135 years ExxonMobil has progressed from a U.S. kerosene marketer operating locally to the greatest publicly-traded fossil fuel and petrochemical enterprise globally.

At present the organization's activities spread throughout most global nations and are most widely known by their recognizable brand names: Exxon, Esso, and Mobil. The corporation develops energy products that lead modern transport, power networks that spread across nations, lubricate manufacturing, construction, and other industries, while supplying petrochemical foundations that generate thousands, if not millions, of consumer commodities.

1870 Rockefeller and his affiliates form the Standard Oil Company (Ohio), with shared establishments representing the most significant refining capacity globally, exceeding any other enterprise. Standard was selected as a brand name to indicate superior and consistent quality.

1882 Standard Oil provides the lubricants for Thomas Edison's primary central generating system. During the same year, the Standard Oil Trust is established and comprises of the Standard Oil Company of New Jersey (Jersey Standard) and the Standard Oil Company of New York (Socony).

1903 Jersey Standard fuel and Mobiloil (Vacuum) lubricants are used for the historic first flight of the Wright brothers, Wilbur and Orville, at Kitty Hawk, North Carolina.

1927 Mobiloil is used in the *Spirit of St. Louis* single-engine monoplane for the first non-stop solo flight from the U.S. to Europe. Its pilot, Charles Lindbergh, was a former U.S. Air Mail pilot, and this flight made him famous overnight.

1928 Mobiloil is used by Amelia Earhart (the first female pilot to cross the Atlantic Ocean), to protect *Friendship*, her historic Fokker tri-motor aircraft (NASA, 2017).

1938 The first industrial production of alkylate globally commences in Baytown, Texas. Alkylation enabled the manufacturing of isooctane (2,2,4-trimethylpentane, C_8H_{18}), a fuel additive used to increase the knock-resistance of the fuel and produce 100-octane aviation gasoline, but also used for cosmetic and personal care products.

1942 The four Jersey Standard petroleum scientists known as the "four horsemen" developed an ameliorated version of the Houdry technique for cat cracking, called Fluid Catalytic Cracking (FCC). The first global fluid catalytic cracker was launched in Louisiana Standard's Baton Rouge refinery, thus establishing the most significant petroleum conversion process. Fluid Catalytic Cracking is considered as the global petrochemical industry standard for developing gasoline, or "the most revolutionary chemical-engineering achievement of the last 50 years," as pronounced by *Fortune* magazine.

1958 Mobil Aviation fuel was used by Pan American Airways in their first trans-Atlantic flight from the U.S. to England.

1982 Exxon's Centennial celebration signifies a century since the foundation of the Standard Oil Trust (1882–1982). This was a productive century with huge leaps of growth from local kerosene refining and trade, to a conglomerate of global caliber, leading the way to oil and gas upstream, midstream, and downstream, and petrochemicals production.

1999 Exxon and Mobil conclude a US$73.7 billion merger agreement and thus become Exxon Mobil Corporation, one of the most influential and pioneering energy conglomerates globally.

2002 ExxonMobil and other stakeholders lead the Global Climate and Energy Project (GCEP) at Stanford University. This is a ground-breaking initiative aiming to develop innovative engineering to mitigate the growing energy demand with significantly reduced greenhouse gas (GHG) pollutants.

2007 A world Offshore Drilling Record is attained by Exxon Mobil Corporation's subsidiary, Exxon Neftegas Limited, by concluding the Z-11 well-drilling operations in Sakhalin island offshore (Russia). This was one of the world's largest offshore projects globally, with the lengthiest calculated draft extended-reach drilling (ERD) well on earth, i.e., a total calculated depth of 37,016 feet (11,282 meters), or exceeding seven miles.

2011 Exxon Mobil Corporation released news on the largest field discoveries in the Gulf of Mexico in the past decade. The three vast deep-water findings include two oil and one natural gas discoveries.

2016 ExxonMobil makes a widespread use of drones (unmanned aerial vehicles) not only to conduct inspections of offshore platforms, refineries, and pipelines, but also to carry out environmental impact assessments. In 2013 the company discovered that the use of aquatic sonar systems can draw whales and other sea mammals towards the site. Subsequently, ExxonMobil used these findings to reverse the process, i.e., by diverting their operations in fish-breeding or passage areas. This is a part of the corporation's strategy to minimize the environmental impact and to protect the aquatic life.

2017 Mr. Rex Tillerson, ExxonMobil's CEO since 2006 and a corporate veteran who joined the company in 1995, assumed office as the 69th U.S. Secretary of State. In the same year, he obtains the Dewhurst Award by the World Petroleum Council for demonstrating "excellence in the petroleum industry."

2017 Mr. Darren W. Woods is elected as Chairman and CEO of ExxonMobil. Mr. Woods, a respected energy veteran of ExxonMobil's executive wing, has already made his mark, with significant global deals, research and environmental initiatives, as well as drilling and refinery investments (ExxonMobil, 2017b).

ExxonMobil retains its title as a premiere "Dividend Aristocrat" in the S&P 500 index among corporations that have amplified their dividend expenditures for 25 successive years or over. The corporation's expansion strategy, known as "Growing the Gulf," encompasses 11 major chemical, refining, liquefied natural gas, and lubricant initiatives in the U.S. Gulf (ExxonMobil, 2017c).

This historic timeline does not only describe the corporation's growth over a century, but also the significance of the energy industry in a nation's growth, security, and wellbeing.

Regulatory Compliance and Corporate Governance

ExxonMobil strives to work strategically in a mode aligned with social responsibility, environmental integrity, and economic prosperity for the regions in which they operate.

Concurrently, they concentrate on safeguarding the health, safety, and security of their global personnel but also their clients, associates, and the public. Every strategic decision, every single project from upstream to midstream and downstream, is accomplished in line with risk management and superior quality and performance (ExxonMobil, 2017d).

Safety in Operations: Control Systems

Management Control Systems (MCS) is an ISO standard and company process that collects and utilizes data to assess the performance of diverse organizational assets, processes, and resources including human, physical, economic and budgetary, but also macro-level data encompassing the entire corporation, its supply chains and global markets.

ExxonMobil's MCS establishes vital principles, guidelines, and techniques that drive their enterprise management and control systems. Their protocol evaluates and quantifies financial control risks, as well as techniques for reducing risks and eliminating vulnerabilities, controlling and monitoring conformity with standards, and communicating such results to the relevant management and operational entities within ExxonMobil (ExxonMobil, 2017d).

Operations Integrity Management System

ExxonMobil's Operations Integrity Management System (OIMS) pinpoints standard objectives for proactively dealing with safety, security, health, environmental, and social risks. The OIMS offers a methodical, structured, and disciplined method of analyzing improvement areas and monitoring dependability and liability throughout lines of authority, and the entire corporate pyramid, including management, operational facilities, and plans. The OIMS Framework is assessed and improved every five years while periodic updates and alterations also take place.

Cogeneration technological innovation encapsulates heat produced by electricity generation to be utilized in development, refining, and chemical development processes. Utilizing cogeneration contributes to reduced greenhouse gas emissions, because of its natural energy efficiency. The corporation's cogeneration plants per se, eliminate the release of roughly 6 million metric tons of greenhouse gases per annum.

The organization pursues around 5,500 megawatts of cogeneration capability in over 100 installations at more than 30 locations globally, and this amount equals the energy consumption of 2.5 million U.S. homes. In the last ten years, an additional 1,000 megawatts of cogeneration capacity and further investment prospects were offered (ExxonMobil, 2017e).

Exxon Mobil Corporation is dedicated to remaining the world's leading petroleum and petrochemical corporation. To sustain this prominent position, they constantly attain outstanding economic and project-related outcomes while concurrently implementing the highest standards in ethics, and social responsibilities (ExxonMobil, 2017f).

ExxonMobil in Space: Aviation and Space Exploration Fuel

Space launch – 1981. NASA used Mobil Jet™ Oil II lubrication for the space shuttle *Columbia* (Orbiter Vehicle Designation OV-102) for its very first mission STS-1 (Synchronous Transport Signal, level 1). The shuttle orbited the earth 36 times and, 55 hours later, safely returned to Edwards Air Force Base (ExxonMobil, 2017g).

Mobil Grease 28 is used on NASA space shuttles (ExxonMobil, 2017h).

In 2000 the company's products were used in NASA's Expedition 1 mission: Lubricants were used to increase safety in life-support backpacks, and the electric power generation units.

ExxonMobil in the Arctic

The company is a primary stakeholder in the Arctic Petroleum Exploration. Exxon Neftegas Limited (ENL) is ExxonMobil's subsidiary operating the Sakhalin-1 consortium involved in offshore exploration and production in Russia, in one of the largest oil and gas projects. This venture entails six among the world's ten largest drilling wells.

The Berkut offshore platform is a distinctive ice-resistant drilling platform, and is designed to operate in the harshest subarctic environments, with below zero temperatures, thick ice, and waves exceeding 60 feet (18 meters), with 6.5 feet (2 meters) ice field in the winter season. It is also the largest offshore platform in the world, weighing 200,000 tons.

The world's longest extended reach well was drilled at Chayvo. The Z-44 Chayvo well has exceeded 40,600 feet (12,376 meters), thus becoming the deepest well globally.

At the same time, the Odoptu and Arkutun-Dagi fields are also part of this ambitious project with vast reserves, whose duration is expected to go past the year 2050 (ExxonMobil, 2017i).

Upstream and Deepwater Exploration

ExxonMobil leads various upstream and deepwater operations such as oil and gas discovery, development, and production. They are among the leading leaseholders in the deepwater Gulf of Mexico with stakes in 339 deepwater blocks (ExxonMobil, 2017j).

Partnerships in Deepwater Technology

The corporation deploys its deepwater systems in Nigeria, in partnership with the Nigerian National Petroleum Corporation. The offshore Erha fields are found at the ocean depth of 4,000 feet (1,219.2 meters). For the past five years the company's annual production increases by about 3.1 billion oil-equivalent barrels (ExxonMobil, 2017k).

Other Innovative Initiatives

A contractual extension with Synthetic Genomics will provide additional time for a market-driven study in creating biofuels out of algae. This method requires that ExxonMobil and SGI researchers design new types of algae that may enhance output of this environmentally friendly energy segment (Woods, 2017).

With regards to industrial procedures, ExxonMobil have announced "cMIST," an innovative technology that generates dehydrated natural gas. The cMIST eliminates the exterior impact for this process by 70%. It furthermore eliminates energy use and air pollution levels.

In the domain of biofuels, ExxonMobil collaborates with the Renewable Energy Group to investigate the creation of biodiesel from farming waste materials.

Envisioning a Sustainable Energy Future

According to Mr. Rex Tillerson, the corporation's former CEO and current Secretary of State, the contemporary energy environment involves three key areas: 1) mitigating the ever-changing operational challenges; 2) the formation of joint ventures among domestic and global oil corporations; and 3) the significance of promising technological innovations (Tillerson, 2016).

The global power sector is estimated to reach an investment of $68 trillion in innovative energy technologies and systems from 2015 to 2040; among which, the allocation of approximately $25 trillion will be required for oil and gas investments. This high volume of investment capital necessitates close public/private sector synergies. Finally, the world still encounters the challenging task of boosting international energy production while at the same time minimizing pollution levels (Tillerson, 2016).

According to Mr. Darren Woods, the corporation's current CEO, mankind shares the common vision of a future of safe, secure, green, and rewarding society. A location where our offspring can grow in anticipation of living standards of good health, contentment, and growth. Cost-effective energy helps humankind meet these goals, while technological innovations materialize the aspirations of financial prosperity whilst minimizing pollutants.

The transformative impact of advanced technologies becomes reality in this corporation dedicated to research and technology. Mr. Darren W. Woods, ExxonMobil's CEO, deeply believes in this and the corporation's strategy strives to its materialization.

Partnerships

Partnerships in the oil and gas industry have historically been pivotal for the growth of the industry. Public/private partnerships, or private–private partnerships, and contracting arrangements ensure the smooth collaboration throughout the entire supply chain: From land leasing to construction, to research and development, to emergency response. It is hard to consider any function in this chapter that is not a by-product of partnerships.

The Organization of the Petroleum Exporting Countries (OPEC) was selected to represent the partnership section, as a long-standing organization that uses global trade as a vehicle to unite oil-producing and exporting nations, but also to define the oil price by controlling the supply–demand equilibria, which in turn are driven by political decisions to produce certain volumes of energy. The role of OPEC in the global energy sector can hardly be exaggerated, and is hereby presented as a Global Energy Partnership offering vital lessons to be learned.

Case Study: OPEC

The Organization of the Petroleum Exporting Countries is a long-established international governmental organization (IGO), established in 1960 and headquartered in Vienna, Austria. Its five founding members are: Iran, Iraq, Kuwait, Saudi Arabia, and Venezuela, which were later accompanied by nine others: Qatar, Indonesia, Libya, United Arab Emirates, Algeria, Nigeria, Ecuador, Angola, and Gabon. Its purpose is to bring together petroleum guidelines and procedures concerning its member states so that they can safeguard reasonable and firm rates for energy corporations; a commercially and financially viable and dependable supply of petroleum products to importing countries; and a equitable return on investment (ROI) to investors and traders. According to the U.S. Energy Information Administration, OPEC represents 42% of the international oil production and 73% of the global "proven" oil reserves, rendering OPEC a significant leverager of global oil prices (EIA, 2015).

The 1960s

OPEC's foundation in September 1960 took place at a period of changeover in the global financial and political scenery, with large scale decolonialization, i.e., the repatriation of indigenous lands and resources, and the declaration of independence of several new states, especially among developing countries. From 1948 through to the late 1960s, crude oil price levels fluctuated between $1.63 and $2.77, which in 2016 inflation-adjusted dollars ranged between $19.80 and $27.73.

At the time, 85% of the global oil reserves was ruled by the "Seven Sisters," i.e., petroleum corporations:

1. Anglo-Persian Oil Company (presently BP),
2. Gulf Oil,
3. Standard Oil of California (presently Chevron),
4. Texaco (subsequently merged with Chevron),
5. Royal Dutch Shell,
6. Standard Oil Company of New York (SOCONY) (stock trading as Mobil, and, presently merged, becoming ExxonMobil),
7. Standard Oil of New Jersey (ESSO/EXXON).

In 1969 OPEC created its joint vision by implementing a *Declaratory Statement of Petroleum Policy in Member Countries*, which underlined the fundamental irrevocable right of nations to exercise sovereign power over their natural resources.

The 1970s

The 1970s was a decade characterized by excessively high prices due to prolonged energy shortages, and long lines at gas stations, out of exaggerated fears of oil production shortages.

Severe financial repercussions were especially felt in the U.S., Canada, Japan, Europe, Australia, and New Zealand.

During the decade two very significant oil crises occurred:

1 In 1973 when the Arab–Israeli war occurred, also known as the "Yom Kippur War" in Arabic, or the "October War" in Hebrew. In 1967 Egypt closed the Suez Canal, and reopened it for navigation eight years later, i.e., in 1975.

   The tensions between the Arab World and Israel rendered the U.S. closer to a nuclear confrontation with the Soviet Union, thus creating a severe geopolitical and economic environment (DOS, 2016).

   In 1975 OPEC proposed a new era of cohesive international relations, to support global financial growth and political stability. In 1976 the OPEC Fund for International Development was founded, where aspiring monetary development programs were launched.

   At the same time, the Suez Canal's closure created serious supply chain disruptions from the oil-rich Middle Eastern countries to their global markets. To accommodate these disruptions, the maritime industry designed the first super tankers that navigated all around Africa, as opposed to many smaller tankers previously using the Suez Canal. These disruptions caused serious energy shortages and economic uncertainty around the world.

2 The 1979 Iranian Revolution brought on oil export disruptions in the Middle East and developed a new map of world coalitions. Despite Iran's financial prosperity, its leader Mohammad Reza Shah encountered strong Shiite opposition lead by Ayatollah Khomeini, who lived in exile. The nation was on the brink of a civil war, and by 1980 the oil production almost discontinued.

   Although the global oil production decreased by 4%, the market uncertainty led to panic and drove the prices far higher than validated by actual supply. This triggered the crude oil prices to double within a year (DOS, 2017).

The 1980s

Although 1980 was a year with record high oil prices ($37.42, or, at the 2015 inflation rate, $109.51), the decade was marked by rapidly declining rates until the 1986 market meltdown ($14.44 – 2015 inflation rate $31.72) due to a global oversupply.

The global economy and oil producers suffered a prolonged economic hardship for two decades, i.e., until 2002. Discussions among OPEC and non-OPEC countries were considerably productive, with mutual aspirations on market security, sustainable supply, and competitive prices. The global energy agenda was now enhanced with ecological concerns.

The 1990s

The markets recovered slowly but steadily during this decade, with a lot of political tensions. The world was transitioning to the post-Soviet era, characterized by increased localism, while satellite systems and other technological advancements greatly empowered globalization. Great advancements were made to promote exporter–importer synergies, while improving OPEC/non-OPEC relations.

The 2000s

Globalization gradually empowered several rapidly developing economies, such as China and India, who enjoyed two-digit GDP growth on almost every year in this decade.

The decade commenced with an impressive recovery, from $16.56 (inflation-adjusted rate $23.89) in 1999 to $27.39 (inflation-adjusted rate $38.29) in 2000, i.e., the price almost doubled within a year.

China's insatiable demand for oil and raw materials drove the market to a peak in early 2008, where the oil prices jumped to record levels, i.e., $91.48 (inflation-adjusted rate $102.00), until their collapse a few months later. The global economic meltdown inspired OPEC and Non-OPEC countries to mutually tackle the commercial challenges and seek common ground.

This era was characterized by speculative investments that gradually replaced the traditional financial houses. The market bubble triggered high-risk ventures to the detriment of energy security or financial security. One year later, i.e., by 2009, the market crashed at $53.48 in 2009 (inflation-adjusted rate $59.93) with rates losing 50% of the previous year's heights (EIA, 2017b).

The market was transformed with misinvestments and depleted cash flow leading companies to reorganization. Liquidation bankruptcies, and mergers and acquisitions grew to become the only options. Traditional banking systems underwent serious restructuring, with large-scale loans becoming the privilege of the few elites (Burns, 2016).

2010 to Date

Geopolitical turmoil in more than half of the oil-producing and oil-importing countries impacted negatively on the oil prices, and on global stability. Despite the steady rates from 2011 to early 2014, oversupply and speculative investments caused yet another energy crisis in 2015 and 2016, with the market seeking for a new balance from 2017 onwards.

As the market struggles to recover, energy professionals need to deal with a lot of market uncertainties, making it difficult to forecast the next market phase and the shift of power within regions, trade blocks, and energy segments.

Special Topic: Future Energy Reserves for the 22nd Century AD: A Case Study on Natural Bitumen and Extra-Heavy Oil

Previous chapters of this book discussed the "Peak Oil" theory and speculations that the oil and gas reserves may be depleted in less than a century. However, based on the new discoveries of bitumen and extra-heavy oil, as well as shale rock formations, the world can still rely on oil and gas beyond the 21st century.

Large deposits of bitumen and extremely heavy crude oil are available all around the globe, however the most substantial reserves are found in the American continent, specifically in Venezuela and Canada. With sizeable advancements currently in place, both of these locations generate a production of approximately 1.5 million barrels (Mb) per day.

Extra-heavy crudes and bitumen belong to the conventional energy classification, and thus play a significant role in supplying the global markets with sustainable energy. Their overall production reaches 4,700 billion barrels (Bb), among which approximately 1,000 billion barrels – the counterpart of the global oil reserves for conventional oil – are technologically, financially, and lawfully feasible to extract. Hence these are defined as proven recoverable reserves.

Both natural bitumen and extra-heavy oil have the footprints of conventional petroleum, as their chemical and textual composition resemble the residuum produced by light oil refinery distillation.

Bitumen and Extra-Heavy Oil

Both bitumen and extra-heavy oil have been produced and deteriorated by organic activity. The difference between bitumen and extra-heavy oil is not an issue of gravity or chemical structure, but of viscosity: Bitumen has increased viscosity compared with extra-heavy oil at normal reservoir environment. Although bitumen is essentially motionless at normal reservoir conditions, extra-heavy petroleum has a certain range of motion.

Heavy Crude Oil

Heavy crude oil is any liquid hydrocarbon with an API gravity below 22.3° and viscosity of at least 100 or 200 centipoise (cp). Its chemical composition is classified from oil that flows on its own, to bitumen, tar sands, and "ultra heavy oil" that is essentially integrated in sand. Due to its high viscosity, the oil cannot pass smoothly from wells to pipelines under typical reservoir environments.

Natural Bitumen (Tar Sands)

The supply chain of shale oil and gas has been duly covered in conjunction with the conventional crude oil and gas, due to the similarities in upstream, midstream, and downstream protocols. This section will briefly cover the bitumen and extra-heavy oil, another promising energy category that positively contributes to national energy independence and security. What makes this category of hydrocarbons so special is their vast volumes and immense output capabilities.

Bitumen is a dark-colored material used in the construction of highways, roofs, and other surfaces. It comprises of tar-like hydrocarbon blends frequently coupled with their non-metal residues that contain carbon, hydrogen, nitrogen, oxygen, calcium, iron, phosphorus, selenium, and sulfur. These weighty compounds become sediments within the well in the sands, once less heavy compounds are discharged right after deterioration.

Bitumen is naturally formed, as deposits of organic bitumen are found in seabed or riverbed environments where prehistoric microorganisms have corroded and have been exposed to pressure and high temperatures.

Found at depths below 75 meters, tar sands are harvested as a solid and not as a liquid. It is typically made up of substantially high particles of asphaltenes; the asphaltenes are chemically modified particles of organic chemical substances, i.e., bitumen with asphalt-like properties. These are found in crude oil together with resins, aromatic hydrocarbons, and saturated hydrocarbons. Their presence in petroleum can significantly impede the production procedure. Consequently, specific asphaltene components will mean the heavy oil has to go through an additional refining procedure known as de-asphalting. The chemical structure of asphaltenes may comprise of portions of sulfur, nitrogen, carbon, oxygen, hydrogen, as well as the heavy metals vanadium and nickel, which are broadly known as dissolvable. Regardless of the high cost and advanced technologies required, energy companies consider this commodity particularly appealing.

When future technologies and processes enhance the efficiency and easiness of upstream and downstream operations, cost-efficiency will also allow the large-scale exploration of these attractive commodities (Attanasi and Meyer, 2010).

Bitumen is typically used for industrial purposes due to its water-resistant and adhesive qualities, but it was furthermore utilized in the healthcare sector as medicine.

The substance is frequently used in road paving, whereby bitumen is known as asphalt, or a blend of bitumen, concrete and other aggregates. Bitumen may be reused, making it a favorable option for engineers as opposed to asphalt roads. Bitumen can be recycled on alternative highway construction projects. Due to its waterproofing qualities, it is additionally utilized to develop roof structures.

Nowadays, bitumen is produced out of crude oil. Its formation by means of distillation eliminates lighter crude oil elements, just like diesel and gasoline, while abandoning the "heavier" bitumen compounds. To create high quality of bitumen, several refining levels are necessary, and the main stages are as follows.

STAGE I: BITUMEN MINING

To generate bitumen, three principal techniques are employed, classified according to the depth of the reserves:

a **Surface mining**: This method is being used for bitumen deposits of low depth (surface) to average depth. It entails the bitumen extraction from a burrow (open pit) where the mineral deposit (the overburden) is removed. Surface mining may entail diverse mining processes, including: Mountaintop removal mining, open-pit mining, and strip mining. If sands are found at small-to-medium depths, the bitumen ore can be retrieved after the removal of a mountaintop, or an extended strip of soil and rock (overburden). Tractors, cranes with grabs, shovels, and other cargo-carrying equipment operate on site to haul the oil sands to a crushing plant. Some of this equipment can carry over 360 tons of blended bitumen and sand. The blend undergoes crushing, grinding or breaking of the ore, and, as the mixture becomes handy to transport, it is subsequently moved to an extraction plant.

b **In situ method**

The in situ recovery method is most suitable for great subterranean depths, e.g., in Alberta, Canada, where the vast majority of oil sands production is located. The two predominant commercial methods of in situ bitumen production are:

i **Cyclic Steam Stimulation (CSS)**, where elevated pressure and elevated temperature (350°C) vapor is inserted into a vertical wellbore in the well, which is cracked by the vapor pressure. As the heavy steam moistures and seeps via the oil sands, the bitumen dissolves and runs through the surface.

ii **Steam Assisted Gravity Drainage (SAGD)** is a technique applied for deeper resources. Over 80% of the reserves require this technique. Sands found at great depths beyond 225 feet (68.58 meters) have significantly less viscosity, hence may be developed by injecting vapor, chemicals, hot air, or less heavy hydrocarbons into the well to further increase viscosity and permit retrieval. Two horizontal well pairs (parallel upper and lower) are drilled, one at the top and one at the bottom of the bitumen reservoir.

At the field plant steam is produced and pumped in the reservoir via the upper injection well. Gradually a vapor chamber is created below ground inside the bitumen area and preserved by injecting more steam. Inside the vapor chamber the bitumen is warmed to ensure that it distinguishes from the adjoining sand and can be transferred. Through gravity, the flowing bitumen moves to the lower reservoir area and is pumped to the top via the producing well.

STAGE II: BITUMEN SEPARATION

Separation pertains to the actual parting of bitumen from sand, water, and other materials, by employing water-based gravitational separation, with the following purposes:

a to enhance the bitumen retrieval from the pulverized solid/liquids mixture of oil sands;
b to create clean bitumen products through froth treatment, i.e., the technique of removing the liquid and solid impurities from the bitumen;
c oil sands tailings with satisfactory settling and consolidation attributes are created by removing solid impurities, without destroying the discharge liquid chemical composition and energy quality of bitumen. This is done by:

   1 **Primary Separation Cell (PSC)** or Primary Separation Vessel (PSV): this is a fundamental retrieval process that includes a gravity segregation cell;
   2 **Main, or Secondary Separation Cell**, which is a supplementary restoration process in which bitumen is retrieved from the middlings flow, usually composed of numerous flotation compounds. The main Separation Cell generates three output flows:

      - **overflow**: a pure bitumen froth solution comprising approximately 60% of bitumen;
      - **middlings**: a high-quality slurry made up of primarily HO_2 and approximately 2–4% bitumen;
      - **underflow**: a rough tailings flow made up of a minimum of 50% solids and sediment bitumen (typically <1%).

STAGE III: BITUMEN UPGRADING

Upgrading is the ultimate phase in which bitumen is transformed into synthetic petroleum, diesel, or other products, ready for the refinery. This final stage might well be transported to an upgrading plant, or carried out on-site. Typically, the extracted bitumen compound is blended with a diluent such as natural gas condensate, or refined naptha, or synthetic crude oil.

References

Attanasi, E. D. and Meyer, R. F. (2010). "Natural bitumen and extra-heavy oil". In *2010 Survey of Energy Resources*, eds J. Trinnaman and A. Clarke. World Energy Council, pp. 123–150. Available at: energy.usgs.gov/portals/0/Rooms/economics/text/WEC10NBEHO.pdf, last accessed: June 3, 2017.

Burns, M. (2014). *Port Management and Operations*. Boca Raton: CRC Press.

Burns, M. (2015). *Logistics and Transportation Security. A Strategic, Tactical, and Operational Guide to Resilience*. Boca Raton: CRC Press.

Burns, M. (2016). "Global Projections for Shipping". NAMEPA – ExxonMobil Conference. Pathways to Sustainability Conference. February 5, ExxonMobil Campus, Texas, USA. Available at: www.namepa.net/past-events-1, last accessed: January 2, 2017.

CIA (2017). *CIA World Factbook*. Available at: www.cia.gov/library/publications/resources/the-world-factbook/geos/in.html, last accessed: June 18, 2017.

Department of Energy (DOE) (2017). *Propane: Liquefied Petroleum Gas (LPG)*. U.S. Department of Energy, Office of Energy Efficiency & Renewable Energy. Available at: www.fueleconomy.gov/feg/lpg.shtml, last accessed: February 12, 2017.

Department of State (DOS) (2016). *1969–1976. Arab-Israeli War*. Office of the Historian, Bureau of Public Affairs, U.S. Department of State. Available at: https://history.state.gov/milestones/1969-1976/arab-israeli-war-1973, last accessed: May 2, 2017.

Department of State (DOS) (2017). *Oil Embargo, 1973–1974*. Office of the Historian, Bureau of Public Affairs, U.S. Department of State. Available at: https://history.state.gov/milestones/1969-1976/oil-embargo, last accessed: May 2, 2017.

Department of Transportation (DOT) (2011). *The Fuel Refinery Process*. U.S. Department of Transportation, Pipeline & Hazardous Materials Safety Administration, Pipeline Safety Stakeholder Communications. Available at: https://primis.phmsa.dot.gov/comm/factsheets/fsrefinery.htm, last accessed: February 8, 2017.

Energy Information Administration (EIA) (2012a). *Crude Oils Have Different Quality Characteristics*. U.S. Energy Information Administration, U.S. Department of Energy. July 16. Available at: www.eia.gov/todayinenergy/detail.php?id=7110, last accessed: January 2, 2017.

Energy Information Administration (EIA) (2012b). *What are Natural Gas Liquids and How Are They Used?* U.S. Energy Information Administration, U.S. Department of Energy. Available at: www.eia.gov/todayinenergy/detail.php?id=5930, last accessed: June 3, 2017.

Energy Information Administration (EIA) (2013). *Coking is a Refinery Process that Produces 19% of Finished Petroleum Product Exports*. U.S. Energy Information Administration, U.S, Department of Energy. January 28. Available at: www.eia.gov/todayinenergy/detail.php?id=9731, last accessed: January 6, 2017.

Energy Information Administration (EIA) (2014). *Shale Oil and Shale Gas Resources are Globally Abundant*. U.S. Energy Information Administration, U.S. Department of Energy. January 2. Available at: www.eia.gov/todayinenergy/detail.php?id=14431, last accessed: June 12, 2017.

Energy Information Administration (EIA) (2015). *International Energy Statistics*. U.S. Energy Information Administration, U.S. Department of Energy. Available at: www.eia.gov/beta/international, last accessed: June 29, 2017.

Energy Information Administration (EIA) (2016a). *Energy Explained*. U.S. Energy Information Administration, U.S. Department of Energy. Available at: www.eia.gov/energyexplained/index.cfm?page, last accessed: June 3, 2017.

Energy Information Administration (EIA) (2016b). *Non Renewable Energy*. U.S. Energy Information Administration, U.S. Department of Energy. Available at: www.eia.gov/energyexplained/?page=nonrenewable_home, last accessed: June 1, 2017.

Energy Information Administration (EIA) (2016c). *Glossary. Residual Fuel Oil*. U.S. Energy Information Administration, U.S. Department of Energy. Available at: www.eia.gov/tools/glossary/index.php?id=Residual%20fuel%20oil, last accessed: January 6, 2017.

Energy Information Administration (EIA) (2016d). *Refining Crude Oil*. U.S. Energy Information Administration, U.S. Department of Energy. Available at: www.eia.gov/energyexplained/index.cfm/index.cfm?page=oil_refining, last access: January 6, 2017.

Energy Information Administration (EIA) (2016e). *Proven Petroleum Reserves 2016*. U.S. Energy Information Administration, U.S. Department of Energy. Available at: www.eia.gov/beta/international/data/browser/#/?pa=0000001001vg0000000000000000000000000000000g&c=410000000200006000000000000g00020000000000000001&tl_id=5-A&vs=INTL.5-2-AFRC-TBPD.A&cy=2015&vo=0&v=H&start=1980&end=2016, last accessed: June 18, 2017.

Energy Information Administration (EIA) (2017a). *Petroleum. Energy Explained.* U.S. Energy Information Administration, U.S. Department of Energy. Available at: www.eia.gov/energyexplained/index.cfm/data/index.cfm?page=oil_home, last accessed: June 3, 2017.

Energy Information Administration (EIA) (2017b). *Gasoline*. U.S. Energy Information Administration, U.S. Department of Energy. Available at: www.eia.gov/energyexplained/index.cfm?page=gasoline_home, last accessed: June 3, 2017.

Energy Information Administration (EIA) (2017c). *What Drives Crude Oil Prices: Supply OPEC.* U.S. Energy Information Administration, U.S. Department of Energy. Available at: www.eia.gov/finance/markets/crudeoil/supply-opec.php, last accessed: June 30, 2017.

Energy Information Administration (EIA) (2017d). *Petroleum Supply Monthly, February 2017, Preliminary Data for 2016.* U.S. Energy Information Administration, U.S. Department of Energy. Available at: www.eia.gov/outlooks/aeo/pdf/0383(2017).pdf, last accessed: June 3, 2017.

Environmental Protection Agency (EPA) (2000). *EPA Office of Compliance, Sector Notebook Project. Profile of the Oil and Gas Extraction Industry*, EPA/310-R-99-006, page 15. Available at: https://archive.epa.gov/sectors/web/pdf/oilgas.pdf, last accessed: June 3, 2017.

Environmental Protection Agency (EPA) (2015). *GHGRP 2015 Supplier of Natural Gas and NGL Supply Chain.* U.S. Environmental Protection Agency. Available at: www.epa.gov/sites/production/files/styles/large/public/2016-10/ghgrp2015suppliersgaswellchart.png, last accessed: June 3, 2017.

ExxonMobil (2017a). *About Us.* ExxonMobil. Available at: http://corporate.exxonmobil.com/en/company/about-us, last accessed: June 25, 2017.

ExxonMobil (2017b). *Management. Darren W. Woods. About Us.* ExxonMobil. Available at: http://corporate.exxonmobil.com/en/company/about-us/management/darren-w-woods, last accessed: June 25, 2017.

ExxonMobil (2017c). *Energy Policy. U.S. Energy Policy.* ExxonMobil. Available at: http://corporate.exxonmobil.com/en/current-issues/energy-policy/united-states-energy-policy/overview, last accessed: June 25, 2017.

ExxonMobil (2017d). *Safety and Health. Process and Management Systems.* ExxonMobil. Available at: http://corporate.exxonmobil.com/en/company/about-us/safety-and-health/process-and-management-systems?parentId=ee28cf94-6de3-4964-9ff3-e67a0b5074cb, last accessed: June 25, 2017.

ExxonMobil (2017e). *Technology. Energy Efficiency.* ExxonMobil. Available at: http://corporate.exxonmobil.com/en/technology/energy-efficiency/cogeneration/overview, last accessed: June 25, 2017.

ExxonMobil (2017f). *Guiding Principles.* ExxonMobil. Available at: http://corporate.exxonmobil.com/en/company/about-us/guiding-principles, last accessed: June 25, 2017.

ExxonMobil (2017g). *About Us. Innovation.* ExxonMobil. Available at: www.exxonmobil.com/en/aviation/about-us/innovation, last accessed: June 25, 2017.

ExxonMobil (2017h). *Worldwide Operations. US Refineries. Beaumont Refinery, Chemical, and Lube Plant.* ExxonMobil. Available at: http://corporate.exxonmobil.com/en/company/worldwide-operations/locations/united-states/us-refineries/beaumont, last accessed: June 25, 2017.

ExxonMobil (2017i). *Our Arctic Presence.* ExxonMobil. Available at: http://corporate.exxonmobil.com/en/current-issues/arctic/presence/our-arctic-presence, last accessed: June 25, 2017.

ExxonMobil (2017j). *Worldwide Operations. Upstream Operations, USA.* ExxonMobil. Available at: http://corporate.exxonmobil.com/en/company/worldwide-operations/locations/united-states/upstream-operations/overview, last accessed: June 25, 2017.

ExxonMobil (2017k). *Technology. Deepwater Drilling.* ExxonMobil. Available at: http://corporate.exxonmobil.com/en/technology/deepwater-drilling, last accessed: June 25, 2017.

Federal Energy Regulatory Commission (FERC, OEIS) (2016). *The U.S the Office of Energy Infrastructure Security (OEIS).* Federal Energy Regulatory Commission, Joseph H. McClelland, Last updated May 31. Available at: www.ferc.gov/about/offices/oeis.asp, last accessed: June 12, 2017.

Hirsch, R., Bezdek, R. and Wendling, R. (2016). *Peaking Of World Oil Production: Impacts, Mitigation, & Risk Management.* U.S. Department of Energy. Available at: www.netl.doe.gov/publications/others/pdf/Oil_Peaking_NETL.pdf, last accessed: January 2, 2017.

NASA (2017). *Earhart Crosses the Atlantic*. Available at: www.nasa.gov/multimedia/imagegallery/image_feature_1112.html, last accessed: June 25, 2017.

National Institutes of Health (NIH) (2017). *Propane*. National Institutes of Health, U.S. National Library of Medicine, National Center for Biotechnology Information. Open Chemistry Database, PubChem. Available at: https://pubchem.ncbi.nlm.nih.gov/compound/propane, last accessed: April 26, 2017.

National Oceanic and Atmospheric Administration (NOAA) (2017). *Kerosene, Chemical Datasheet*. National Oceanic and Atmospheric Administration, U.S. Department of Commerce. Available at: https://cameochemicals.noaa.gov/chemical/960, last accessed: May 6, 2017.

Occupational Safety and Health Administration (OSHA) (2017). *Petroleum Refining Process*, Section IV: Chapter 2. Occupational Safety and Health Administration, U.S. Department of Labor. Available at: www.osha.gov/dts/osta/otm/otm_iv/otm_iv_2.html, last accessed: February 8, 2017.

Railroad Commission of Texas (RRC.TX) (2017). *Barnett Shale Information*. Railroad Commission of Texas. Available at: www.rrc.state.tx.us/oil-gas/major-oil-gas-formations/barnett-shale-information, last accessed: May 6, 2017.

Tillerson, R. W. (2016). "The value of partnerships in delivering energy for the future". Rex W. Tillerson, Chairman and Chief Executive Officer, ExxonMobil. Abu Dhabi International Petroleum Exhibition and Conference, Nov. 7, Abu Dhabi, United Arab Emirates. Available at: http://corporate.exxonmobil.com/en/company/news-and-updates/speeches/the-value-of-partnerships-in-delivering-energy-for-the-future, last accessed: June 14, 2017.

Woods, D. W. (2017). *CERA Week Keynote Speech*. Darren W. Woods, Chairman and Chief Executive Officer, March 6, Houston, Texas. Available at: http://corporate.exxonmobil.com/en/company/news-and-updates/speeches/growing-the-gulf, last accessed: June 5, 2017.

5 Coal Energy Security (Fossil Fuels, Non-Renewable)

> Coal . . . We may well call it black diamonds.
>
> Every basket is power and civilization; for coal is a portable climate. . . .
>
> Watt and Stephenson whispered in the ear of mankind their secret, that a half-ounce of coal will draw two tons a mile, and coal carries coal, by rail and by boat, to make Canada as warm as Calcutta, and with its comforts bring its industrial power.
>
> Ralph Waldo Emerson (1803–1882), "The Conduct of Life" (1860), collected in *Emerson's Complete Works* (1892), Vol. 6, 86

There are four main forms of carbon, all of which are widely used in the energy industry. These include coal, graphite, graphine, and diamonds. This chapter will primary explore the use of coal, which is one of the oldest and most prevailing energy sources globally. In addition, this chapter will include concise references to the applications of graphite and graphine, two carbon-based materials that will play an important role in the future of energy.

Diamonds, on the other hand, are widely used in the geothermal energy industry, and their applications will be duly discussed in Chapter 8.

Furthermore, the conversion of graphite into diamond has applications in the nuclear energy segment, and will be duly discussed in Chapter 6 of this book.

Coal: An Overview

Coal is a combustible sedimentary mineral rich in carbon and hydrocarbons, and carbon represents more than half of its mass. It is considered as a non-renewable energy source since its formation requires millennia. It represents one of the most substantial and abundant power sources worldwide. It is also one of the most ancient energy sources, utilized at least since 4000 BC for cooking and home heating purposes. For the past two centuries coal has been used as a principal inorganic energy source necessary for a commercial procedure.

Globally, 6.9 billion tons of hard coal are generated, with the major coal-producing nations found in different continents; the top hard coal producers being China, the U.S., India, Australia, and Indonesia. The U.S. possesses over a quarter of the verified global coal reserves. While a large portion of the coal produced is consumed in its country of origin, aggregate coal exports have at least doubled since 2000 (WCA, 2017a).

Energy security, and the coal segment in particular, is a significant component of homeland security. Coal-rich nations possess a significant "energy buffer" to mitigate any energy disruptions, shortage or high oil price fluctuations. Furthermore, the buffer provides a time extension until alternative renewable energy technologies become increasingly sustainable.

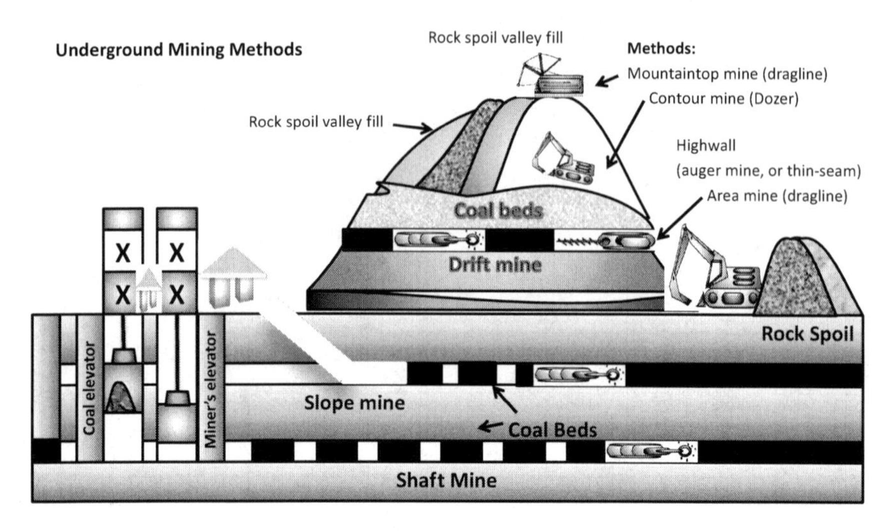

Figure 5.1 Coal mining methods: Underground and surface mining.

Source: the Author, adapted from: EPA, 2017. *Basic Information about Surface Coal Mining in Appalachia*, U.S. Environmental Protection Agency. Available at: www.epa.gov/sc-mining/basic-information-about-surface-coal-mining-appalachia(public domain)2670, last accessed: June 12, 2017.

The Stages of Coal Formation

Mineral coal is a fossil fuel, or organic rock, which is high in carbon content and typically tinted black, brown, or grey. Its formation takes hundreds of millions of years and follows these stages.

1 Generation of Peat or Turf

Coal derives from peat or turf, i.e., a heterogeneous mixture that is generated out of 30% or higher ratio of decomposed flora and fauna. Peat is the initial stage towards the formation of coal. It is created when this sedentarily gathered bulk has been depleted of oxygen and has been amassed in a wetland-saturated atmosphere. Decomposition is precipitated in higher temperatures, pressures, and when the peat is buried in great depths. This is a mass of moderately carbonized organic debris that will be transformed into coal within a geologic time period.

Coal formation can be autochthonous or allochthonous, and takes approximately 300 million years.

a **Autochthonous or indigenous organic matter** was formed as dense vegetation grew on earth and eventually died, and large tree trunks fell into wetlands, i.e., swamp waters and moors, thus creating an organic plant matter (phytoclasts).

b **Allochthonous organic matter** has been moved from its preliminary region of formation to a new location.

Chemical decomposition of the organic particles, i.e., the flora and fauna residues, was triggered by oxygen found in the water and the atmosphere. This is how peat was created.

2 Peat Pressure by Sedimentary Rocks

Geological alterations over the millennia buried the peat even deeper into the ground. Sedimentary rocks formed on top of the peat, pressed, dried out and buried the peat even deeper into the ground.

3 The Transformation of Peat into Coal

Peat, a saturated dense material, is the preliminary stage in the creation of coal. It will evolve into one of the above four grades of coal, i.e., lignite, bituminous, subbituminous or anthracite, depending on its preservation over the millennia (EIA, 2011). The element of time, temperature during burial and depth of burial will determine the peat's conversion into a particular grade.

Peat is transformed into lignite when preserved in high temperatures and great depths. The longer the time preserved, and the higher the temperatures, will transform lignite into superior classes of coal.

Types of Coal

The four main coal categories are lignite, subbituminous, bituminous, and anthracite. Just like any mineral, coal classification and chemical characteristics are determined by geomorphology, i.e., the geological structure, the depth where coal is found, the coal formation time and the overall configuration of physical formation and development of coal-rich landforms. Coal classifications are determined according to the chemical composition and percentage of carbon, hydrogen, and oxygen as well as lower content of sulfur, nitrogen, ash, and undesirable chemical compounds such as sodium and chlorine. Different regions may produce coal of different chemical composition. The four principal categories of coal are as follows:

1 **Lignite:** This is the most common type among global coal reserves and has the lowest quality among the coal classifications. Lignite, or brown coal, is a soft, dark-colored mineral. This coal category was stored at the lowest pressure and temperature so, for this reason, it contains the highest carbon content, i.e., about 60–70%, and the lowest energy composition among the other types, i.e., 25% to 35% (DOE, 2013; EIA, 2011).
2 **Subbituminous:** Subbituminous coal is also known as black lignite or black/brown coal. It is a plain-colored type of coal that contains a relatively low carbon content, i.e., about 35% to 45%. When consumed, it emits a relatively higher amount of energy, i.e., calorific heat, compared with lignite. One of its benefits is its abundance and relatively low production costs due to the relative ease of surface mining.
3 **Bituminous:** Also known as "soft coal." The highest energy content is present in this type of coal.
4 **Anthracite:** Anthracite represents the finest quality of coal, representing only 2% of the global production. It has the highest carbon and energy content, is the most dense and compact type of coal, and its superb quality mainly depends on high pressure and temperature during burial.

Table 5.1 demonstrates the chemical qualities and uses of all coal types.

Coal is also classified according to its volatility. As depicted in Table 5.2, coal possessing high heat content is less volatile compared with coal of low heat content.

Table 5.1 Coal types: Chemical qualities and uses.

Types	*Peat*	*Lignite*	*Subbituminous*	*Bituminous*	*Anthracite*
Other names	Turf	Brown coal	Black lignite, black or brown coal	Steam coal, rock coal, and steinkohle (German)	Hard coal
Quality	Lowest	Low	Intermediate	Higher	Highest
Classifications	Time preserved, and temperature will determine peat's transformation into a particular class of coal	A and B	A, B, and C	1 Per volatility: i A: high-volatile; ii B: medium-volatile; iii C: low-volatile 2 Per chemical composition: i Thermal (steam coal) ii Metallurgical (coking coal)	i Semi-anthracite ii Anthracite iii Meta-anthracite
Hydrocarbon creation	Premature gas		Oil & gas	Wet gas	Dry gas
Carbon content		25%–41%	42%–45%	45%–86%	87%–98%
Energy content (heat value) Mj/kg		14.7–19.2	19.3–24.4	24.4–33	33–38
Calorific value BTU/pound	5,000–6,100	6,200– 8,300	8,300–10,500	10,500–14,000	14,000–15,000
Moisture	>60%	35%– 75%	20%–34%	8%–19%	2%–7%
Volatile matter	>50%	50%	45%	15%–35%	2%–14%
Uses	Subject to transformation into a particular class of coal	Power generation	i Power generation ii Industrial applications iii Cement production	a Thermal (steam coal) uses: i Power generation ii Industrial applications iii Cement production b Metallurgical (coking coal) uses: steel & iron production	Industrial and domestic use

Source: the Author, based on data from: EIA, 2011. *Subbituminous and Bituminous Coal Dominate U.S. Coal Production*, August 16. U.S. Energy Information Administration. Available at: www.eia.gov/todayinenergy/detail.php?id=2670, last accessed: July 1, 2017.

Table 5.2 Coal types: Volatility and heat content.

Coal type	*Volatility %*	*Heat content kj/kg*
Anthracite	7–12	35,380
Non-baking coal	10–14	
Forge coal	14–19	
Fat coal	19–28	
Gas coal	28–35	34,960
Gas flame coal	35–40	33,910
Flame coal	40–45	32,870
Lignite	45–65	28

Sources: the Author, based on data from EIA, 2011. *Subbituminous and Bituminous Coal Dominate U.S. Coal Production*, August 16. U.S. Energy Information Administration. Available at: www.eia.gov/todayinenergy/detail.php?id=2670, last accessed: June 12, 2017.

Calorific value measured in kcal/kg is a particular property to measure the energy contained in coal or other forms of energy. It is verified by estimating the heat generated via combustion. The higher the age and grade of coal, the higher its calorific value and the composition of volatile compounds.

The substances that produce non-renewable fuels differ significantly depending on the depth of the soil that it covers underground. Because of these variants and the coal formation time, various kinds of coal were produced. Based on its chemical properties, each specific coal type and grade has different uses.

Uses of Coal

Coal is called the backbone of the electric system: At least 40% of global electricity is generated from coal. One third of global energy is generated by coal, thus making coal the second most popular energy source after petroleum. Despite its high demand, coal is an affordable and reliable energy form. It plays a significant role in the global energy production, 70% of the world's steel production, metallurgy, cement production, and much more (National Coal Council, 2017).

Coal production includes coking coal, lignite, anthracite, bituminous, and sub-bituminous coal. The uses of all these types of coal are diverse and entail a myriad of applications, as reflected in Figure 5.2. The global steel and non-recycled iron production rely on coking or metallurgical coal. Although coal is found in prolific quantities and qualitative variations, its transportation and storage are rather straightforward with limited operational, logistics, or supply-chain-related complexities.

According to the World Coal Association, the uses of coal entail the following:

- About 50% of global electricity is presently fueled by coal-fired power plants.
- 70% of steel produced today uses coal.
- 200 kg of coal is needed to produce 1 ton of cement.
- 300–400 kg of cement is needed to produce 1 cubic meter of concrete.
- 50% of the energy used to produce aluminum comes from coal. (WCA, 2017a; WCA, 2017b; WCA, 2017c).

Electric power generation is a significant source of coal utilization. Once coal is extracted, it is moved to an energy-producing station, compressed to a minuscule compound size, and subsequently heated. High temperature from the combusting coal is utilized to generate vapor,

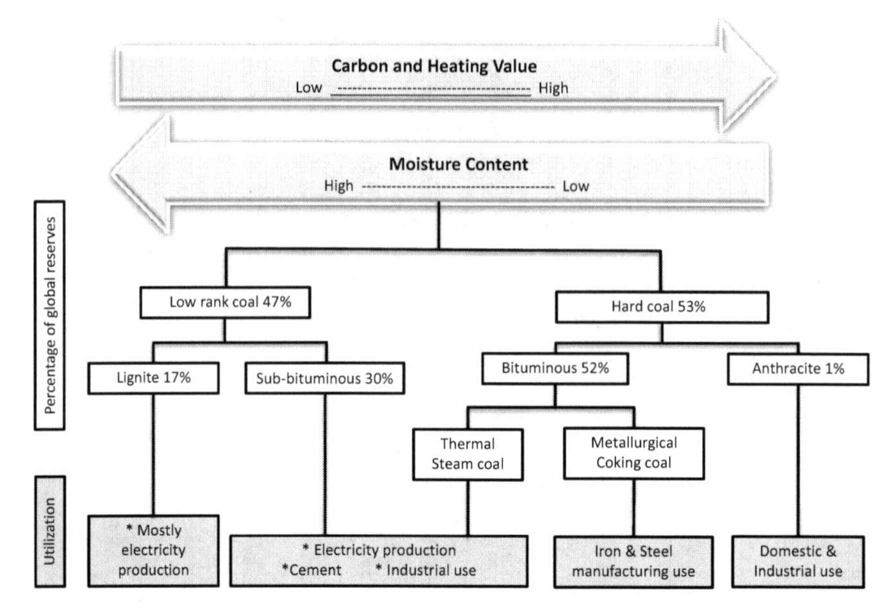

Figure 5.2 Uses and estimated percentage of the global coal reserves per coal type.
Source: World Coal Association, 2017a, 2017b.

which transforms a power generator to generate electric power. The majority of the electricity absorbed, i.e., 40% of global electricity production, is produced by burning up coal.

Coals of different chemical structure are employed as flammable fossil fuels for producing usable energy and generating steel across the globe. Coal is the fastest developing power source throughout the world (EIA, 2017a; IEA, 2017).

Coal products and byproducts are utilized by almost every industry, including other energy sectors.

As a first stage, coal is moved to power plants or refineries for pulverization. Its subsequent processing will be determined by the demand of specific products or byproducts:

- **Electricity**: It is moved to electric companies in order to be utilized by coal-powered electric power plants.
- **Heat** used for the paper and concrete industries.
- **Coal coke** (or coking coal).

Coal processing will generate the following coal products and byproducts:

1 **Thermal coal** is used for electricity generation;
2 **Coking coal, also called metallurgical coal** has multiple applications:

   a The construction industry, which encompasses land and sea structures and deep offshore structures, the use in coke ovens, etc.
   b Steel production: The global crude steel production exceeds 1.6 billion tons and coking coal is used to produce over 70% of steel (National Coal Council, 2017; WCA, 2017a; WCA, 2017b; WCA, 2017c). Hence, coal becomes a significant element of the industry.

   c Cement production.
   d Coal or BBQ bricks.

3 **Silicon metal** is utilized in the production of silanes and silicones:

   a Silanes are used to stabilize the composite substances, enhance the adhesion in polymers and among dissimilar materials, protect from corrosion;
   b Silanes and silicones are used in the solar power and photovoltaic industry;
   c Security threats: Due to its reactive nature, increasing safety measures are required to mitigate the threat of explosive silicon or silane products.

4 **Coal-powder** is used for chemicals, medicines, and cosmetics.

   a Kidney dialysis machines use activated carbon;
   b Dentistry (silanes and silicones dental porcelain for fillings, veneers, crowns and bridges);
   c Water filters and air purifiers use activated carbon.

5 **Activated carbon** is considered as a widely used adsorbent, as it will adsorb most gaseous products and other odors and pollutants (WCA, 2017e; 2017f). It is used in the food and beverage industry as a purifier.
6 **Carbon adsorbents** are used to eliminate perfluoroalkyl acids (PFOA) from potable reuse systems.
7 **Coal tar** is a by-product of bituminous coal distillation.

   a It is used in the medical industry. According to the U.S. Food and Drug Administration (FDA), it is classified as Category I, i.e., a safe and effective over the counter (OTC) medicine ingredient for the treatment of psoriasis, dandruff, and seborrhea (Cosmetic Ingredient Review (CIR) Expert Panel, 2008).
   b It is utilized in the food and cosmetics industry (personal care items). Coal tar dye is used for textiles.

History

The Significance of Coal Throughout Historical Eras

Since the Industrial Revolution and the use of coal, the world has associated national security with energy independence.

Coal was the prevailing energy mode during the 19th century, and greatly contributed to the technological evolution during critical eras in the history of humankind.

- **The Industrial Revolution** needed a robust, reliable, abundant energy source like coal in order to be materialized.
- During the **OPEC oil crisis** in 1974 and subsequent years, the significance of coal was re-established, as industries and nations required an alternative energy segment to ensure power sustainability.
- **Globalization** and several eras of technological and commercial growth used coal among other energy sources, both renewables and non-renewables.
- **21st century energy independence and energy security:** It is worth noting that

   a not all nations are rich in oil and gas reserves;
   b energy-importing nations need a leveraging point to withstand oil price fluctuations, and oil supply disruptions;
   c even nations that are rich in oil and gas need to plan ahead for a future time when their reserves will be depleted.

Based on the above assumptions, nations can use coal, among other energy types, to mitigate domestic energy demand, but also to reach their goals of energy independence and energy security (EIA, 2017a; EIA, 2017b).

4300 BC–800 BC: British flint axes entrenched in coal have been discovered, dating back to the Later Neolithic and Bronze Ages. This evidence suggests that the British extracted coal well prior to the Roman invasion.

3490 BC: Evidence of coal surface mining and domestic use of coal has been found in prehistoric China.

371–287 BC: The Greek scientist Theophrastus in his book "On Stones" (Lap. 16), referred to the geological formation of coal, its flammable nature, and recommended applications in metallurgy.

43 AD: After the Romans invaded England, they utilized British coal, mined in several British locations, in order to produce iron tools and armaments to protect the Roman Empire.

1300s: The Hopi Indians in America utilized coal to bake clay pottery and for heating and cooking purposes (EIA, 2017a).

1760–1840: The Industrial Revolution commenced in England, with several technological innovations. Large-scale coal mining was developed, with coal used in manufacturing, the military and defense, iron production, and transportation.

1769–1781: The Scottish inventor and engineer James Watt developed and patented the Watt steam engine, which created constant revolving motion. The engine could use coal, wood fuel or water to generate steam.

1800s: Coal became the prevailing energy source used to power locomotives, and subsequently steamships and homes. Coal was the primary energy source for street lighting.

1880s: Coal is no longer mined by hand. Coal-cutting machines are used for underground mining.

1882: Thomas Edison created the first coal-powered plan for electricity generation. New York City was the first city to use electricity for domestic lights globally.

1920s: Coal-handling technologies gradually replaced the use of animals in coal mines. Rail and cargo-handling equipment boosted productivity. The use of horses, mules, and dogs was gradually eradicated.

1930: Hard hats were first used by mine workers.

1950s: Coal became the prevailing energy source for locomotive and maritime transportation, manufacturing and domestic use.

1973–1974: A global energy crisis caused by the OPEC-driven petroleum embargo generated increased demand for coal (EIA, 2017a; EIA, 2017b; California Energy Commission, 2007).

Coal: Commercial and Economic Development

Several global regions contribute to the production of coal. As reflected in Figure 5.3, the annual production rates have been growing for the past decade or so. This growth can be attributed to green coal technologies, combined with the world's growing needs for reliable energy. Another interesting observation pertains to the non-OECD economies and the world's developing and less-developed economies using more coal compared with the developed economies.

Chapter 3 of this book duly explained a strong positive correlation between energy consumption, population growth, and economic growth, and Figure 5.3 verifies this correlation.

Coal mines are available in every continent, yet different grades and products (i.e., coking coal, lignite, anthracite, bituminous, and subbituminous coal) are found in different regions.

China is by far the major global producer of anthracite (88%) and bituminous coal (72%). Aruba leads the metallurgical coal production with 75% of the global ratio.Armenia leads on subbituminous coal with 52% of the global production, followed by Aruba with 30%. Finally, lignite is produced in China (31%), Germany (21%), the U.S. (8%), Russia (8%), Poland (7%), Australia (7%), Greece (6%), and other regions (EIA, 2000–2016).

The Asia Pacific region leads the global demand, with 73% of the world coal consumption according to Figure 5.4.

The top coal-consuming nations include China, the U.S., India, Germany, and Russia, whereas the world's top coal reserves are found in China, the U.S., India, Australia, Indonesia, and other nations, as shown in Figures 5.5. and 5.6. The same figures illustrate a strong positive correlation between China's annual coal consumption and the global coal reserves. China appears to lead by far both consumption and reserves' indices, thus surpassing the rest of the world.

China

With an annual coal consumption exceeding 3.2 tons per annum, China is a major coal producer, consumer, and importer. As of 2017, China produces most of its electric power from imported coal and is dependent on fossil fuel imports.

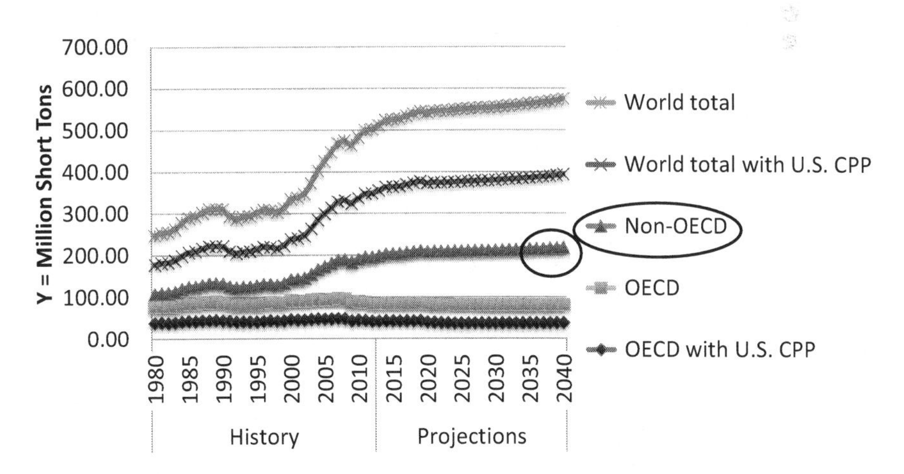

Figure 5.3 World coal production by region 2012–2040 (million short tons).

Source: World Coal Association, 2017a, 2017b.

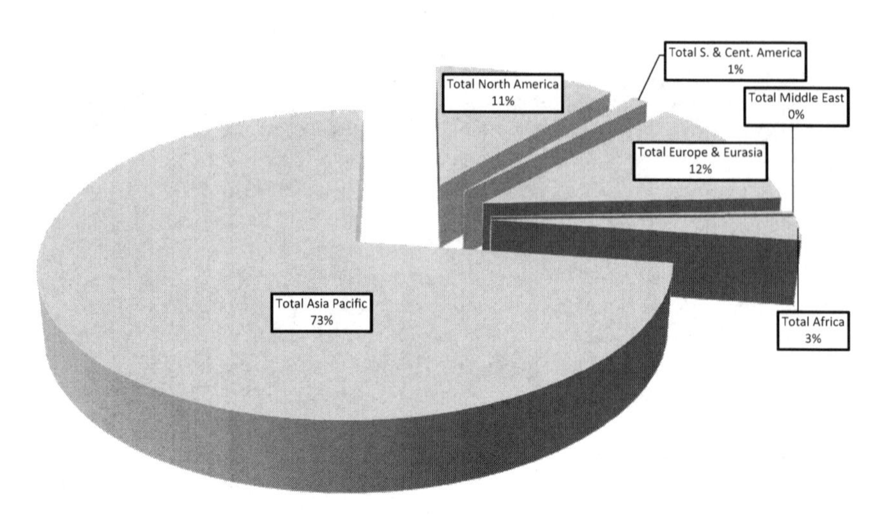

Figure 5.4 Coal consumption by region, 2015.

Source: the Author, based on data from EIA, 2016. *International Energy Outlook 2016*.

Release Date: May 11, Report Number: DOE/EIA-0484(2016). U.S. Energy Information Administration. Available at: www.eia.gov/outlooks/ieo/coal.php, last accessed: May 12, 2017.

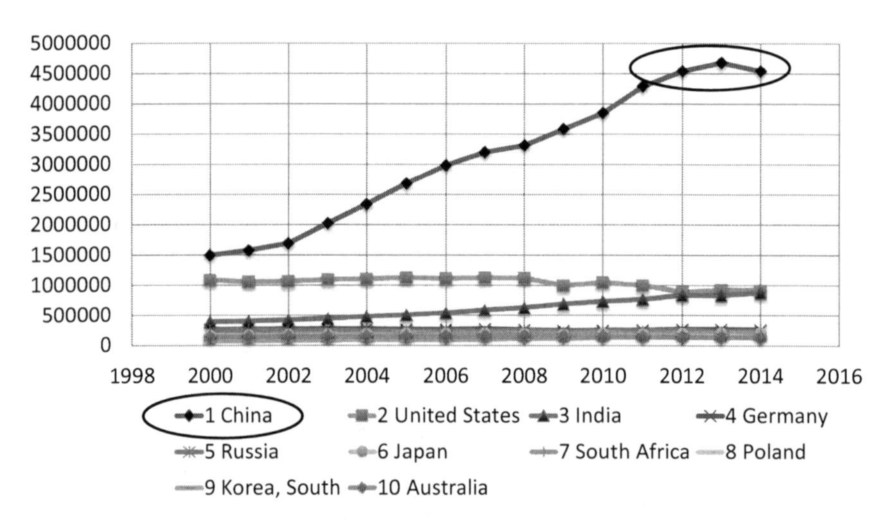

Figure 5.5 Consumption: Total primary coal (1,000 ST), 2000–2014.

Source: EIA, 2000–2016.

To meet the energy demand of China's increasing population and industrial growth, China has to develop a contingency plan for decades ahead, when its coal reserves will be depleted (World Energy Council, 2019).

The last five-year plan stipulates China's radical shift into investments for the development of renewable energy, in an effort to attain its energy independence and green energy goals.

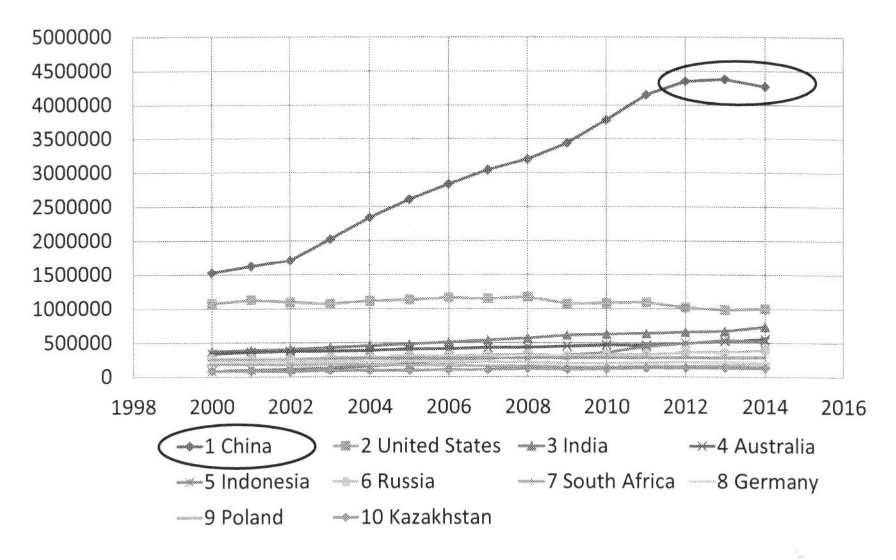

Figure 5.6 Reserves: Primary coal (1,000 ST), 2000–2014.

Source: EIA, 2000–2016.

India

India imports about 70% of its petroleum needs, mainly from the Middle East, Sudan, Nigeria, Syria, and the Caspian Sea. However, it is rich in coal reserves. Furthermore, it has the option to invest in green coal technology and other energy segments, and thus ensure energy independence. Otherwise, if no sufficient plan is in place, India's dependence will increase to 92%.

U.S.

In the U.S., coal production exceeds other non-renewable energy sources, due to the profusion of coal reserves, surpassing 25% of worldwide reserves. In fact, over 25 among the 50 States of America are rich in coal. Almost half of America's electricity production relies on coal (DOE FE, 2017a).

UK

The UK has a centuries-old tradition with the use of coal, and the Industrial Revolution was greatly enabled by coal. Britain's coalfields include Durham and Northumberland, North and South Wales, Yorkshire, Scotland, Lancashire, the East and West Midlands and Kent.

Case Study: South Africa

The history of coal is reflected in the history of South Africa, a pioneer in development of coal liquefaction from 1948 to 1994, and it has continued to lead the global industry ever since. The triggers of South Africa's innovations in coal liquefaction included the fact that the nation is depleted of oil and gas reserves, combined with the commercial and political isolation from its global partners due to the Apartheid system (meaning "racial segregation"). In 1948, in the aftermath of the Great Depression and World War II, the country was ruled by the Afrikaner National Party which introduced a political system based on racial discrimination, with benefits for the white minority, and diminished social and political freedom for the people of color or mixed descent, mainly of African and Asian origins. Under a climate of political isolation and energy dependency, indirect liquefaction was the nation's optimum solution to better utilize its coal mines.

The Supply Chain of Coal

When comparing the entire supply chain of coal, there are two particularities:

1 there is a great variety of transportation methods entailing the haulage of coal; and
2 the cost of long-distance transport is higher compared with the actual mining expenditures. For this reason, key players within the coal supply chain, especially upstream and midstream, are strategically located in premises close to mines.

Mining Management: The Supply Chain of Coal Mining

For the sake of consistency, we will classify the coal site management activities and all the supply chain procedures into upstream, midstream, and downstream:

1 Upstream: Exploration and Production (E&P)

a **Proactive planning – Stage 1**: Focus on commercial, financial, and feasibility issues:

- mine exploration – discovery and evaluation of coal reserves;
- feasibility studies;
- cost-benefit analysis;
- obtaining permits for mining from local/state, tribal or federal government: "Land Acquisition," "Mining Lease," "Mining Concession Agreement" (MCA), "Grant of Rights."

b **Proactive planning – Stage 2**: A risk management and regulatory compliance approach:

- risk management, resource management and contingency planning;
- regulatory compliance – social responsibilities; safety, security and environmental compliance.

c **Site preparedness**:

- land clearing, dewatering;
- preliminary terrain work;
- mine engineering and groundwork planning.

d **Coal mine exploration**:

- rock and soil drilling;
- use of bulldozers and other equipment for soil and cargo purposes;
- underground coal mining – roof bolting and collapsible drill steel enclosure;

- cargo handling procedures – loading and unloading;
- transportation from mine to site storage areas;
- soil leveling and site restoration.

COAL MINING METHODS

The coal mining process requires the utilization of advanced technologies and equipment to extract coal from seams (i.e., coal deposits or coal beds). Table 5.3 demonstrates the prevailing surface mining and underground mining methods, whereby the appropriate extraction method is selected based on the location and depth of coal mines.

Modern extraction technologies and working rates are much faster compared with the technologies and work processes of the past decades. There are three main coal mining methods: Underground mining, mountaintop removal, and surface mining.

a **Surface mining or strip mining** is the method selected when coal is found at a shallow ground level, i.e., less than 200 feet from the surface of the earth. Explosives may be used to remove the mountain tops. Large-sized equipment is used to eliminate the topsoil and overburden (i.e., rock layers) in order to uncover coal layers (Figure 5.7). In the U.S., approximately 66.6% of coal requires surface mining.

b **Open pit mining** is a variation of surface mining, where ore, waste, soil and trees and other unwanted materials are removed, thus generating a hole where the mining operations take place. This is depicted in Figure 5.8.

c **Mountaintop removal** is a variation of surface mining. Explosives are used to remove the soil and rocks off the mountaintops in order to uncover coal layers. When the coal reserves at the particular mine are depleted, the forestation procedure takes place: Topsoil is used to reconstruct the mountain soil and trees and plants are re-planted.

d **Underground mining or deep mining** is applied when coal is found at a deeper level, i.e., from 200 feet to several miles deep. The coal workers use elevators to move in different levels and carry the coal. Also, rail wagons, slopes, and advanced technologies are used in order to extract the coal and bring it to the surface (Figure 5.9).

Table 5.3 Coal mining methods.

Surface mining methods	*Underground mining methods*
Strip mining	Longwall mining
Removing overburden in rectangular strips	(half of the underground mining)
Contour mining	Continuous mining
Removing overburden around a mountainside or along a ridge	Use of a "board and pillar" system
Mountaintop removal	Room and pillar mining
Removing overburden through combining contour and area mining	
X	Blast mining or conventional mining
X	Short-wall mining
X	Retreat mining

Source: the Author, adapted from World Coal Institute, 2009. Available at: /web.archive.org/web/20090428202846/http://www.worldcoal.org/pages/content/index.asp?PageID=92, last accessed: July 21, 2017.

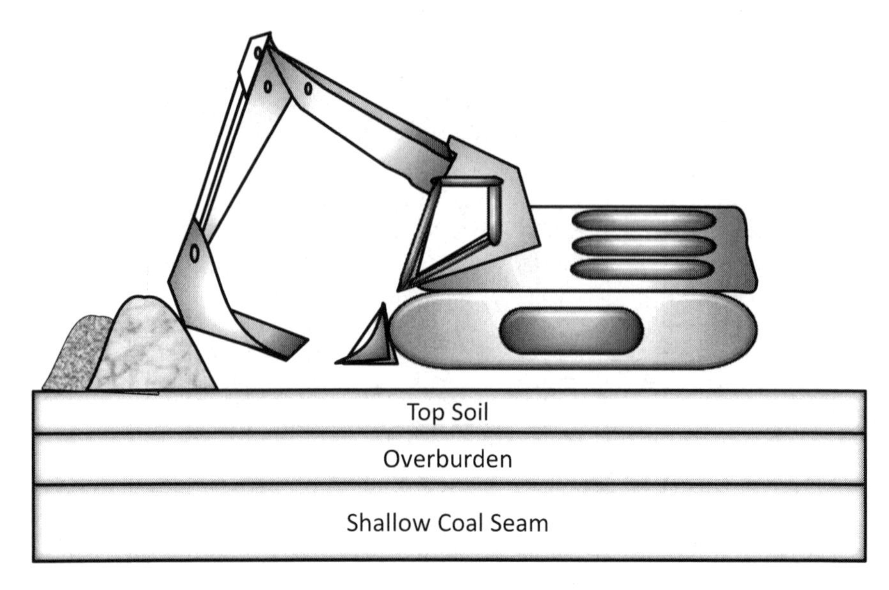

Figure 5.7 Surface coal mining: Removing topsoil and overburden.

Source: the Author, adapted from EIA, 2017d. *Coal Explained. Coal Mining and Transportation*. U.S. Energy Information Administration, last reviewed by EIA: September 7, 2016. Available at: www.eia.gov/energyexplained/index.cfm?page=coal_mining, last accessed: January 20, 2019.

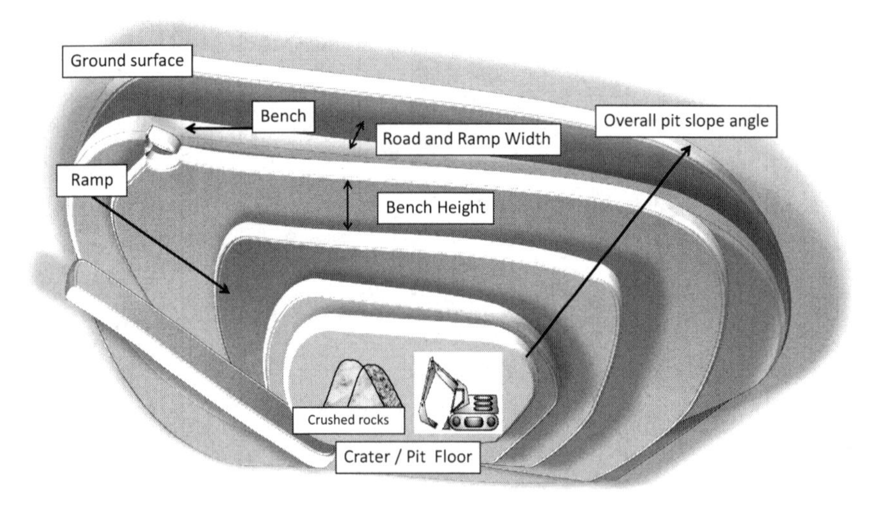

Figure 5.8 Open pit mining.

Source: the Author, adapted from EIA, 2017d. *Coal Explained. Coal Mining and Transportation*. U.S. Energy Information Administration, last reviewed by EIA: September 7, 2016. Available at: www.eia.gov/energyexplained/index.cfm?page=coal_mining, last accessed: January 20, 2019.

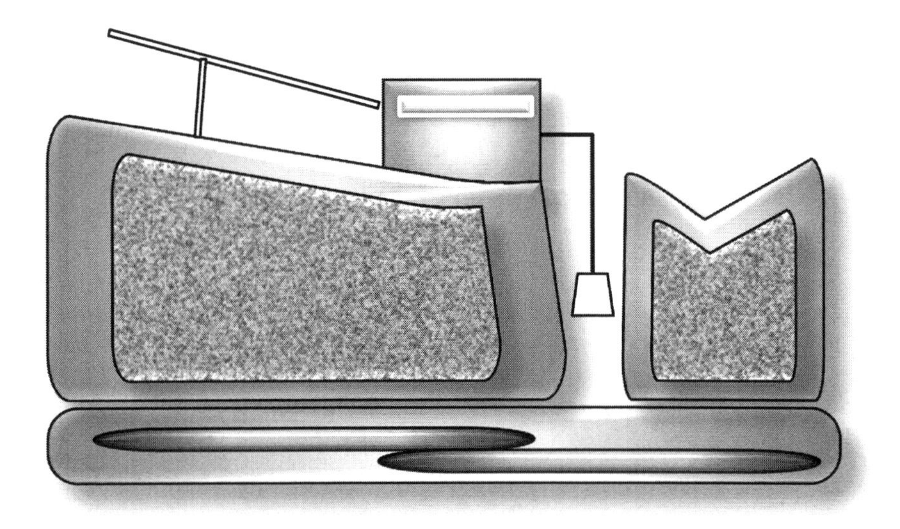

Figure 5.9 Coal underground mining.

Source: the Author, adapted from EIA, 2017d. *Coal Explained. Coal Mining and Transportation*. U.S. Energy Information Administration, last reviewed by EIA: September 7, 2016. Available at: www.eia.gov/energyexplained/index.cfm?page=coal_mining, last accessed: January 20, 2019.

COAL HANDLING AND PREPARATION PLANT (CHPP)

Once coal has been extracted via any of the above three methods, the coal is transported to a CHPP for further cleaning and processing of the coal, with the purpose to enhance the coal's heating value and quality. The CHPP plant is typically located in the vicinity of the mine, for two reasons:

a in order to eliminate the high transportation costs – the greater the distance, the higher the operational and transportation costs;
b in order to create a more reliable supply chain segment – the greater the distance, the higher the security, safety and commercial risks.

2 Midstream: Processing, Logistics and Transportation Networks

- Transporting coal to storage and power plants/processing facilities.
- Preliminary energy processing.
- Transporting energy products to distribution centers and global markets.

3 Downstream: Processed Energy Commodities Distributed to Global Markets

- Further energy processing: Meeting the market demand.
- The majority of coal-powered plants are situated in coal-producing regions, and this strategic decision impacts the trade routes, transportation and supply chain networks.
- Transportation of energy products to global distribution centers and markets.

Technologies, Operations, and Processes

This section covers the most prevailing coal-processing technologies. The three main coal conversion methods entail:

1 Pyrolysis (pulverization).
2 Liquefaction.
3 Gasification.

Furthermore, some of the modern Clean Tech methods that are producing increasingly cleaner energy-efficient products are also discussed.

It is worth noting that these methods are also applied in renewable energy, in particular for the conversion of biomass into biofuel (Chapter 7 of this book).

Pyrolysis and Pulverized Coal Systems

Pulverization or pyrolysis is defined as the generation of thermal energy through the burning of coal in a boiler. The main purpose of pulverization is to grind coal into a fine powder before combustion in the boiler of power plants, to make sure that all carbon particles are exposed to oxygen, and hence burn efficiently.

Modern industrial activities require high volumes of coal, and the sheer quantities deprive the oxygen that is available in the atmosphere, preventing it from penetrating and burning all the coal particles so that heat is discharged through combustion. When coal particles remain unburned they are released in the form of emissions, i.e., particulate matter, and pollute the atmosphere via excessive CO_2 and NOx values. The inability to fully utilize 100% of this energy portion is considered as energy inefficiency.

Effective pulverization offers environmental benefits by reducing harmful emissions released from unburned particles. At the same time, pulverization will increase energy efficiency by capturing more energy, and therefore by improving the combustion performance. Pulverized coal power plants surpass 95% of the international coal-fired capacity.

While good quantity of coal depends on the source and chemical composition of the raw material, pulverization aims to ensure optimum quality of coal, sustainable calorific value, low ash content, and benign chemical composition.

Types of Pulverized Plants

There are four main types of pulverized plants, their main difference being the combustion pressures and temperatures.

1 Subcritical Pulverized (SubCPC)

The majority of active PC energy plants belong to this category, i.e., they use boilers that generate steam of temperature 1,022°F (550°C), pressure around 22 MPa, and obtain power conversion efficiencies around 33% to 37% HHV.

2 Supercritical Pulverized (SCPC)

The inception of SCPC energy plants took place in the U.S. in the 1960s. These plants use boilers that generate steam of temperature 1,049°F (565°C), pressure around 24.3–24.7 MPa, and obtain power conversion efficiencies around 37% to 40% HHV.

These plants were popular in the 1960s and 1970s, however over the decades their popularity reduced due to certain structural and performance issues encountered with the boiler components.

3 Ultra-Supercritical Pulverized (USCPC)

4 Advanced Ultra-Supercritical (AUSCPC)

These are the most advanced and energy-efficient technologies where high temperature water is used to generate extremely high-pressure steam without being boiled. Increased unit efficiency is achieved, which means that less energy is used, thus eliminating environmental pollution, water use, solid waste, and operating costs (AEP, 2017).

Table 5.4 demonstrates the technological advancements among the types of coal pulverized plants. There is a strong positive correlation between the level of advancement (output) and the increase of operating efficiency, temperature, and pressure rates (input).

Pyrolysis, Biochar and Syngas

Pyrolysis is a highly valuable low-cost method of producing electricity from biomaterial. Pyrolysis has the capacity to significantly lower transportation expenses, provided that:

- **Biochar** is recycled and submerged in the ground; and
- The **syngas flow** is used as energy for these methods.

Biochar is a type of charcoal that increases soil fertility. It is carbon-rich solid, generated from biomass throughout the pyrolysis process. It is sustainable and secure. It has the ability to store high volumes of greenhouse gases in the soil for centuries, hence eliminates environmental pollution, e.g., of carbon dioxide, methane, nitrous oxide, and so on.

It is considered as a promising method of enhancing soil fertility and geological stability. For this reason it has potential applications in long-term sequestration because of its carbon-negative impact. By eliminating high amounts of carbon dioxide (CO_2) from the environment, long-term sequestration is enabled.

Coal Liquefaction

Coal liquefaction, or Coal to Liquid Fuels (CTL), Coal Conversion or Coal To X, is a method of transforming coal into petrochemicals and hydrocarbons. Coal conversion is based on two basic chemical processes, liquefaction and gasification, as described below. The main benefits of these methods include:

- alternative energy enhances a region's energy security and energy independence;
- ample global reserves of coal, i.e., its raw material;
- high demand in the production of transportation fuels;
- reliable, tested technologies, infrastructures, and superstructures available.

There are two methods of liquefaction.

1 Direct Coal Liquefaction (DCL)

The DCL method consists of two options: The hydrogenation (Bergius method) and carbonization methods. The hydrogenation or Bergius method is named after Friedrich Bergius who won the Nobel Prize in Chemistry in 1931 "for the invention and development of high pressure methods" (Nobel Prize, 2017). The process entails the generation of synthetic gas via the hydrogenation of volatile coal at high pressure and temperature.

Table 5.4 Types of coal pulverized plants (efficiency, temperature, and pressure rates shown in approximation).

	Subcritical (SubCPC)	*Supercritical (SCPC)*	*Ultra-Supercritical (USCPC)*	*Future: Advanced Ultra-Supercritical (AUSCPC)*
Operating efficiency	<37%	40%–42%	42%	50%
Operating (main stream) pressures	Below the critical point of water 170–220 Bar 22–22.5 MPa	Above the critical point of water 250 Bar 24.3–24.7 MPa	Above the critical point of water 300 Bar 27–30 MPa	Above the critical point of water 350 Bar
Operating temperatures	below the critical point of water	Above the critical point of water	Above the critical point of water	Above the critical point of water
Main steam temp	1,022°F, 540°C	1,049°F, 565°C	1,157°F, 625°C	1,292°F, 700°C
Reheat steam temp	Single	Single	Double	Double +

Source: the Author.

Figure 5.10 demonstrates how atomic hydrogen (H), a free radical consisting of one electron and one proton, is converted into molecular Hydrogen (H_2), comprising of two electrons and two protons.

The carbonization method, or pyrolysis or destructive distillation, entails the generation of carbon through thermochemical conversion, i.e., the breakdown of composite organic substances into carbon.

Figure 5.11 exhibits the conversion of carbon monoxide (CO) into carbon dioxide (CO_2) comprising of two electrons and two protons. The formation of covalent bonding, or molecular bonding (in contrast to ionic bonding), provides chemical stability deriving out of electron pairs, since atoms have the same electronegativity (same attractive and repulsive bonds).

2 Indirect Coal Liquefaction (ICL)

Coal gasification is accomplished with oxygen and steam to carbon monoxide and hydrogen (synthetic gas), in order to generate syngas (i.e., synthetic gas). Subsequently it is purified from acid gases, dust, and tar.

The Fischer-Tropsch method entails a series of chemical reactions that convert the carbon monoxide gases and hydrogen gases into kerosene, gasoline, and other liquid hydrocarbons.

Coal Gasification: IGCC, Pre-Combustion, Post-Combustion, and Oxy-Combustion Technologies

Integrated Gasification Combined Cycle (IGCC)

The Integrated Gasification Combined Cycle is a well-known process in which coal is transformed into a synthetic gas under pressure and temperature. The syngas is processed to remove impurities such as sulfur and particulates. The cleaned syngas is fired in a

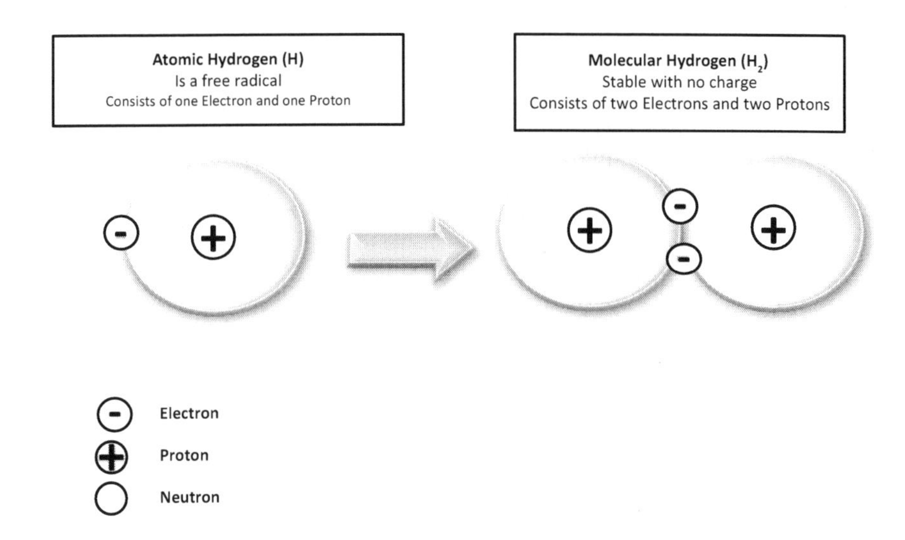

Figure 5.10 The conversion of atomic hydrogen (H) into molecular hydrogen (H_2).

Source: the Author.

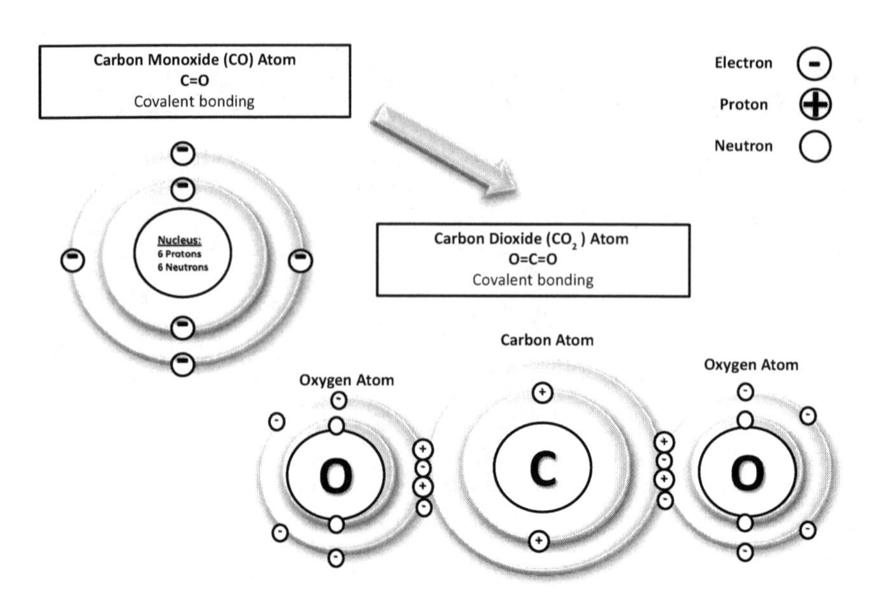

Figure 5.11 The conversion of carbon monoxide (CO) into carbon dioxide (CO_2).
Source: the Author.

combustion turbine that drives a generator to produce electricity. The hot exhaust from the turbine is passed through a heat recovery steam generator (HRSG) to produce steam used to drive a second turbine-generator set.

While resembling the Natural Gas Combined Cycle (NGCC) as this method utilizes gas and vapor turbines to produce electricity, syngas is used in this instance. Essentially coal is the typical component, however the gasification of any carbon-based feedstock is also possible, including the use of diverse components like biomass, petroleum coke, residues, municipal waste, and so on. This method is not only low-cost, but also enables the burning of coal along with several other compounds like biomass and municipal waste. This flexibility helps the industry meet with environmental regulations and eliminate greenhouse gases, and it can be used to experiment by combining the burning of coal and other fossil fuels with renewable energy and innovative chemicals (DOE, 2013; NETL, 2017; DOE FE, 2017a).

Gasification and pre-combustion, post-combustion, and oxy-combustion technologies are efficiently incorporated into an IGCC system to eliminate air emissions and lower the operational costs.

a **Pre-combustion capture (PCC)** pertains to the preliminary transformation of coal into a blend of hydrogen and carbon dioxide method, achieved via gasification or chemical reforming (DOE, 2013).

b **Post-combustion capture (PCC)** entails the segregation and storage of CO_2 and other feedstock from gaseous releases emitted from coal combustion in order to produce synthetic gas. Carbon dioxide generated from this process is segregated, hauled, and finally stored.

For instance, in gasification techniques, coal and other feedstock is fractionally oxidized in vapor and oxygen/air under extreme pressure and temperature in order to develop syngas (DOE FE, 2017b).

Oxy-Fuel Combustion

Coal and other fossil fuels are combusted in a compound of oxygen and recirculated flue gas and thus facilitate optimum cost-effective CO_2 encapsulation. Low quality coals typically contain higher humidity ratio, diminished energy, and higher pollutant releases. Green technologies eliminate humidity ratio before the burning process, and, as a consequence, eliminate pollutants by almost half.

Coal Gasification and the Production of Syngas

Synthetic fuel or syngas is defined as an energy segment generated by coal, natural gas or biomass feedstocks. It is utilized in the production of hydrogen and electric power. According to the U.S. Energy Information Administration, synthetic fuel is any fuel "produced from coal, natural gas or biomass feedstocks" generated via the chemical process of conversion. Synthetic fuels are generated by warming coal in large boilers, via chemical conversion into synthetic liquid or synthetic crude products and byproducts, often mixed with a combination of industrial and municipal waste (EIA, 2007; EIA, 2017a; EIA, 2017c).

Synthetic fuel generated from coal is a technique effectively implemented for the first time by World War II Germany. The outcome of WWII depended, to a great extent, on each nation's oil reserves, i.e., an aggregate sum of each nation's domestic production and imports from allied nations. In the beginning of WWII, the U.S. had available 1 billion oil barrels, Russia had 183 million, England 76 million, and Germany only 44 million barrels, most of which was imported. Germany therefore employed its leading scientists to experiment with synthetic fuel (Becker, 1981).

Synthetic fuel products and methods include:

1 **Benzol**: **A** byproduct of coking coal mixed with gasoline.
2 **Lignite coal**: The heating of soft brown coal, **which is** distilled into fuel oil and generates 10% gasoline and 90% lower-grade fuel.
3 **Coal, diesel oil and other fuels via the Fischer-Tropsch method**: Oil molecules are generated when coal is compacted into gas and blended with hydrogen.

A main benefit of synthetic fuel is the limited carbon footprint during the production or combustion process, as air emissions are diminished compared with the direct combustion of coal.

Clean Coal Technology (CCT)

Clean Coal Technology encompasses several technological advancements created to alleviate environmental damage caused by coal production or consumption, while also eliminating harmful emissions through the capturing of carbon. The CCTs are being designed and improved to alleviate coal pollution in response to green energy, greenhouse gases, and climate change goals, both at a global level and at a government level through the Environmental Protection Agency. Traditionally, CCT seeks to eradicate high concentrations of gaseous sulfur dioxide (SO_2), nitric oxides (NOx), mercury, particulate matters, and other air pollutants.

Clean Coal Technologies have greatly contributed to the reduction of hazardous emissions by 60% over the past 50 years. Hence, although coal production has increased two-fold, since then the emissions of CO_2, ozone, particulate matter, SO_2, NO_2, etc., have been greatly reduced.

The CCTs comprise of several methods discussed in this chapter pertaining to chemically cleansing mineral substances and harmful particles from fossil fuels; draining reduced coal grades in order to optimize their heating value, and consequently, the performance of their transformation into electrical power. Also, CCTs contribute to the processing of combustion exhaust gases, also known as flue gases, in order to eradicate contaminants to progressively more rigorous grades and at increased functionality, including seizing and storing systems that obtain the CO_2 from the combustion exhaust gases, and so on[1] (EIA, 2017a; EIA, 2017c; DOE, 2017b; DOE, 2017c; DOE FE, 2017c).

Bio-Energy with Carbon Capture and Storage (BECCS)

Bio-Energy with Carbon Capture and Storage is a promising greenhouse gas purging method that releases negative carbon dioxide. It is considered as the leading technology to attain the global goals of low carbon emissions. Its utilization is based on a) the emission of CO_2 by crops and timber, b) the process of carbon capture and storage through the burial of CO_2 in underground development, and c) the biomass application in energy plants and industries. It is used for both biomass energy, i.e., bioenergy, and geologic carbon and storage methods.

Carbon Capture, Utilization and Sequestration or Storage (CCUS)

Carbon Capture, Utilization and Sequestration or Storage is a cutting-edge technology of extracting, sequestration in porous rock formations, storage, and re-utilizing carbon dioxide. The U.S. Department of Energy has invested $6.5 billion in technological research and development (DOE, 2017a; DOE, 2017b).

Circulating Fluidized Bed (CFB)

These are technologies pertaining to an increasingly popular coal combustion process that aims to attain reduced air pollution. A CFB boiler is used to fluidize ground coal, typically blended with pounded limestone, through transferring upwards high-temperature burning oxygen by way of a supplier tray. The fluidized area over the distribution dish is the area where the combustion takes place. The coal's sulfur and limestone inter-react and eliminate sulfur oxide pollutants.

The process enhances the operational performance as it utilizes increased oxygen streams to circulate the bed substance with surrounding high-volume high-temperature cyclone dividers. The gas proceeding to the temperature exchanger is now filtered and purified, thus positively contributing to the feed layout, improving the connection among flue gas and sorbent. At the same time, it diminishes the pipeline corrosion of the heat exchanger, while ameliorating the combustion and overall performance of the SO_2 capture.

The Claus Process

The Claus Process eliminates air pollution by removing particulate matters (PM), mercury, sulfur dioxide (SO_2) and often carbon dioxide (CO_2).

The synthetic gas is subsequently processed through the water-gas switch solution to transform carbon monoxide (CO) into carbon dioxide (CO_2), and atomic hydrogen (H) into molecular hydrogen (H_2). Hence, the generated gaseous compound is now carbon dioxide (CO_2) and molecular hydrogen (H_2) respectively. From this mixture, the carbon dioxide is segregated, hauled, and eventually isolated, whereas the molecular hydrogen gas is combusted (DOE FE, 2017d).

Benefits, Risks, and Recommendations

The technological capabilities within the coal industry have greatly developed over the past decades or so, thus minimizing the environmental impacts while optimizing the energy-efficiency impacts. Coal is a cost-efficient fuel with high energy content compared with other non-renewable energy sources. This combination makes it an attractive option with significant advantages.

Eliminating Environmental Effects Through Technologies

Green Energy Technologies

Coal remains the "fuel of choice" for electricity generation, with innovative steam turbine technologies using less combusting coal to produce higher volumes of thermal energy. Innovative technologies have been established to monitor and control environmental emissions: Modern plants utilize innovative technologies and techniques to diminish about 99% of particulate matters, and 95% of sulfur oxide (SOx) and nitrogen oxide (NOx) emissions (EIA, 2017b; National Coal Council, 2017).

LAND MANAGEMENT

Modern coal management techniques minimize land and environmental impact, and utilize minimum land acreage. Land management eliminates unwanted effects such as: a) disturbance and change of geological strata in coal mine locations, b) delamination, i.e., geological foundering or soil instability caused by coal mining. Land management processes ensure that the vertical, horizontal, and lateral motion of the soil does not lead to leakage or contamination of the coal seams (deposits) that actually serve as aquifers which filter and store the water.

WATER MANAGEMENT

Proper water management/land management and water drainage operations are required to prevent acid rain and rainwater runoff occurring in coal mine locations.

OTHER FUELS PRESENT IN COAL MINING LOCATIONS

Coalbed methane (CBM), or coal seam gas (CSG), or coalbed gas, or traces of mercury, may be present in the vicinity of coal mining operations. In case it is present in subversive coal mines, the fuel must be monitored and controlled to ensure mine safety and avoid leakage, and subsequent explosion.

Eliminating Environmental Effects Through Mine Management

Geological and Pre-Production Planning

Mine subsidence is defined as the horizontal or vertical motion of the soil level due to limited pre-planning or restoration measures, causing the overburden to recede. The concavities and channels formed not only affect the geological formation with long-term effects, but may impact the safety of the mining operations. Geological and engineering studies are required to examine region-specific, project-specific, technology-specific subsistence effects. Based on the particular mining arrangement and blueprint, the soil subsidence will be duly monitored and controlled.

Water Management Contingency Planning

Water management contingency planning is needed for rainwater runoff separation. Clear water runoff must be filtered and segregated, and subsequently released in neighboring rivers or lakes. Water containing salt or soil deposits from mine operations should be recycled and filtered so that it may be reutilized through:

- soil leveling and inhibition processes; and
- coal operational facilities.

Dust Control

Dust control is a method of subduing and leveling dust generated during coal mine drilling, grinding operations, and due to weather, especially during heavy winds. It is achieved through:

- sprinkling water on land, conveyors, and stockpiles;
- investing in drilling technologies with dust collectors;
- dedicating a region in the periphery of the coal mine to serve as a mine management site.

Eliminating Noise and Visual Disturbance

To eliminate noise and visual disturbance during mining operations:

- tree planting in the region surrounding the coal mines;
- invest in exploration equipment with dust collection technologies;
- post-mining: Reforestation and optimum land uses (Matetic et al., 2017).

Mitigating the Environmental Impact

The global regulations striving for clean air and green energy initiatives have eliminated greenhouse gases and environmental pollution for many nations.

Coal pollution mitigation or clean coal technologies have been developed and funded as a result of environmental regulations seeking to protect the environment.

The Environmental Protection Agency utilized the Clean Air Act in order to impose a ceiling on aggregate sulfur dioxide (SO_2) emissions from energy plants, distributing commerce-friendly emission permits to diverse polluting entities.

Taxation and regulatory frameworks will promote the use of green technologies that ameliorate the coal-powered plants' efficiency. High investment requirements may impose delays or restrictions on their expansion (National Coal Council, 2017).

Moreover, there is ongoing research to further improve clean coal technologies, which will yield energy efficiency and reduce environmental impact.

Coal Transportation Methods

Coal is a relatively easy energy cargo to store and transfer, with very basic safety knowledge required. Overall, the longer the voyage, the more difficult it is for the shipper, carrier, and receiver to monitor and control cargo quality and quantity. Coal cargoes are prone to contamination when holds are unclean. Increased safety measures are required when coal is exposed to humid weather. High risk of fire is present when exposed to oxygen.

The most prevailing methods for coal transportation throughout its supply chain include:

1 **Belt conveyor systems, in particular three main types:**

   a **Underground mines**. This mining method requires higher safety measures due to the confined space of mining and transportation processes. There is oxygen depletion combined with flammable cargo that increases the risk of fire and human health issues. To mitigate the risk it is necessary to use non-flammable material for cargo handling. Another risk entails the time lost and diminished productivity when the belt conveyor needs to be maintained or repaired.

   b **Above-ground mines** typically require longer-distance belt systems. Two main threats involving this type of belt include: i) the threat of conveyor stretching (elongation), that may lead to belt breakdown or high maintenance costs. To mitigate this risk, robust materials should be used for these belts. ii) the threat of cargo contamination is another high-impact, high-likelihood risk. Sturdy belt materials should also be used to mitigate this risk in order to minimize cargo damage and pertinent delays.

2 **Slurry pipelines** are the ideal method to carry high volumes of coal (several million tons per annum) through great distances (several dozens of miles). These are haul compounds of water and pulverized coal, i.e., a fine powdered type of coal also called coal dust. A mixture of water and pulverized coal is used to eliminate cargo residues (settling). This is the most economic, reliable, and efficient method to carry coal through long distances.

   Risks include: a) cargo transport interruptions leading to cargo settlement and sealing the pipeline, b) prone to sabotage, c) prone to corrosion, d) prone to cargo theft, especially in remote areas, e) time and resources needed to install and maintain the pipelines.

3 **Haul trucks and trams** for concise parcels of coal cargoes.

4 **Trains**: Coal unit train systems may include 100 or more rail cars, carrying a 100 tons of coal per wagon.

5 **Vessels and barges:** Ideal for international voyages and long-distance voyages respectively. In both cases, risk is entailed in cargo handling.

Recommendations and Conclusions on the Risks of Coal

The global coal strategy focuses on enhancing energy output and eliminating the high investment and maintenance expenditure of CO_2 capture and storage, both in existing and newly developed coal production facilities. Modernization and green tech will enable coal to retain its position as a key global fuel that is established as an environmentally friendly fuel.

Coal is energy efficient; a reliable fuel in terms of calorific value. Abundant coal supports national energy security. New plants are of a better condition in terms of environmental and technological performance. The sustainable development and commercialization of Green Coal Technologies contributes to environmental protection, higher efficiency, lower cost, regional development, financial growth, and so on.

The global coal industry has implemented security measures to enhance energy efficiency and clean coal results, including:

- exercising due diligence since the pre-planning stage;
- pollution control protocols;
- monitoring and controlling environmental pollution impact including:
    - filtering emissions,
    - restoring mined lands,
    - social responsibility measures.

Case Study: World Coal Association

The World Coal Association (WCA) is the international network representing the coal industry that was established by significant global coal producers and stakeholders. The strategy of its members plays a critical part in defining the future of the global coal industry.

The WCA is the only global organization that represents the entire coal industry, and, as such, its goal is to showcase and obtain recognition for the worldwide significance of coal in accomplishing an energy-efficient low-carbon future.

The WCA leadership has a keen understanding of the regulatory, technological, and market developments over time, and utilizes this knowledge to impact policy-making at the utmost status for at least 30 years.

Membership is available to global coal stakeholders from the corporate and not-for-profit sectors, and member enterprises are represented at Chief Executive Officer or Chairman level (WCA, 2017a; WCA, 2017b; WCA, 2017c).

WCA and the Significance of the Coal Industry

Coal plays a crucial role in the development and construction of contemporary societies by enabling accessibility to baseload electrical power, and it is an essential foundation for several industries, and any construction, infrastructure, and superstructure raised globally. As an example, metallurgical coal is an important component of steel and a substantial part of the globe's cement is made out of coal, both essential components in advancing civilization. While this book encompasses the theme of energy security, it is worth verifying the significance of coal in global security, as structural integrity is another significant element of national and industrial security.

The incorporation of these sustainable growth requirements also need to be aligned with the goals of ecological mandates stipulated in the Paris Agreement. The WCA purports that the route to zero emissions from coal commences with the implementation of "high efficiency, low emission" coal technologies (HELE) and advances to "carbon capture use and storage" (CCUS). Breakthrough approaches to environmental and climate change issues will only occur by means of technological innovations and utilization of low emission engineering, hence the use of CCUS is of paramount significance.

The WCA designs and facilitates workshops that emphasize industry best practices, but also undertakes research concentrated on the environmental benefits and global community rewards of the broader implementation of HELE coal technologies. Furthermore, the WCA has set up a global Platform for Accelerating Coal Efficiency (PACE); PACE envisions for coal to be used in nations around the world, in an advanced technological environment where innovative, state-of-the-art techniques are integrated in order to ascertain the worldwide trajectory to zero emissions (WCA, 2017d; WCA, 2017e; WCA, 2017f).

WCA and the UN Global Compact

In May 2017 the WCA was officially endorsed as a signatory to the UN Global Compact (UNGC), which represents a global platform of leaders for the advance, application, and sharing of responsible corporate applications. Established in 2000, it is the most prominent corporate sustainability action globally, with over 9,000 companies and 3,000 non-business participants located in over 160 nations and in over 70 local networks.

The WCA endorses and substantiates the ten doctrines of the UN Global Compact pertaining to human rights, labor, environment, and anti-corruption. These are as follows:

Human Rights

Principle 1: Businesses should support and defer the safety of globally proclaimed human rights; and

Principle 2: Ensure that they are not involved with human rights abuses.

Labor

Principle 3: Companies should maintain the freedom of association and the actual acknowledgment of the right to collective bargaining;

Principle 4: The abolition of all kinds of involuntary and compulsory labor;

Principle 5: The elimination of child labor; and

Principle 6: The eradication of discrimination pertaining to employment and profession.

Environment

Principle 7: Companies should sustain a preventive approach to environmental issues;

Principle 8: Undertake initiatives to support greater environmental accountability; and

Principle 9: Boost the growth and dissemination of environmentally friendly technologies.

Anti-Corruption

> Principle 10: Companies should operate against corruption in all its kinds, including but not limited to bribery or blackmail.

The WCA is dedicated to backing up the UNGC in implementing and proceeding these principles and to applying them as a vital instrument of WCA's strategy and culture (WCA, 2017b).

The WCA has four core strategic goals that shape their mission.

1. **Influencing and engaging**: Reinforce their stature and leverage by attracting world leaders, strategic thinkers, influencers, and policy makers in coherent, evidence-based argument to position the coal industry as accountable and progressive.
2. **Powering economies**: Establish that coal plays an essential role in eliminating global energy poverty, enabling urbanization and transporting cost-efficient energy to sustain contemporary economies within developed and developing nations.
3. **Meeting environmental challenges**: Establish that world climate goals can only be attained with substantial global support for cleaner coal technologies.
4. **Building sustainable societies**: Determine that coal production and utilization plays a significant role to the progress of secure and sustainable global communities (WCA, 2017a; WCA, 2017b; WCA, 2017c; WCA, 2017d).

Partnerships: Carbon Sequestration and Leadership Forum (CSLF)

The Carbon Sequestration Leadership Forum (CSLF) is a government-level effort dedicated to the continuing development of ameliorated cost-effective systems for carbon capture and storage (CCS). These technologies entail the collection of carbon dioxide (CO_2) and other environmentally harmful gases from power stations and other establishments, and their subsequent storage, so that they are not released into the atmosphere (DOE FE, 2017a, 2017b; CSLF, 2017).

The CSLF serves as an advocate of these green technologies by stimulating global familiarization and understanding of the pertinent laws, policies, economics, and other governmental and administrative frameworks. The CSLF consists of 26 member governments (25 countries as well as the European Commission), which translates into 60% of the global population, i.e., 3.5 billion people (DOE FE, 2017c).

Its mission pertains to promoting the advancement and implementation of CCS technologies by way of collaborative initiatives that tackle key technical, financial, and environmental obstacles. Some of its major goals include:

- Pinpointing key challenges to attaining enhanced technological capacity.
- Determining potential aspects of research and multilateral partnerships on CCS technologies.
- Coordinating partnershipatory initiatives with all areas of the global research community, such as global enterprises, universities, government, and non-government organizations (CSLF, 2017).

In 2017 the CSLF selected and recognized two U.S. Department of Energy (DOE) innovative initiatives driven by the Office of Fossil Energy's National Energy Technology Laboratory (NETL). The chosen initiatives are the National Risk Assessment Partnership

(NRAP) and the Carbon Capture Stimulation Initiative (CCSI) together with its second phase, Carbon Capture Simulation for Industry Impact (CCSI2).

The NRAP is a multi-lab collaborative effort on creating a defensible science-based technique and model for risk assessment and measurement at carbon storage sites. Its findings determine decision-making and risk-management processes. Those endeavors led to the development of the worldwide use of the NRAP Toolset, a unique suite of computational methods to promote and precipitate the advancement, scale-up, and commercialization of carbon-capture technologies.

The CCSI entails an inter-governmental partnership between national laboratories, enterprise, and universities, involved in the use of the CCSI Toolset. The CCSI2 will apply the advanced simulation tools developed by CCSI to accelerate the commercialization of carbon capture technologies.

These distinguished DOE projects will comprise the list of CSLF's globally recognized and implemented projects. Most important, the best practices and lessons learned will drive future initiatives for knowledge sharing, and the acceleration of CCS technology implementation (DOE FE, 2017d).

Special Topics: Graphite and Graphene

Graphite and graphene are two other forms of carbon that relate to coal and are anticipated to play a key role in the future of the energy industry, both in renewable and non-renewable energy

Graphite

Graphite is a crystalline carbon (carbon allotrope) whose chemical composition is exclusively carbon (C). It is generated naturally in igneous and metamorphic processes of coal seams, when carbon is exposed to high temperatures and high pressures.

Graphitization is the practice of converting carbon into graphite, hence commercially producing graphite out of petroleum coke or simply coke. Coal distillation generates coke, among other derivatives.

The industrial production of graphite derives from the increasing high demand for graphite. Due to its very unique chemical composition, graphite has a multitude of uses: It is a good electricity conductor, can withstand high temperatures, is extremely soft and flexible, yet extremely durable. An effective lubricant, it is also utilized in the steel industry (graphite crucibles), molds in foundries and brake linings.

Graphite holds the future of energy-storing structures: Artificial diamond batteries that are generated from radioactive graphite absorb nuclear waste and convert it into small-scale electric currents that can last for thousands of years (University of Bristol, 2016).

Graphene

Graphene is a conductive carbon allotrope, i.e., created from carbon atoms. It is considered as a two-dimensional crystal because its single atom structure consists of two dimensions (width and length), yet is lacking height as the third dimension. The unique properties of 2D crystals extend graphene's applications in the energy industry in the form of nanomaterials, sensors, electronics (semiconductors, transistors, circuits), bio-devices, optics (solar cells, touch screens, flat light-emitting technologies like OLEDs (Organic Light Emitting Diodes) and LCDs (Liquid-Crystal Display)) and much more.

Conclusion

Coal will retain its leadership role among the world's top energy sources as long as government and industry segments invest in innovative technologies to ensure that coal turns into a considerably cleaner power source in the future markets. The private and public sectors need to re-establish coal as the green fuel of the future. They need to develop new environmentally friendly technologies to eliminate emissions and improve energy efficiency in order to disprove ongoing considerations regarding environmental pollution and emission of greenhouse gases that ingest and release thermal radiation.

In conclusion, the multiple usage and applications of coal products and by-products are not widely known to the public. Modern technologies eliminating the environmental and health impact are not widely known either. Finally, social and environmentalist groups favoring renewable energy fail to pinpoint that coal products are actually used in the alternative energy infrastructure and superstructure.

Note

1 Energy.gov, 2017, *Clean Coal Technology* (www.energy.gov.coal).

References

American Electric Power (AEP) (2017). *Pulverized Coal Technologies*. American Electric Power. Available at: www.aep.com/about/IssuesAndPositions/Generation/Technologies/PulverizedCoal.aspx, last accessed: March 25, 2017.

Becker, P. W. (1981). "The role of synthetic fuel in World War II Germany. Implications for today?" *Air University Review*, July–August 1981. US Air Force. Available at: www.airpower.maxwell.af.mil/airchronicles/aureview/1981/jul-aug/becker.htm, last accessed: March 25, 2017.

California Energy Commission (2007). *Energy Quest*. California Energy Commission, October. Available at: www.energyquest.ca.gov/time_machine/index.php, last accessed: March 25, 2017.

Carbon Sequestration and Leadership Forum (CSLF) (2017) *Overview*. Carbon Sequestration and Leadership Forum. Available at: https://www.cslforum.org/cslf, last accessed: July 1, 2017.

Cosmetic Ingredient Review (CIR) Expert Panel (2008). "Final safety assessment of Coal Tar as used in cosmetics". *International Journal of Toxicology 27*(2): 1–24. Available at: www.ncbi.nlm.nih.gov/pubmed/18830861, last accessed: December 24, 2016.

Department of Energy (DOE) (2013). *Fossil Energy Study Guide: Coal*. U.S. Department of Energy. Available at: https://energy.gov/sites/prod/files/2013/04/f0/HS_Coal_Studyguide_draft1.pdf, last accessed: July 12, 2017.

Department of Energy (DOE) (2017a). *Pre-Combustion Carbon Capture Research*. U.S. Department of Energy. Available at: https://energy.gov/fe/science-innovation/carbon-capture-and-storage-research/carbon-capture-rd/pre-combustion-carbon, last accessed: July 5, 2017.

Department of Energy (DOE) (2017b). *Carbon Capture, Utilization & Storage*. U.S. Department of Energy, Office of Fossil Energy. Available at: www.energy.gov/carbon-capture-utilization-storage, last accessed: July 5, 2017.

Department of Energy (DOE) (2017c). *Coal. Clean Coal Technology*. U.S. Department of Energy. Available at: www.energy.gov.coal, last accessed: July 5, 2017.

Department of Energy: Fossil Energy (DOE FE) (2017a). *Coal. Clean Coal Technology*. Office of Fossil Energy, U.S. Department of Energy. Available at: https://energy.gov/fe/science-innovation/clean-coal-research/hydrogen-coal, last accessed: July 5, 2017.

Department of Energy: Fossil Energy (DOE FE) (2017b). *DOE Projects Recognized by International Carbon Sequestration Group*. Office of Fossil Energy, U.S. Department of Energy, June 20. Available at: https://energy.gov/fe/articles/doe-projects-recognized-international-carbon-sequestration-group, last accessed: July 5, 2017.

Department of Energy: Fossil Energy (DOE FE) (2017c). *Coal: Our Most Abundant Fuel*, Fossil Energy Study Guides. Office of Fossil Energy, U.S. Department of Energy. Available at: www.fe.doe.gov/education/energylessons/coal/gen_coal.html, last accessed: July 5, 2017.

Department of Energy: Fossil Energy (DOE FE) (2017d). *Pre-Combustion Carbon Capture Research*. Office of Fossil Energy, U.S. Department of Energy. Available at: https://energy.gov/fe/science-innovation/carbon-capture-and-storage-research/carbon-capture-rd/pre-combustion-carbon, last accessed: June 6, 2017.

Energy Information Administration (EIA) (2007). "Synthetic fuels". *Annual Energy Outlook 2007*. U.S. Energy Information Administration. Available at: www.eia.gov/oiaf/archive/aeo07/pdf/0383(2007).pdf, last accessed: March 25, 2017.

Energy Information Administration (EIA) (2011). "Subbituminous and bituminous coal dominate U.S. coal production". *Today in Energy*. U.S. Energy Information Administration. August 16. Available at: www.eia.gov/todayinenergy/detail.php?id=2670, last accessed: May 25, 2017.

Energy Information Administration (EIA) (2000–2016). *International Data, 2000–2014*. U.S. Energy Information Administration. Available at: www.eia.gov/beta/international/data/browser/#/?pa=00000000000000000000000000000000000jg088000000000p&c=ruvvvvvfvtvnvv1urvvvvfvvvvvvfvvvou20evvvvvvvvvnvvuvo&ct=0&tl_id=1-A&vs=INTL.7-6-AFG-MST.A&vo=0&v=H&start=2000&end=2014, last accessed: July 1, 2017.

Energy Information Administration EIA (2016). *World Coal Production by Region, 2012–2040 (Million Short Tons)*. U.S. Energy Information Administration. Available at: www.eia.gov/outlooks/ieo/pdf/coal.pdf, last accessed: July 1, 2017.

Energy Information Administration (EIA) (2017a). "Coal". *Today In Energy*. U.S. Energy Information Administration. Available at: www.eia.gov/today in energy, last accessed: March 21, 2017.

Energy Information Administration (EIA) (2017b). *History of Energy, Energy Timelines*. U.S. Energy Information Administration. Available at: www.eia.gov/kids/energy.cfm?page=timelines, last accessed: July 1, 2017.

Energy Information Administration (EIA) (2017c). "World coal production by region, 2012–40 (million short tons)". *Coal Energy Outlook*. U.S. Energy Information Administration. Available at: www.eia.gov/outlooks/ieo/pdf/coal.pdf, last accessed: July 1, 2017.

Energy Information Administration (EIA) (2017d). *Coal Explained. Coal Mining and Transportation*. U.S. Energy Information Administration, last reviewed by EIA: September 7, 2016. Available at: www.eia.gov/energyexplained/index.cfm?page=coal_mining, last accessed: January 20, 2019.

Environmental Protection Agency (EPA) (2017). *Basic Information about Surface Coal Mining in Appalachia*. U.S. Environmental Protection Agency. Available at: www.epa.gov/sc-mining/basic-information-about-surface-coal-mining-appalachia (public domain)2670, last accessed: July 1, 2017.

International Energy Association (IEA) (2017). *The Use of Coal*. International Energy Association. Available at: www.iea.org/topics/coal, last accessed: May 9, 2017.

Matetic, R., Kovalchik P., Peterson S. and Alcorn L. (2017). *A Noise Control For A Roof Bolting Machine: Collapsible Drill Steel Enclosure*. National Institute for Occupational Safety and Health, Pittsburgh Research Laboratory, USA Centers for Disease Control and Prevention. Available at: www.cdc.gov/niosh/mining/userfiles/works/pdfs/ancfa.pdf, last accessed: January 1, 2017.

NASA (2017). *Coal Strip Mine in Germany*. NASA Earth Observatory. Available at: https://earthobservatory.nasa.gov/IOTD/view.php?id=90477, last accessed: July 1, 2017.

National Coal Council (2017). *The National Coal Council, A Federal Advisory Committee to the U.S. Secretary of Energy*. The National Coal Council. Available at: www.nationalcoalcouncil.org/Documents/Energy-Education/1-Coal-Past-Present-and-Future-Final.pdf, last accessed: May 25, 2017.

National Energy Technology Laboratory (NETL) (2017). *Coal Gasification. Energy Systems. Commercial Power Production Based On Gasification*. National Energy Technology Laboratory, USA. Available at: www.netl.doe.gov/research/coal/energy-systems/gasification/gasifipedia/igcc, last accessed: July 20, 2017.

Nobel Prize (2017). *The Nobel Prize in Chemistry 1931, Carl Bosch, Friedrich Bergius*. Available at: www.nobelprize.org/nobel_prizes/chemistry/laureates/1931/bergius-bio.html, last accessed: December 25, 2016.

University of Bristol (2016). "'Diamond-age' of power generation as nuclear batteries developed". Press release issued: November 25. Cabot Institute, University of Bristol, UK. Available at: www.bristol.ac.uk/news/2016/november/diamond-power.html, last accessed: May 25, 2017.

World Coal Association (WCA) (2017a). *Basic Coal Facts*. World Coal Association. Available at: www.worldcoal.org/basic-coal-facts, last accessed: July 20, 2017.

World Coal Association (WCA) (2017b). *What WCA Does*. World Coal Association. Available at: www.worldcoal.org/about-wca-0/what-wca-does, last accessed: July 20, 2017.

World Coal Association (WCA) (2017c). *About WCA*. World Coal Association. Available at: www.worldcoal.org/about-wca-0, last accessed: July 20, 2017.

World Coal Association (WCA) (2017d). *WCA and UN Global Compact*. World Coal Association. Available at: www.worldcoal.org/sustainable-societies/wca-and-un-global-compact, last accessed: July 21, 2017.

World Coal Association (WCA) (2017e). *How is Steel Produced?* World Coal Association. Available at: www.worldcoal.org/coal/uses-coal/how-steel-produced, last accessed: July 21, 2017.

World Coal Association (WCA) (2017f). *Uses of Coal*. World Coal Association. Available at: www.worldcoal.org/coal/uses-coal, last accessed: July 21, 2017.

World Coal Institute (2009). *The Coal Resource. A Comprehensive Overview of Coal*. World Coal Institute. Available at: www.worldcoal.org/file_validate.php?file=coal_resource_overview_of_coal_report%252803_06_2009%2529.pdf, last accessed: January 20, 2019.

World Energy Council (2019). *Coal in China*. World Energy Council. Available at: www.worldenergy.org/data/resources/country/china/coal, last accessed: January 20, 2019.

6 Nuclear Energy Security

(Non-Renewable) Uranium, Plutonium, Thorium

> If you really care about this environment that we live in—and I think the vast majority of the people in the country and the world do—then you need to be a supporter of nuclear energy, this amazingly clean, resilient, safe, reliable source of energy.
>
> U.S. Secretary of Energy Rick Perry advocates for Nuclear Power (NEI, 2018)

Nuclear Energy: An Overview

Not all non-renewable energy sources are fossil fuels. Nuclear energy or atomic energy is a non-renewable, yet inexhaustible, energy segment. Its unique features make it an indispensable ally in every nation's security strategy and energy mix. As scientific and technological breakthroughs enhance our understanding of this unique universal power, humankind will be able to better harness nuclear energy and reduce the impact of an accident. The great benefits of this fuel have been misunderstood (or misrepresented), hence the public opinion may not have a clear understanding of this truly remarkable energy source and its significance in a nation's energy portfolio and our daily lives (McGinnis, 2017).

Nuclear energy pertains to an atom's nucleus or core. Atoms are the small constituents within molecules that compose solid, liquid, and gaseous products. The nucleus is the core of an atom and consists of protons and neutrons that are encircled by electrons. Protons hold a positive electrical charge, neutrons have no charge, whereas electrons hold a negative charge (EIA, 2017a).

The nuclear industry consists of two different types of reactions, i.e., fission, the traditional atom-splitting technique, and fusion energy, which uses plasma to fuse atoms and thus produce energy. Fusion, the so-called energy of the future, promises inexhaustible power to be used on planet earth, and in space exploration and space colonization.

Nuclear Fission and Nuclear Fusion

Every atom comprises a nucleus, which in turn is made up of protons and neutrons. To generate nuclear energy a nuclear reaction is induced to change the number of protons and/or neutrons, and consequently change the nucleus of an atom. The energy launched by atomic nuclei reactions (also called radioactive processes) derives from nuclear fission or fusion.

- a In nuclear fission a reaction or a radioactive decay process is induced in which an atomic nucleus is split in two to produce smaller nuclei. This process generates energy.
- b In nuclear fusion a reaction is induced in which two atomic nuclei are united (fused) to produce a larger nucleus. Again, this process generates energy.

Nuclear Fission

Nuclear fission or splitting takes place in all power plants, most of which utilize uranium, plutonium, or thorium atoms to discharge radiation or thermal energy. As the atomic bonds, e.g., of uranium-235 or plutonium-239, are broken via nuclear fission, an enormous amount of energy is released, i.e., nuclear energy. The fission process divides the nucleus into two radioactive nuclei and thus releases neutrons. The energy that is produced thereby can generate electricity or nuclear weapons. Figure 6.1 demonstrates nuclear fission and how the process splits the uranium 235 atom to generate energy.

The nuclear fission method has been implemented for at least half a century, yet over the past few years new techniques and trends have increased performance and safety. The possibilities of this energy segment would greatly unfold should governments establish policies that attract and involve private funds (IEA, 2017). The nuclear energy segment has now set new goals, ranging from the expansion of the nuclear plants' commercial life to the building of both larger and smaller plants suitable enough to fit the grid of particular locations.

The element of security is pivotal for innovations of dual use. Nuclear technologies have been invented by the military for military use, yet their energy applications are immense. As scientific and technological innovations generate a wide array of nuclear energy uses, new laws and commercialization policies must be in place to prevent any illegitimate use of nuclear energy.

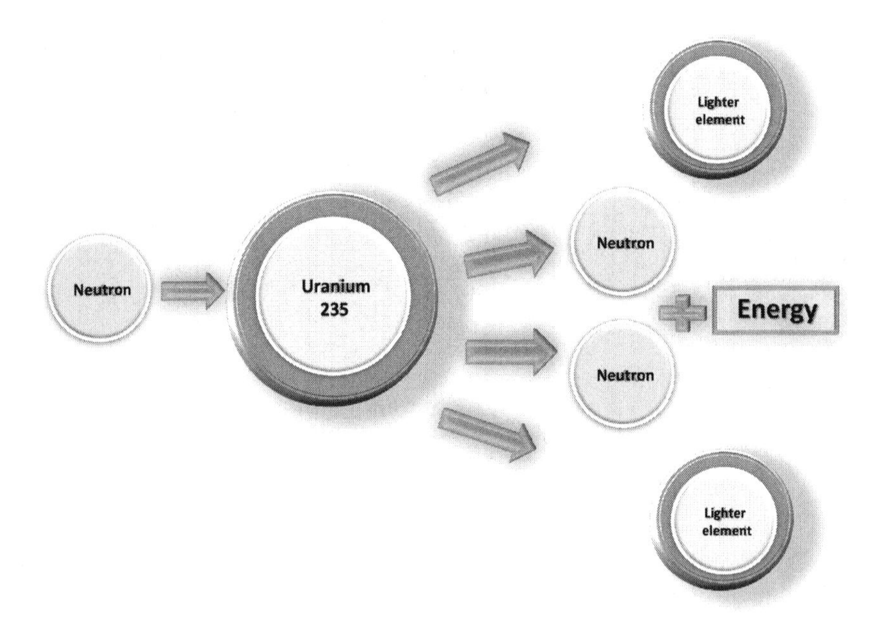

Figure 6.1 Nuclear fission. How fission splits the uranium atom.

Source: the Author, adapted from EIA, 2017. *How Fission Splits the Uranium Atom*. EIA and the National Energy Education Development Project (public domain). Available at: www.eia.gov/energyexplained/images/atom.jpg, last accessed: April 12, 2017.

FISSION AND NUCLEAR CHAIN REACTION

The nuclear chain reaction process pertains to the consecutive atomic fission. An extra fission process is generated in another nucleus, which in turn creates more neutrons (EIA, 2017b). Hence, atoms are divided in a repetitive consecutive manner, and are capable of producing more energy as an output compared with the input of energy used.

Controlled chain reactions or simply chain reactions are a vital part of the operations of a nuclear reactor. Typically such reactions take place with uranium-235 and other heavy isotopes, due to the sustainable discharge and absorption of neutrons. In the case of uranium-235 for example, a single atom enthralls a neutron and breeds two fission fragments (two atoms), thus producing three neutrons and vast energy.

Nuclear Fusion

Nuclear fusion is the converse progression of nuclear fission. The process entails the atomic fusion of lighter elements such as heavy hydrogens obtained from sea water. Once pressurized, it can release vast amounts of energy. This is a natural process observed in the universe and its planets to generate energy, e.g., when two hydrogen isotopes fuse to produce helium and release vast amounts of energy. In physics, fusion is defined as the process of fusing or uniting atoms to generate a weightier atom, such as heavy hydrogens, with a larger mass compared with the common hydrogens comprising of a single proton. For example: Deuterium is a heavy hydrogen comprising of one proton and one neutron, with atomic weight 2.014u and excess energy: 13135.720± 0.001 keV. Tritium is another radioactive hydrogen isotope that comprises of one proton and two neutrons, hence their atomic weight is 3.016u.

Figure 6.2 demonstrates nuclear fusion and shows how the process unites two isotopes of hydrogen (deuterium and tritium) to create helium and hence generate energy.

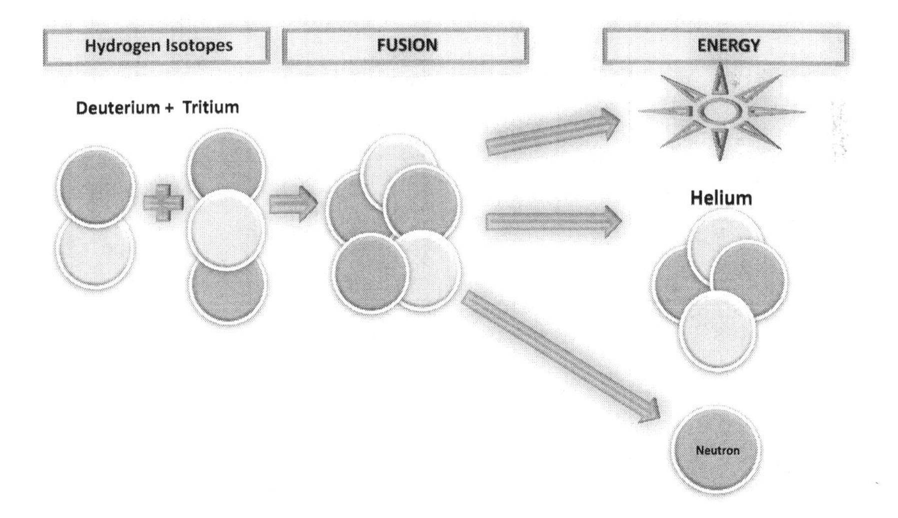

Figure 6.2 Nuclear fusion. How fusion unites two isotopes of hydrogen (deuterium and tritium) to generate helium.

Source: the Author.

NUCLEAR FUSION AND PLASMA: THE FOURTH STATE OF MATTER

There are four principal states of matter: Liquid, solid, gas, and plasma. While the first three states are well known and understood by the majority of the population, few people are familiar with the plasma state despite its abundant presence on earth. Plasma is an electrically conducting assembly of positive and negative particles, whose powers of attraction and repulsion will determine its outlet and behavior. Heavy ion fusion or ionization is the process of creating plasma by the electrical charging of gaseous products to the point where they generate positively charged ions and negatively charged electrons. Electric lightning is a natural demonstration of plasma energy, yet the man-made achievements involving plasma technologies are most impressive. Plasma technology or "engineered plasmas" have brought nuclear fusion to a new level.

Some of the most common uses of plasma technologies include plasma television screens, neon and fluorescent lights, etc. However, there are state-of-the-art plasma technologies that are not widely known though they hold tremendous potential, such as:

a thermal treatment and plasma blasting hazardous waste, i.e., dangerous goods' waste disposal;
b plasma gasification for the destruction of chemical weapons – this excludes elementary waste such as heavy metals or nuclear material like plutonium;
c medical uses: e.g., plasma scalpels and surgery lightsabers, used from cold plasma, are used for surgical skin closure to make surgeries more precise and less bloody.

FUSION TESTING WITH TEMPERATURES HOTTER THAN THE SUN

The benefits of harnessing the fusion process combine the infinite energy volume produced and the elimination of nuclear waste. While aspirational at present, synergistic research initiatives from several nations aim to simulate this natural process in a controlled scientific laboratory or nuclear plant setting as a fusion reaction is too powerful to be harnessed with the existing scientific and technological know-how.

Existing scientific knowledge suggests that the fusion process takes place at extremely high temperatures. As of 2018, several fusion labs globally (e.g., in the U.S., China, the UK, Europe, etc.) experiment with different temperatures at different trial times.

The Chinese experiment with 90 million degrees Fahrenheit for an 102-second duration blast. The British Tokamak Fusion Reactor experiments with 180 million degrees Fahrenheit (82 million degrees Celsius), whereas the International Thermonuclear Experimental Reactor (ITER) project (a global initiative consisting of 35 nations and a $20 billion funding pool) has pushed the bar up to 212 million degrees Fahrenheit (100 million degrees Celsius). In these high temperatures nuclei are divided from electrons and plasma is created that is capable of disseminating a considerable amount of energy.

History

> Nuclear energy is playing a vital role in the life of every man, woman, and child in the U.S. today. In the years ahead it will affect increasingly all the peoples of the earth. It is essential that all Americans gain an understanding of this vital force if they are to discharge thoughtfully their responsibilities as citizens and if they are to realize fully the myriad benefits that nuclear energy offers them.
>
> Edward J. Brunenkant, The United States Atomic Energy Commission

Nuclear energy is a remarkable energy segment that has drawn many enthusiastic scholars. Around 500 BC, Democritus of Abdera (the Greek philosopher and scientist) was the first scholar to name and define atoms as the smallest particles that are "undivided". Almost 2,500 years later, science would prove that atoms can in fact be divided and produce energy therefrom.

From 1789 to the early 20th century, several European and American pioneers worked tirelessly to better understand the atom and the manner in which electrons are positioned around the nucleus. These include, to name but a few, Martin Klaproth, Wilhelm Rontgen, Henri Beckquerel, Piere and Marie Curie, Samuel Prescott, Ernest Rutherford, and Frederick Soddy.

In 1802 the British chemist John Dalton (1766–1844) stated in his book *A New System of Chemical Philosophy* that elements are formed from certain combinations of atoms and all atoms of the same element (e.g., uranium atoms) can be identified because they have identical chemical composition. Over the next decades this theory was confirmed and extended as our understanding of the electrons, neutrons, and protons, mass number, etc., evolved. Namely, the atomic number (Z) of a neutral atom in each element equals the number of protons in the nucleus, and this helps us to recognize the different elements. The mass number of the atom (M) is the sum of protons and neutrons in the nucleus.

In 1905 Einstein developed the Special Theory of Relativity and the famous energy equation $E = m c^2$ where E is rest energy, m is rest mass (or matter) and c represents the speed of light. In his work Einstein purported that energy and matter are the same thing taking different forms: Energy can be transformed into mass (matter) and vice versa, and this conversion occurs instantly, at the speed of light. Einstein's Theory of Relativity was actually proven by the first nuclear bomb experiments, and later by the first nuclear energy plants: The transformation of mass can yield energy, with applications in the production of electricity or for military purposes.

In the 1930s, scientists like James Chadwick, Cockroft and Walton, Enrico Fermi, Otto Hahn, Leo Szilard, Fritz Strassmann, Lise Meitner, and Otto Frisch conducted experiments on diverse aspects of nuclear fission, from the discovery of neutrons to the use of accelerated protons to generate energy.

How the Manhattan Project Brought about the End of World War II

The Manhattan Project (1939–1946) was a top-secret scientific project during WWII, steered by selected political and scientific leaders in the U.S., U.K., and Canada, for the development of the world's first atomic bomb. The notable scientists Albert Einstein and Enrico Fermi had immigrated to the U.S. to get away from Europe and the Germano-Italian-led Axis powers, and they informed the U.S. President Roosevelt that scientific discoveries on nuclear energy could potentially be used by Nazi Germany. President Roosevelt, supported by the British Prime Minister Winston Churchill and Canada's Prime Minister William Lyon Mackenzie King, supported the development of the world's first atomic bomb.

Scientists from Columbia University, the University of Chicago, and the University of California at Berkeley participated in this highly confidential project. Uranium from Canada's Eldorado mine was supplied for all testing and the final development of the atomic bomb (NPS, 2017; DOE, 2017a).

- In **July 1945** the U.S. Army's Manhattan Engineer District (MED) conducted trials for the world's first atomic bomb detonation at Trinity Site, New Mexico, USA. The bomb was given the code name "Gadget," while the project had the code name "Manhattan Project."

- In **August 1945** the world's first atomic bomb, with the code name "Little Boy," dropped on Hiroshima, Japan, and the second bomb, with the code name "Fat Man," was dropped on Nagasaki, Japan. The Axis powers (Japan, Germany, Italy) were defeated, thus bringing an end to World War II.
- In **August 1946** the Atomic Energy Act (McMahon Act) determined that the U.S. would monitor, manage, and control the innovative nuclear framework they had created with England and Canada, its WWII allies, through the inception of the Atomic Energy Commission (AEC).
- In **1953** the U.S. President Eisenhower gave the "Atoms for Peace" speech, introducing nuclear energy at a time of global peace as an instrument that offered infinite electricity in energy-deprived regions. His speech inspired the inception of the International Atomic Energy Agency (IAEA).

In the same year (1953), the first commissioned nuclear-powered submarine U.S.S. *Nautilus* was constructed by General Dynamics Corporation. Figure 6.3 depicts its launching. During the 25 years of its glorious career, USS *Nautilus* broke all previous records on speed and distance, and successfully made the most perilous naval operations, from "Mission Sunshine" in Pearl Harbor, Hawaii, to the North Pole, under the Arctic polar ice pack (U.S. Navy NHH, 2017).

In the 1950s–1960s the U.S. operated the first commercial nuclear plants, employing both Fast Breeder Reactors (FBR) and Light Water Reactors (LWR), while new technologically improved versions such as the Boiling Water Reactor (BWR) and the Pressurized Water Reactor (PWR) were commercialized (Argonne National Laboratory, 2017).

Figure 6.3 Launching of USS *Nautilus* (SSN 571), January 21, 1954, built by General Dynamics Corp., USA.

Source: U.S. Navy NHH. 2017. *U.S.S. Nautilus*. Naval History and Heritage, U.S. Navy (public domain). Available at: www.history.navy.mil/browse-by-topic/ships/uss-nautilus.html, last accessed: April 12, 2017.

In 1984, coal and nuclear power became the two prevailing electricity sources in the U.S. (DOE, 2017b).

By 1991, 32 nations possessed nuclear power plant technology, which was either commercially operating or in the stage of construction (DOE, 2017b).

Commercial and Economic Development

As discussed in the history section, for the past 60 years the U.S. remains a pioneer in the development of nuclear energy and the design and configuration of nuclear plants. To date, nuclear fission has been utilized to produce thermal and electric energy. In fact, about 20% of U.S. electricity is made using nuclear energy (McGinnis, 2017).

Nuclear power plants are very robust structures that, by design and construction, are very difficult to penetrate. In the U.S. for example, there are 62 nuclear plant sites comprising of 100 reactors and safeguarded by about 9,000 well-trained, heavily armed security officers, equipped with the state-of-the-art detection and surveillance technologies (NEI, 2017a).

According to the World Nuclear Association, over 30 nations operate nuclear plants and another 20 nations are considering initiating nuclear power programs (WNA, 2017a). While in the past only developed countries would launch such ambitious projects, there are increasingly more developing and less developed countries, of diverse economic, demographical, and geographical location, that seek to develop their nuclear energy sector. Recent crises like the 2008 global economic meltdown and the 2015–2016 oil price collapse initiated a renewed interest in nuclear energy from over 50 nations worldwide. Consequently, many leading nuclear nations are establishing partnerships and "mentoring" agreements with nations that seek to invest in nuclear energy.

Establishing a nuclear energy plant requires years of planning and designing the national and regional infrastructure. Nations should be determined to finance such long-term endeavors and evaluate the quality, capacity, and the structural and technological integrity of their national grid.

Taking into consideration the aging grid in most countries, a reasonable appraisal should estimate the partial upgrading, if not total replacement, of the grid, bearing in mind that large-scale upgrades may be almost as costly as the development of a new grid.

Nuclear energy is a most sustainable and reliable energy resource, as nuclear plants have the capacity to function for at least 18 months, 24/7, with limited interruptions or need to replenish their sources. At a global level, the U.S. remains the pioneer for safe, secure, constructive use of atomic energy, with well-designed policies and a regulatory framework on health, safety, security, quality systems and the environment (McGinnis, 2017).

Global nuclear capacity grows with about 60 new reactors currently being built in 15 nations, including Asia and Russia. The U.S. invests in upgrading or retrofitting that prolongs the life span from 40 to 60 years (WNA, 2017a).

At the same time, other developed nations like France, China, Russia, South Korea, Canada, and others have also developed nuclear technologies, and are also part of global scientific initiatives that strive towards industrial growth and the next stage of nuclear plants.

Table 6.1 demonstrates the top global nuclear-producing nations' (2016) production in billion kWh (NEI, 2017a). It is worth noting that while Iran's statistics are not published, the nation is very active in the development of nuclear technologies both for energy production and for the development of nuclear weapons.

Table 6.1 Top global nuclear-producing nations (2016) billion kWh.

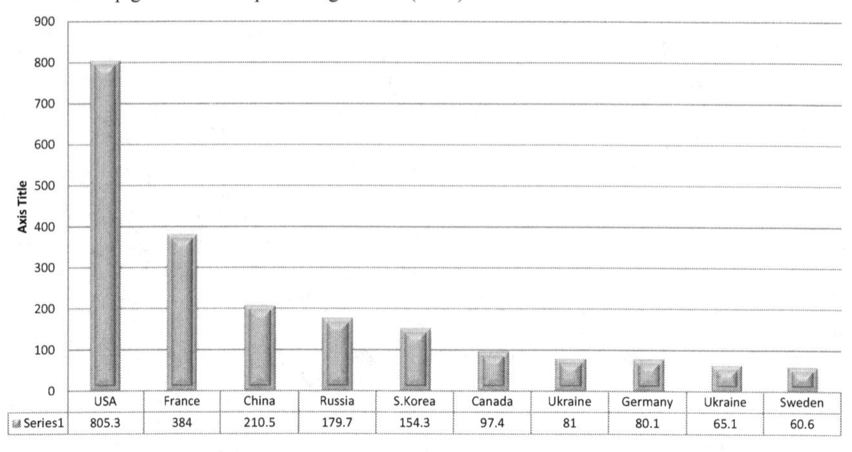

Source: NEI, 2017a. *Top Global Nuclear Generating Countries (2016) Billion kWh.* Nuclear Energy Institute. Available at: www.nei.org/Knowledge-Center/Nuclear-Statistics/World-Statistics/Top-10-Nuclear-Generating-Countries, last accessed: August 15, 2017.

Commercial Nuclear Production Today

According to the World Nuclear Association, there are at least 440 commercial nuclear energy reactors globally, functioning in 31 nations, with an aggregate capacity of at least 390,000 MWe. These reactors supply at least 11.5% of the global electric grid. Civil nuclear energy can offer a cumulative of 17,000 reactor years of expertise. One-quarter of the world's reactors have load factors of more than 90% (WNA, 2017a; WNA, 2017b; WNA, 2017c). This is a sustainable green energy segment that is reliable enough to meet the base-load demand.

Baseload Power

Nuclear energy is environmentally friendly, dependable, and consistent, and offers on-demand baseload. Baseload power, or baseload supply, pertains to the minimum energy volume that needs to be produced by a particular plant or energy segment in order to satisfy the regional energy needs (NEI, 2017b).

Reliability is key for a power plant to satisfy the regional markets, and, as discussed in previous chapters of this book, non-renewable energy segments are technologically mature enough to provide sustainable energy. Nuclear energy also belongs to this category of reliable sustainable energy.

Research and Military Activities

The global map of nuclear energy is growing rapidly, with new energy players in the commercial, research, and the military areas contributing to the global nuclear capacity.

- Research on nuclear technologies is also expanding, with 55 nations operating nearly 250 research reactors and multi-billion dollar investment deals tied to these research activities.
- In addition, 180 nuclear reactors power department of defense activities with about 140 nuclear aircraft carriers, ships, and submarines (WNA, 2017b; WNA, 2017c).

Dual Purpose Nuclear Plants: Ocean Water Desalination and Electricity Production

In addition to the use of nuclear energy for feeding the electric grid, it is also used in desalination plants that utilize the excess heat from nuclear plants to remove the salt from ocean water. Through the methods of reverse osmosis, ultrafiltration, and multi-stage flash distillation, these plants condense the pure water and evaporate salt water.

There are currently approximately 15,000 desalination plants globally, mostly in the Middle East and North Africa. The major nuclear desalination plants are found in India, Japan, and Kazakhstan, but also in the U.S. and Israel. The latest key players include Argentina, China, and South Korea, which have industrialized small modular designs of nuclear reactors that combine both electricity and fresh water generation (NEI, 2017b; NEI, 2017c).

The benefits of these operations are immense: According to World Health Organization, one in three humans globally is impacted by water scarcity (WHO, 2011). The world population has exceeded 7.5 billion, and, according to the United Nations, by 2100 it will exceed 11.2 billion (UN, 2015; US Census Bureau, 2017). Another parameter to the water scarcity equation is the impact on animals, plants, and the environment.

New Commercial Plants, New Global Players

The global map of nuclear energy is growing rapidly, with new energy players and nations that are currently educating themselves on the possibilities of nuclear energy.

- From a commercial standpoint, over 60 reactors are being built, with Southeast Asia leading, and anticipated to further enhance the global nuclear energy capacity. Several existing plants are being upgraded and therefore increase the global capacity. For example, U.S. programs are intended to extend plant life and capacity.
- At least 45 nations are actively considering embarking upon nuclear power programs. These range from sophisticated economies to developing nations. The front runners are UAE, Turkey, Belarus, and Poland (WNA, 2017b; WNA, 2017c).

Case Study: Measuring and Eliminating Risks in Nuclear Plants

The nuclear energy segment combines many noteworthy characteristics that inspire admiration or even a "reverential awe" about this unique energy that reflects the secrets of the universe. Nuclear plants operate with innovative technologies, require a high degree of scientific aptitude, training, and specialization, yet they offer reliable sustainable energy of high calorific value.

The U.S. and other leading nations in the global nuclear energy production have established a regulatory framework where nuclear power plant operators need to establish a risk management strategy. Companies need to confirm that their sites can endure design-basis security threats, and that they have a strategy for resource management along with the capability to respond to beyond-design-basis threats.

The Energy Risk Triangle

The energy risk triangle entails human factors and partnerships, technologies, and regulations, as identified in Figure 6.4 (Burns, 2014). These conclusions apply for nuclear energy as well as all energy types.

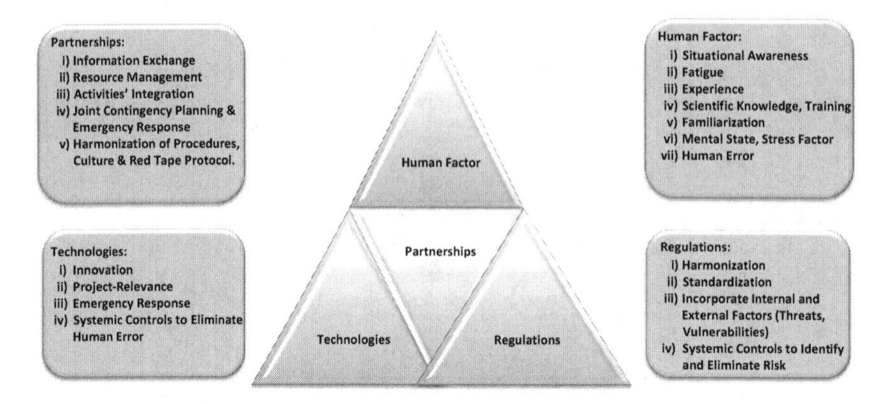

Figure 6.4 The energy risk triangle.

Source: the Author, based on Burns, 2014. "Supply chain security: a strategy to resilience, through policies, partnerships and technologies". DHS Supply Chain Security Workshop, Houston, April 23. Organized by the U.S. Department of Homeland Security.

Technologies and the Supply Chain

The Supply Chain of Nuclear Energy

In previous chapters of this book reference was made to the upstream, midstream, and downstream processes of the supply chain. Since nuclear energy uses minerals like uranium, plutonium, and thorium as fuel, the preliminary part of the mineral extraction resembles the stages of coal mining or shale gas upstream process.

The Raw Materials: Uranium, Plutonium, Thorium

Like every other supply chain, the nuclear power process commences with the mining, enrichment processing, and the fuel cycles' protocol of particular materials, typically uranium, plutonium, and thorium, and their isotopes, that are being used in a nuclear plant in order to generate energy.

Fissile vs Fertile Materials

The term fissile applies to materials that can successfully withstand a neutron chain reaction and the fission process. Fissile elements pertain to low-energy slow nuclei, usable in a thermal neutron reactor, but also transuranic elements with an odd number of neutrons. The most significant fissile elements in nuclear power are plutonium-239, (i.e., an isotope of plutonium) and uranium-235 (^{235}U, i.e., an isotope of uranium that is a primordial nuclide, i.e., is naturally available and mined).

The term fertile pertains to natural elements that are not per se fissionable (or fissile via thermal neutrons), yet can be transformed into fissile materials by the method of irradiation in a reactor.

The two main fertile materials are uranium-238 and thorium-232. For instance, thorium-232 can be transformed into uranium-233, uranium-234 can be transformed into uranium-235, and uranium-238 can be transformed into plutonium-239 (NRC, 2017b).

1 **Uranium**

Uranium is a light silver-colored element with the chemical symbol U and atomic number 92 in the periodic table of elements, i.e., its atom consists of 92 protons and 92 electrons, among which six are valence electrons (DOE, 2017c).

Uranium is the most prevailing mineral used in nuclear reactors to induce fission and is considered as a fissile element. It is enhanced with the uranium-235 isotope (^{235}U) that is scarcely available in nature, i.e., represents about 0.72% of the naturally available and extracted uranium. Pursuant to its mining, it is converted into pellets of nuclear rods and utilized in nuclear plants (EIA, 2017a). In particular, U-235 is the most preferred isotope due to the unchallenging division of its atoms (WNA, 2017a).

Among the advantages of uranium is the high energy density it possesses, i.e., one ton of uranium contains energy comparable to 14,000–23,000 tons of coal, and its availability in many geographical locations, thus facilitating its extraction and transportation to and from different sites. Contemporary improvements in the mining methods and operational protocols have improved the health and environmental risks (CDC, 2017).

Uranium is considered a heavy metal, i.e., it is one of the heaviest natural elements, meaning its diffusion process is slower. Enriched uranium is transformed from its ore form into solid ceramic fuel pellets that are packaged in sealed into zirconium alloy tubes, and thereby used as fuel in nuclear reactors. The U-235 isotope can be concentrated in a process called "enrichment," making it "fissile" and suitable for use in nuclear reactors or weapons. Over 99% of the uranium extracted is available in the U-238 form. U-234 comprises less than 1% of all natural uranium forms, but it is much more radioactive, representing about 50% of radioactivity from all forms of uranium found in the environment (EPA, 2017a).

2 **Plutonium**

Plutonium is a radioactive metallic element with the chemical symbol Pu and atomic number 94. It is the second transuranium element among the actinide series.

Plutonium represents at least one-third of the energy generated in the nuclear power plants globally, however most of the plutonium used in the nuclear plants is man-made and this takes place in a reactor as a by-product, as uranium atoms enthrall neutrons.

Plutonium is found in abundance in various nuclear fuel storage facilities globally. To make it usable, a method known as reprocessing is applied to isolate plutonium from uranium and other fission products through the utilization of chemicals (NRC, 2017c).

Back in World War II, plutonium was generated as part of the Manhattan Project for the "Trinity Test" conducted in New Mexico. Since then, it has assumed a prominent position within the transuranium elements because its use is elementary, both as an explosive constituent in nuclear weapons and in nuclear power production (LANL, 2017b).

Among the isotopes of plutonium, five are considered as fissile elements, i.e., their atomic nucleus can be divided effortlessly once hit by a neutron. These include: Pu-238, Pu-239, Pu-240, Pu-241, and Pu-242. Plutonium isotopes, just like other radioactive isotopes (radionuclides), transmute as they degenerate. They could either transform into different plutonium isotopes or into other elements like uranium or neptunium (NRC, 2017c).

3 Thorium

Thorium is another radionuclide or radioactive metal created naturally, with the chemical symbol Th and atomic number 90. Under natural atmospheric conditions thorium is a solid. It is generated both naturally and man-made in a laboratory setting, yet all of its forms and isotopes are radioactive. Its trace elements are found in the environment, i.e., on the ground, in aquatic and rock formations, and in flora and fauna.

Thorium represents a most suitable nuclear fuel substitute when uranium becomes scarce and finely depleted, due to the vast reserves found naturally. One of its benefits entails its ability to breed fissile fuel (^{233}U) from natural thorium (^{232}Th). Possible scenarios for using thorium in the nuclear fuel cycle include use in different nuclear reactor types (light water, high temperature gas cooled, fast spectrum sodium, molten salt, etc.), advanced accelerator-driven systems, or even fission–fusion hybrid systems. The most likely near-term application of thorium in the United States is in currently operating light water reactors (LWRs) (NRC, 2014–2016).

As a general rule, naturally produced thorium exists in the forms of Th-232, Th-230 or Th-228 isotopes.

The thorium fuel cycle is a nuclear fuel cycle that uses an isotope of thorium, 232-Th, as the fertile material. In the reactor, 232-Th is transmuted into the fissile artificial uranium isotope 233-U which is the nuclear fuel. Unlike natural uranium, natural thorium contains only trace amounts of fissile material (EPA, 2017a).

Thorium consumption worldwide is small compared with that of most other mineral commodities. Thorium is used in a variety of catalysts, ceramics, optics, and metal applications. In catalysts, ThO_2 is used in petroleum cracking and in the production of nitric and sulfuric acids. Interest in thorium as a nuclear fuel continues, in part, owing to its high abundance relative to uranium (USGS, 2014).

Table 6.2 presents the chemical properties of the three most significant nuclear fuels, uranium, plutonium, and thorium.

Table 6.2 Chemical properties of uranium, plutonium, and thorium.

Uranium			
Atomic Number:	92	Atomic Radius:	240 pm (Van der Waals)
Atomic Symbol:	U	Melting Point:	1135°C
Atomic Weight:	238	Boiling Point:	4131°C
Electron Configuration:	$[Rn]7s^2 5f^3 6d^1$	Oxidation States:	6, 5, 4, 3, 0

Source: Los Alamos National Laboratory (LANL) (2017a). *Uranium, Periodic Table of Elements.* Los Alamos National Laboratory. Available at: http://periodic.lanl.gov/92.shtml, last accessed: May 15, 2017.

Plutonium			
Atomic Number:	94	Atomic Radius:	243 pm (Van der Waals)
Atomic Symbol:	Pu	Melting Point:	640°C
Atomic Weight:	244	Boiling Point:	3228°C
Electron Configuration:	$[Rn]7s^2 5f^6$	Oxidation States:	7, 6, 5, 4, 3, 0

Source: Los Alamos National Laboratory (LANL) (2017b). *Plutonium, Periodic Table of Elements.* Los Alamos National Laboratory. Available at: http://periodic.lanl.gov/94.shtml, last accessed: May 15, 2017.

Thorium			
Atomic Number:	90	Atomic Radius:	237 pm (Van der Waals)
Atomic Symbol:	Th	Melting Point:	1750°C
Atomic Weight:	232	Boiling Point:	4788°C
Electron Configuration:	$[Rn]7s^2 6d^2$	Oxidation States:	4, 0

Source: Los Alamos National Laboratory (LANL) (2017c). *Thorium, Periodic Table of Elements*. Los Alamos National Laboratory. Available at: http://periodic.lanl.gov/90.shtml, last accessed: May 15, 2017.

Uranium Ore Mining Methods

The process of extracting uranium ore entails the main techniques used in coal extraction (Chapter 5), i.e., open pit, underground mining, and in situ methods:

a **Open pit mining** (also known as open cast or open cut) is a surface mining method applicable when uranium ore is found close to the soil surface. The minerals are excavated from a borrow or an open pit.

b **Underground mining** (also known as "hard rock" or "soft rock" mining) is applicable when excavating minerals, e.g., ores, in deeper sedimentary deposits.

c **In-situ mining** pertains to the extraction of uranium and other ores from the soil via boreholes and further processing it. In-situ leaching ("ISL") or solution mining entails the injection of liquids (also known as "leaching liquors") into the ground via the leaching method in order to extract the particular minerals. Alkaline solutions are used to retain the uranium in solution. This is the prevailing method as it involves minor geological disruption, is not as labor intensive, and supports a rapid environmental recovery.

How Nuclear Energy Differentiates from Other Energy Sources

The nuclear energy supply chain has the following particularities that make it stand out from other energy sources:

a the high-tech configuration of a nuclear plant, requiring the highest safety and security measures;

b The uranium "enrichment" process – typically via the isotope separation method – where input fuel must be enriched or chemically enhanced to yield a higher calorific value, or a higher breeding ratio;

c the production of high-yield energy that has various utilities, from electricity to military weapons, to aerospace fuel, to ocean water desalination.

d Highly innovative designs and patents, and military affiliations, are the reason why nuclear energy is under the auspices of governments and the Department of Defense, but also explains why certain countries adopt this particular energy segment for security, aeronautics and space missions.

e Waste management is a significant part of nuclear energy security. Radiological waste is the secondary outcome of nuclear energy production, and decisions must be made as to i) radioactive waste disposal, ii) radioactive waste cleanup, and iii) reprocessing protocols to produce nuclear weapons. National regulations will define the protocols, i.e., in determining how different types of waste will be treated, recycled, or disposed of.

A Nuclear Plant Configuration

A nuclear plant consists of numerous reactors and supporting facilities, and the following description is a rather concise version of the principal configuration, facilities, and operations, which is also reflected in Figure 6.5.

1. The turbine building is a confined metallic structure that encompasses:
   a. The electrical power generator, turbines, cooling, and lubrication structures. The power transformers supply electric power to the plant and to the electric grid.
   b. The condensate and feed-water systems that condense exhaust steam from the turbine and supply water to and from the generators.
   c. A control room monitors and controls the operating and technical performance of all equipment in the plant.
2. The containment building is a gas-tight dome-shaped shell where the nuclear reactor is confined. To ensure the fission output will not be escaping into the environment during an accident, it is constructed of several-feet-thick walls made of steel or lead-plated concrete.
3. The cooling tower: As the turbines in a nuclear plant release steam, the cooling tower ensures that the exhaust heat is released into the air and not in the water.

The reactor structure generates heat, which is transported from the primary cooling loop to the steam generator and into a secondary cooling loop, where heat vaporizes the water and generates steam. Subsequently, steam is thrusted upon the turbine and, as it moves, electricity is produced (NRC, 2017a).

Nuclear Reactor Generations

The commercialization of nuclear reactors commenced in the 1950s. In 1956 the U.S. launched the first commercial nuclear plants, employing both Fast Breeder Reactors (FBR) and Light Water Reactors (LWR). In 1960, two versions of the Light Water Reactor (LWR) were developed in the U.S. by the Argonne National Laboratory and commercialized: Namely, the world's first Boiling Water Reactor (BWR) was commercialized by General Electric and the Pressurized Water Reactor (PWR) was commercialized by Westinghouse (Argonne National Laboratory, 2017).

Since then, the nuclear energy industry has grown in terms of cost efficiency, thermal efficiency, technological advancements, commercial life, and automated functions that reduce risk due to the elimination of human error.

Table 6.3 shows the generations of commercial nuclear reactors, with their key characteristics.

Nuclear Plants: Commercial Life Expansion from 40 to 60+ Years

It is worth noting that most of the Generation I and several Generation II plants have concluded their commercial life and are now shut down, whereas many Generation

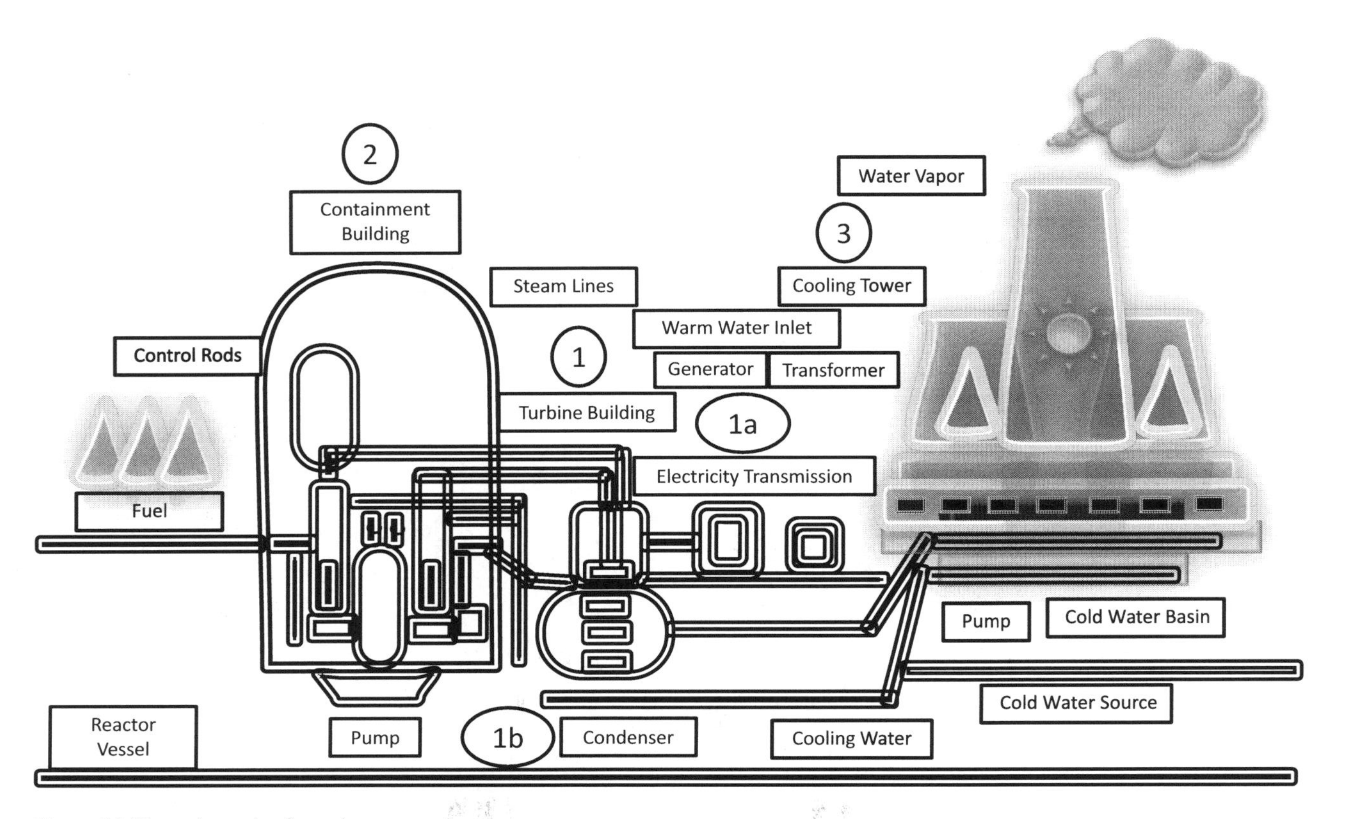

Figure 6.5 The major parts of a nuclear power plant.

Source: the Author, based on: EPA, 2017. Major Parts of a Nuclear Power Plant. U.S. Environmental Protection Agency. Available at: https://archive.epa.gov/climatechange/kids/solutions/technologies/nuclear.html, last accessed: April 12, 2017.

II and III plants have expanded their commercial life span from 40 to 60+ years, for three reasons:

1 due to innovative technologies that facilitate efficient Upgrading and Retrofitting (U&R);
2 high capital investment requirements and ongoing demand for nuclear-powered electricity forced the industry to retrofit as opposed to shut down and rebuild;
3 the Generation IV International Forum and their scientific team have been working on new designs and technologies for both large scale and small modular technologies. As the commercial life of Generation II and II has been prolonged, Generation IV will bring a new era in nuclear power.

In addition to the generational classifications, nuclear energy has also expanded in terms of geographical coverage and the variety of technologies, sizes, and methods. Table 6.4 demonstrates the main commercial reactor types globally.

As demonstrated in Tables 6.3 and 6.4, there are numerous classifications of nuclear reactors, with several subcategories, that would merit a separate publication to reflect on their technological capabilities, energy efficiency, and operational performance. As the present book focuses on the security of all energy segments, this section will briefly cover the most popular reactor categories globally.

Table 6.3 The commercial nuclear reactor generations (1950s–2020s).

Generation/ timeline	*Classification & models*	*Characteristics & performance features*
Generation I 1950s–1970s	Class: Fast Breeder Reactors (FBR) Light Water Reactors (LWR) Models: Shippingport, Dresden, Magnox/UNGG	• Prototypes of commercial use.
Generation II 1970s–1990s	Class: Boiling Water Reactor (BWR) Pressurized Water Reactor (PWR) Models: CANDU (Canada, Heavy Water) RBMK (Russian, Graphite) AGR (England, Graphite) WWER (Russian, Pressurized Water, also used in Europe, China, India, etc.)	• Improved safety and lower cost. • Technological improvements: increased monitoring and controlling, thus minimizing human error.
Generation III 1990s–2010s Generation III+ 2010s–2020s	Class Advanced Boiling Water Reactors (Japan, UAE) Generation III+ 100 MWe-class Westinghouse AP1000 (USA) ESBWR, GE Hitachi (USA, Japan) VVER-1200/392M (Russia)	• New multinational partnerships are the norm. New technologies improve surveillance for safety, security, and performance monitoring. • Improved thermal efficiency. • Retrofitting and maintenance systems prolong commercial life from 40 to 60+ years, thus enhancing the return on investment ratio.

Generation IV 2020s–2030s	The Generation IV International Forum monitors and evaluates testing on the following categories: Class: 1 Gas-cooled fast reactors (Helium-cooled Hydrogen Production) 2 Lead-cooled fast reactors (large and small modular sizes) 3 Molten Salt Reactors: i The Molten Salt Fast Neutron Reactor (MSFR), and ii the Advanced High-Temperature Reactor (AHTR) 4 Sodium-cooled fast reactors (SFR) 5 Supercritical water-cooled reactors (SCWR) 6 Very High-Temperature Gas Reactors (graphite-moderated, helium-cooled).	• Large scale and small modular technologies. • Many new plants under construction. New designs and licensing arrangements. • Harmonization of regulatory framework encompasses technologies and operational processes. Improved performance and security through a holistic appraisal. Efforts to further expand the commercial life of plants via the replacement of obsolete technologies.

Source: the Author, incorporating the Generation IV designs (2020s–2030s), as presented in the following sources:

1 DOE, 2005. DOE Office of Nuclear Energy, Science and Technology, Office of Advanced Nuclear Research, U.S. Department of Energy. Available at: http://nuclear.inl.gov/deliverables/docs/gen-iv-10-yr-program-plan.pdf, last accessed: May 21, 2017.
2 DOE, 2017. *Observations on A Technology Roadmap for Generation IV Nuclear Energy Systems*. U.S. Department of Energy. Available at: https://energy.gov/ne/downloads/observations-technology-roadmap-generation-iv-nuclear-energy-systems, last accessed: May 21, 2017.
3 WNA, 2017d. *Generation IV Nuclear Reactors*. World Nuclear Association. Available at: www.world-nuclear.org/information-library/nuclear-fuel-cycle/nuclear-power-reactors/generation-iv-nuclear-reactors.aspx, last accessed: May 21, 2017.
4 WNA, 2017e. *Advanced Nuclear Power Reactors*. World Nuclear Association. Available at: www.world-nuclear.org/information-library/nuclear-fuel-cycle/nuclear-power-reactors/advanced-nuclear-power-reactors.aspx, last accessed: May 21, 2017.

Light-Water Reactor (LWR)

The light-water reactor (LWR) is one of the most common types of nuclear reactors. It belongs to the thermal-neutron reactor family, where heat is generated by controlling the nuclear fission.

Conventional water that comprises protium (a hydrogen-1 isotope) is used for its cooling and neutron moderation operations, i.e., the process of inhibiting the velocity of fast-moving neutrons.

There are three main types of LWR:

- pressurized-water reactor (PWR);
- boiling-water reactor (BWR);
- supercritical-water reactor (SCWR).

Pressurized-Water Reactor (PWR)

The PWR is the most common nuclear reactor used globally, especially by the U.S., France, Japan, Russia, and China. Water is heated to very high temperatures, yet the water

Table 6.4 Categories of commercial nuclear reactor.

Reactor type	*Fuel*	*Countries*	*Moderator*	*Coolant*	*Number*
Pressurized water reactor (PWR)	Enriched UO_2	US, France, Japan, Russia, China	Water	Water	290
Boiling water reactor (BWR)	Enriched UO_2	US, Japan, Sweden	Water	Water	78
Pressurized heavy water reactor (PHWR)	Natural UO_2	Canada, India	Heavy water	Heavy water	47
Light water graphite reactor (LWGR)	Enriched UO_2	UK	Graphite	Water	15
Gas-cooled reactor (GCR)	Natural U, enriched UO_2	Russia	Graphite	Carbon dioxide	14
Fast breeder reactor (FBR)	PuO_2 and UO_2	Russia	None	Liquid sodium	3

Source: WNA, 2017f. *Nuclear Fuel Cycle. Nuclear Power Reactors*. World Nuclear Association. Available at: www.world-nuclear.org/information-library/nuclear-fuel-cycle/nuclear-power-reactors/nuclear-power-reactors.aspx, last accessed: June 21, 2017.

is pressurized in order to inhibit its boiling stage with the purpose of retaining the nuclear energy yield in the reactor, while separating the steam deriving from the steam generator. The procedure is depicted in Figure 6.6.

The reactor's core includes about 150–200 fuel assemblies that are cooled by water disseminated by the use of pumps. In normal conditions the entire operational system is powered via the electric grid, yet auxiliary generators are also available to supply power in case of an emergency (NRC, 2017d).

Boiling-Water Reactor (BWR)

The BWR signifies approximately 15% of the global reactors and they are widely used in the U.S., Japan, and Sweden. The coolant is retained at a lower pressure compared with the PWR, enabling it to reach a boiling point. As the water is heated, it evaporates and rotates the main turbine, which turns the generator, thus creating electrical power.

Unlike the PWR design, this design does not have a steam generator, thus increasing the risk of turbine radioactivity (CNA, 2017).

Supercritical-Water Reactor (SCWR)

The SCWR is a hybrid technology that combines a simplified version of the LWR and the supercritical coal-fired boiler design. It attains higher thermal efficiency compared with the LWRs, due to the higher operating temperatures and pressures they develop. Research findings in 2003 presented certain challenges related to thermal-hydraulic and thermal-nuclear vulnerabilities resulting in the total loss of feedwater heating. A corrective action entails the installation of a "high-capacity, high-pressure, fast-acting auxiliary feedwater system." However, its design would impose safety, operational, and financial risks related to requirements for additional investment, resources, and technological acumen (Buongiorno and MacDonald, 2003).

More than a decade later, the Generation IV researchers recognize the model's strengths, i.e., optimum fuel utilization, increased thermal efficiency (44%, contrasted to about 35% for other reactors), and are working on improving the design while eliminating certain technical vulnerabilities.

Light Water Graphite Reactor (LWGR)

This is a graphite-moderated, pressurized water-cooled reactor that is used in Russia. It is a hybrid of the advanced gas-cooled reactor and the steam-generating heavy water reactor. A simple circuit configuration is offered in this direct steam cycle, where steam is produced directly in the reactor and subsequently passes into the external steam containers where it is segregated from water. After the Chernobyl accident of 1986, this model has improved its safety features.

Pressurized Heavy Water Reactor (PHWR)

The PHWR uses natural uranium in its unenriched form as fuel. It is mainly used in Canada and India, and represents about 12% of reaction worldwide. Heavy water, i.e., a type of water that comprises deuterium, a hydrogen isotope, is used for its cooling and neutron-moderation operations. One of its benefits is the ability to continue operations and fuel replenishment operations without the need to be shut down.

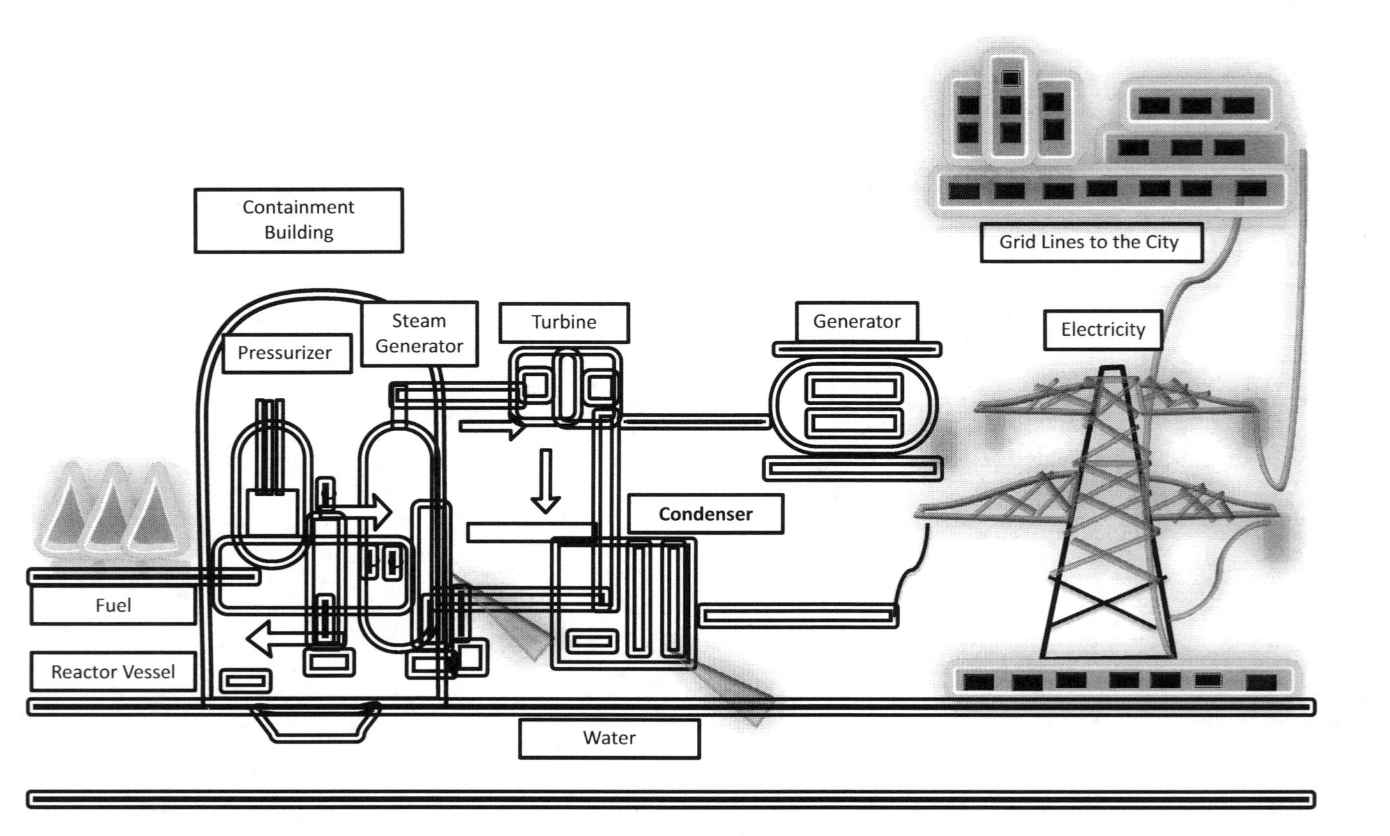

Figure 6.6 The pressurized-water reactor (PWR).

Source: the Author, based on NRC, 2017d. *Pressurized Water Reactors*. U.S. Nuclear Regulatory Commission. Available at: www.nrc.gov/reactors/pwrs.html, last accessed: April 12, 2017.

Despite its higher cost, heavy water contributes to neutron economy in two ways:

1 fuel enrichment processes are not needed for the nuclear reactor to function;
2 the reactor can still carry out alternate fuel cycles.

Gas-Cooled Reactor (GCR)

This type of reactor is a continuation of the original nuclear reactor designs. Widely used in England, its two main representatives are the advanced gas-cooled reactor (AGR), where enriched uranium is the fuel, and the Magnox, where uranium is used as fuel. Both designs use graphite as a moderator, and carbon dioxide, helium, or another inert gas as the coolant.

Breeder Reactors: Thermal (TBR), Intermediate (IBR), and Fast (FBR) Breeders

Breeder reactors pertain to the nuclear reactors that, though they resemble the conventional nuclear energy reactors, have the capability to generate more fuel than the fuel used as input through the fission process.

Only a few isotopes available in quantity are capable of sustaining the fission process. These isotopes are uranium-233, uranium-235, plutonium-239, and plutonium-241, and the term fissionable substance or material refers to these four isotopes. All four undergo radioactive decay: ^{235}U is the only one that exists in any quantity in nature; the other three may have existed on earth billions of years ago but they have now entirely decayed. Consequently, if we wish to produce ^{233}U, ^{239}Pu, or ^{241}Pu, they must be generated artificially in a laboratory setting (OSTI, 1971).

The breeding process can be organized in two key methods:

1 The fertile and fissile input is treated in the same place, in a manner in which the breeding and splitting process happens concurrently in the same facilities.
2 The fissile core-and-breeding blanket process separates the fertile from the fissile input: At the first stage, a fissile core produces the heat and neutrons while a separate blanket does all the breeding.

The breeding ratio in a breeder reactor is higher than 1, suggesting that such reactors produce more fissionable fuel than they consume, as, for each neutron that is enthralled by an atom, more than one new atom of fissionable substance is produced.

Breeding Doubling Time pertains to the breeder reactor's time needed to breed surplus matter, with a ratio higher than 1, and thus refuel a subsequent reaction. A common doubling time for standard nuclear plants is approximately ten years, meaning that the nuclear reaction output of a particular reactor could be sufficient to generate energy for a decade, yet there is sufficient surplus energy generated over this time that a different reactor can be fed for another decade.

There are three main types of breeder reactors, thermal (TBR), intermediate (IBR) and fast (FBR), classified according to their breeding ratio (>1), and according to their breeding doubling time (years required to double aggregate output).

THERMAL BREEDER REACTORS (TBR)

Thermal breeder reactors (TBR) use thermal neutrons such as thorium-232 as fertile material to breed fissile uranium-233. Thermal neutrons represent the category of slower neutrons since they retain the temperature levels of the adjacent coolant. The thermal breeding process favors the use of thorium fuel in order to diminish the accumulation of heavier transuranic matter. The risk in using low energy neutrons is that the breeding process may transform them into neutrons that are non-fissile. For example, uranium-235 is transformed into uranium-236, and plutonium-239 is transformed into plutonium-240.

INTERMEDIATE BREEDER REACTORS (IBR)

Intermediate breeder reactors (IBR) pertain to intermediate speed of the neutrons, with the use of less neutron moderators compared with the thermal reactors. When the reactor functions without a moderator it yields higher calorific value.

FAST BREEDER REACTORS (FBR)

Fast breeder reactors (FBR) pertain to the faster speed of the neutrons, with the use of fewer or no neutron moderators compared with the thermal and intermediate reactors. They utilize fertile neutron material as input fuel (typically uranium-238 or thorium-232) to generate fissile plutonium and other high transuranic materials, i.e., synthetic radioactive elements with an atomic number greater than uranium (92). The use of coolants (like liquid sodium) in the FBR process compromises the fast neutrons' efficiency in igniting fission, yet they attain a high calorific value and the capability to multiply, i.e., produce more energy than their original input.

Benefits, Risks, and Recommendations

The Industry at a Glance

The global nuclear energy community shares identical strategic goals for the particular energy segment that aim to eliminate the common risks and to further explore the benefits and opportunities of the new era in the nuclear sector.

The U.S. Energy Policy Act of 1992 has reflected on the key risks and opportunities, which are attuned with the most recent industry goals and include, among others:

1. The preservation of safety and design standards, with a focus on radiation and reactor safety standards, facility and site management, environmental protection, and so on.
2. The maximizing of regulatory compliance and the minimizing of regulatory risks.
3. The elimination of financial risks (DOE, 2017b).

The nuclear sector is high profit, high risk: Its strategic geopolitical and socioeconomic significance make nuclear plants popular targets for security attackers. At the same time, the nuclear energy industry's security strategy is regulated and monitored by the government, thus making the facilities safe and secure. The sturdy construction and configuration of nuclear facilities, in combination with the advanced surveillance and defense systems, make nuclear targets hard to penetrate.

While nuclear energy was originally considered as a non-renewable source, breeding techniques, as well as the development of synthetic fuel to feed the nuclear reactors, have transformed this segment into a renewable one. Longevity is an appealing feature for governments that are constantly exploring optimum solutions for improving the energy security, energy independence, and the energy mix.

Regulatory Framework for the Nuclear Energy

Over the past decades governments have been developing a standardized set of policies to ensure safety, health, and environmental integrity. Government and state regulatory frameworks aim to diminish safety- or pollution-related risks deriving both from naturally occurring radioactive material and from technologically enhanced radioactive material deriving from nuclear-related activities.

Chapter 2 of this book includes information on particular government departments in charge of energy, mining, transportation, and environmental regulations, and their areas of responsibility should be studied in conjunction with the regulatory framework covered in this chapter.

In the U.S., the legal framework for nuclear facilities and waste is stipulated by the following set of acts and regulations:

1 The Department of Energy, the Environmental Protection Agency and the Department of Transportation Hazardous Materials (Dangerous Goods) Regulations.
2 The Nuclear Regulatory Commission (Predecessor: United States Atomic Energy Commission).
3 The Resource Conservation and Recovery Act (RCRA) of 1976, which established a framework for the appropriate management of hazardous waste.
4 The Nuclear Waste Policy Act of 1982, which is an all-inclusive federal program for the long-term storage of nuclear waste.

Classification of Radioactive Materials

There are many ways of classifying the hazardous impact of toxic waste, such as the nine HazMat/Dangerous Goods classifications.[1]

The Environmental Protection Agency in most nations monitors, controls, and regulates the two main classifications of radioactive material, both "NORM" and TENORM". According to the U.S. EPA, these are defined as follows:

- **Naturally Occurring Radioactive Material (NORM)** is defined as, "Materials which may contain any of the primordial radionuclides or radioactive elements as they occur in nature, such as radium, uranium, thorium, potassium, and their radioactive decay products, that are undisturbed as a result of human activities."
- **Technologically Enhanced Naturally Occurring Radioactive Material (TENORM)** is defined as, "Naturally occurring radioactive materials that have been concentrated or exposed to the accessible environment as a result of human activities such as manufacturing, mineral extraction, or water processing" (EPA, 2017c; EPA 2017d).

For example, the mining operations of uranium ores generate different forms and levels of bulk waste material, which may comprise radionuclides of uranium, radium, and thorium: a) excavated top soil, b) overburden, c) weakly uranium-enriched waste stones, d) subgrade ores, e) evaporation pond sludge and scales (EPA, 2017c; EPA 2017d).

Radioactive Waste, Storage, and Other Environmental Considerations

Nuclear energy represents a sustainable reliable energy segment. Previous sections discussed the improved operational processes throughout the nuclear supply chain, from the extraction of uranium to the fission protocols. Technological advancements have been made to ensure that health and the environment are duly protected from nuclear plant activities.

Radioactive waste, i.e., the by-product of nuclear fission for energy generation or medical and research purposes, is not dispersed to the ecosystem but is treated internally as part of the facility's waste management plan, whose framework is defined by the government.

Over the past decades, significant advancements have been made to maximize energy efficiency and to reduce the quantity, and the health and environmental impact of radioactive waste. Among the diverse sources of toxic and hazardous materials (dangerous goods) produced globally, radioactive waste represents no more than 1%. It is also worth noting that the nuclear industry is regulated and monitored in a manner so that all of its activities, especially its toxic waste, are recorded and accounted for. Long-term storage is currently the safest and most viable solution.

Half-Life of Radioactive Isotopes

The half-life of a radioactive substance is the time interval necessary for its atomic nuclei to diminish by one-half. The other half decays as its particles and energy are released into the atmosphere.

Radioactive isotopes have a prolonged half-life that lasts for thousands or even millions of years. For example, plutonium-239 has a half-life of 24,000 years, uranium-234 of 25,000 years, uranium-235 of about 700 million years, and uranium-238 of about 4.5 billion years (EAD.ANL.gov[2]; NRC, 2017e[3]).

Consequently, a regulatory framework is necessary to implement safety measures and preventive action in order to protect life and the environment.

Waste Levels: Disposal Solutions

Safe disposal of radionuclides entails a designated location where the radioactive waste is permanently disposed of to make certain that there is no radioactive impact on life and the environment. The waste volume combined with the degree of radioactivity will determine the selection of the optimum disposal method and the storage protocols, i.e., selection of method, depth of the disposal, etc. Table 6.5 demonstrates the basic waste types, storage, and disposal options.

Table 6.5 Waste types, storage, and disposal options.

Waste type	*Storage and disposal options*
• Low-level waste (LLW): disposal is both safe and uncomplicated. • Intermediate-level waste (ILW): disposal requires increasing safety measures. • Long-lived intermediate-level waste (ILW). • High-level waste (HLW): disposal requires high safety measures.	• At-reactor storage and near-surface disposal: The vast majority of nuclear power plants store nuclear waste in designated locations within their facilities, or in near-surface locations, in specially designed steel and concrete casks. • Under water storage: It typically takes five years for used fuel to remain under water, and subsequently it is moved to a dry storage location. • Deep geological disposal is appropriate for highly radioactive waste.

Source: the Author.

Nuclear Waste Policies Revisited and Research on Future Disposal Solutions

The United Nations, international treaties, governments, and nuclear scientists are seeking for solutions to the compelling need to dispose of nuclear waste. Numerous nations invest in research projects with the purpose of making the nuclear industry safer, greener, and with optimum benefits for the global community.

Some of the ongoing research on potential disposal solutions include: Sea disposal, under seabed disposal, under ice disposal, deep well injection, subduction zone disposal, disposal in outer space, etc.

However, it is worth noting that current international agreements, treaties, and environmental regulations may be in conflict with the above disposal methods. Policies will need to be revisited as the scientific findings will be conclusive: An alignment will be sought between the agreements, the technological and scientific capabilities, and the need for long-term disposal solutions.

Upon observing various scientific proposals that still need to materialize, it seems that the decision-making process has its own supply chain protocol. Let us call this "the long journey from scientific research to political ratification," and distinguish the tactical process from the strategic steps. Figure 6.7 demonstrates how an energy industry challenge, i.e., waste disposal, passes from the scientific realm into the world of politics.

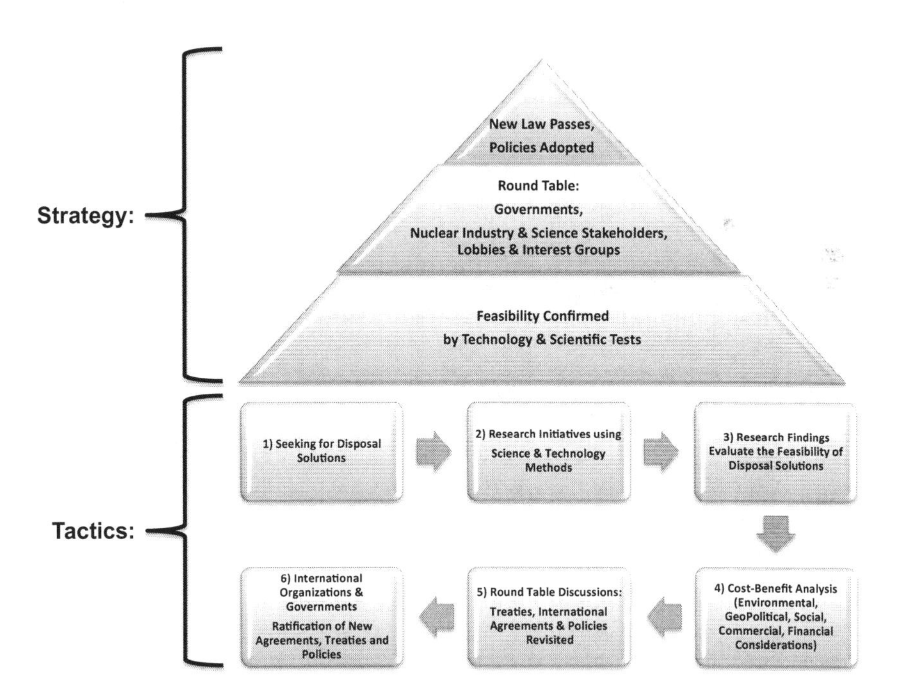

Figure 6.7 Energy policy: The long journey from science to politics.

Source: the Author.

The Future of Nuclear Energy

> Half the people that were out there saying 'No nukes' and 'Shut down the power plants' are now realizing that, some would say, nuclear is the best way to go for energy for the future.
>
> So I think it's natural to reexamine your beliefs as you age up.
>
> Robert Downey Jr.
> (*Robert Downey Jr. Talks Nuclear During a Marvel Interview*. NEI, 2018)

Operational and Maintenance Considerations

Dynamic and planned maintenance pertain to the need to preserve safety and performance throughout a plant's commercial life.

Natural wear and tear, corrosion, or material degradation (e.g., caused by neutron irradiation and high temperatures) will impact the structural integrity, causing safety and security vulnerabilities while also impacting efficiency. Nuclear plants contain many nuclear reactors and cost billions of dollars. Among the intricate infrastructure and superstructure, the steam generators are costly and critical for a plant's safety.

A plant's strategic maintenance system will determine if certain components will be replaced, retrofitted, or simply maintained, throughout the plant's commercial life.

Obsolete instrumentation and control (I&C) technology is another critical area for consideration, and this includes the transition from analogue to digital convertors.

New Generation Plants

For the past two decades or so the global nuclear energy industry has benefited from the innovative technologies and the implementation of standardization protocols. It can demonstrate a proven track record of safety, security, and environmental integrity, with the exception of specific nuclear plant disasters that were partially triggered by natural disasters and exogenous factors.

The Generation IV of nuclear plants is a fact: Innovative designs gradually kindle the construction of new reactors, characterized by fuel cycle efficiency, time and resource efficiency, and proliferation resistance. Technological novelties are not only adopted in the construction of new plants, but also in the upgrading and retrofitting of the existing plants.

Early plant retirement has been a challenge for the past decade or so. Yet this is a new and promising era with increased optimism in the nuclear industry. The extension of a plant's commercial life from 40 to 60+ years enhances profitability and the return-on-investment ratio, while operational efficiencies drive down the cost per unit of the nuclear-based electricity.

Countries like the U.S., France, Russia, Canada, Japan, etc., with experience in the nuclear energy production, offer support and technological expertise to newest members such as China, India, and several developing and less developed economies that are investing in new nuclear plants.

An example of this transnational collaboration is the U.S.–Indian civil nuclear cooperation agreement, where the two nations aim to strengthen the global nonproliferation regime while enhancing energy security, economic and technological development, and retaining environmental integrity.

Knowledge Transfer: Safely Passing the Baton

"Passing the baton" is an interesting notion that refers to the transfer of scientific knowledge and hands-on expertise from one nation to another, and from one generation to another. This transition is especially critical in terms of security, especially for the nuclear industry, for two reasons:

1. **Global trade agreements: When innovation falls into the wrong hands**

   International agreements on knowledge transfer may entail risks where nuclear technologies are not used for purely "peaceful nuclear cooperation." To quote the ancient Greeks once more, "today's friends may become tomorrow's foes." Even if a change in diplomatic relations is not the case, the history of political science has shown us that politics are governed by the laws of supply chains: Knowledge transfer may pass from one nation to another through an intricate pipeline of trade agreements.

   As an example, over the past year North Korea has partnered with over 80 countries, most of which are key players in nuclear and hydrogen energy production (i.e., China, Russia, India, Pakistan, France, Iran, and so on). Upon observing North Korea's ability to develop nuclear and hydrogen bombs, one can assume that this high level of expertise is the product of knowledge transfer in an intricate supply chain where innovative technology somehow strayed and went in the wrong hands.

2. **Mind the "generation" gap**

   Nuclear reactors have a commercial life span of about 40–60 years. As new technologies tend to further expand this timeframe, the human factor plays a pivotal role.

   A well-known challenge (and opportunity) in the nuclear industry (and the energy industry as a whole) is the operational transition from one technological generation to another. For example, the professionals accustomed to operating the Generation II nuclear reactors underwent rigorous training in order to become familiar with Generation III systems. The transition from analogue to digital convertors has been another example where the generation gap had to be filled in.

   Therefore, nuclear scientists and highly specialized personnel will need to pass on the baton to the younger generation in a manner that does not compromise security, and yet ensures the industry's well-kept secrets are not lost through the generation gap.

Notes

1 For a tutorial on hazardous materials (i.e., Dangerous Goods) see Burns, M. (2016). *Logistics and Transportation Security: A Strategic, Tactical and Operational Guide to Resilience*. London: CRC Press, Taylor & Francis. In particular, for the HazMat Classifications, see pages 151–185.

2 EAD.ANL (2017). *Is Uranium Radioactive?* Environmental Science Division of Argonne National Laboratory for the United States Department of Energy (DOE), Office of Environmental Management (EM). Available at: http://web.ead.anl.gov/uranium/faq/uproperties/faq5.cfm.

3 NRC (2017d). *Backgrounder on Radioactive Waste*. U.S. Nuclear Regulatory Commission. Available at: www.nrc.gov/reading-rm/doc-collections/fact-sheets/radwaste.html, last accessed: April 30, 2017.

References

Argonne National Laboratory (2017). *Reactors Designed by Argonne National Laboratory*. Nuclear Engineering Division, Argonne National Laboratory. Available at: www.ne.anl.gov/About/reactors/lwr3.shtml, last accessed: April 12, 2017.

Brunenkant, E. J. (1971). *Understanding the Atom Breeder Reactors*. The United States Atomic Energy Commission, U.S. Department of Energy, Office of Science (OSTI). Available at: www.osti.gov/includes/opennet/includes/Understanding%20the%20Atom/Breeder%20Reactors.pdf, last accessed: April 12, 2017.

Buongiorno J. and MacDonald P.E. (2003). *Supercritical Water Reactora. (SCWR), Progress Report for the FY-03 Generation-IV. R&D Activities for the Development of the SCWR in the U.S.* Idaho National Engineering and Environmental Laboratory, INEEL/EXT-03-01210, Idaho, USA, September. Available at: http://nuclear.inel.gov/gen4/docs/scwr_annual_progress_report_gen-iv_fy-03.pdf, last accessed: April 30, 2017.

Burns, M. (2014). "Supply chain security: A strategy to resilience, through policies, partnerships and technologies". DHS Supply Chain Security Workshop, Houston, April 23. Organized by the U.S. Department of Homeland Security.

Burns, M. (2016). *Logistics and Transportation Security: A Strategic, Tactical and Operational Guide to Resilience*. London: CRC Press, Taylor & Francis.

Canada Nuclear Association (CNA) (2017). *Types of Reactors*. Canada Nuclear Association. Available at: https://cna.ca/technology/energy/types-of-reactors, last accessed: April 30, 2017.

Centers for Disease Control and Prevention (CDC) (2017). *Uranium. What Form Is It In?* U.S. Centers for Disease Control and Prevention. Available at: https://emergency.cdc.gov/radiation/isotopes/uranium.asp, last accessed: April 30, 2017.

Department of Energy (DOE) (2005). *Generation IV Nuclear Energy Systems Ten-Year Program Plan*. DOE Office of Nuclear Energy, Science and Technology, Office of Advanced Nuclear Research, U.S. Department of Energy. Available at: http://nuclear.inl.gov/deliverables/docs/gen-iv-10-yr-program-plan.pdf, last accessed: May 21, 2017.

Department of Energy (DOE) (2017a). *Manhattan Project National Historical Park*. U.S. Department of Energy. Available at: https://energy.gov/management/office-management/operational-management/history/manhattan-project/manhattan-project-0, last accessed: April 12, 2017.

Department of Energy (DOE) (2017b). *The History of Nuclear Energy*. The U.S. Department of Energy, Office of Nuclear Energy, Science and Technology, Washington, D.C. 20585. Available at: https://energy.gov/sites/prod/files/The%20History%20of%20Nuclear%20Energy_0.pdf, last accessed: April 30, 2017.

Department of Energy (DOE) (2017c). *Nuclear Fuel Facts: Uranium*. U.S. Department of Energy. Available at: www.energy.gov/ne/nuclear-fuel-facts-uranium, last accessed: April 30, 2017.

Department of Energy (DOE) (2017d). *Observations on a Technology Roadmap for Generation IV Nuclear Energy Systems*. U.S. Department of Energy. Available at: https://energy.gov/ne/downloads/observations-technology-roadmap-generation-iv-nuclear-energy-systems, last accessed: May 21, 2017.

Energy Information Administration (EIA) (2017a). *Energy Explained, Nuclear, How it Works*. Available at: www.eia.gov/energyexplained/index.cfm?page=nuclear_home, last accessed: April 12, 2017.

Energy Information Administration (EIA) (2017b). *How Fission Splits the Uranium Atom*. Energy Information Administration and the National Energy Education Development Project (public domain). Available at: www.eia.gov/energyexplained/images/atom.jpg, last accessed: April 12, 2017.

Environmental Protection Agency (EPA) (2017a). *Radiation Protection. Radionuclide Basics: Uranium*. U.S. Environmental Protection Agency. Available at: www.epa.gov/radiation/radionuclide-basics-uranium, last accessed: July 1, 2017.

Environmental Protection Agency (EPA) (2017b). *Major Parts of a Nuclear Power Plant*. U.S. Environmental Protection Agency. Available at: https://archive.epa.gov/climatechange/kids/solutions/technologies/nuclear.html, last accessed: April 12, 2017.

Environmental Protection Agency (EPA) (2017c). *Radiation Protection. TENORM: Uranium Mining Wastes*. U.S. Environmental Protection Agency. Available at: www.epa.gov/radiation/tenorm-uranium-mining-wastes, last accessed: July 1, 2017.

Environmental Protection Agency (EPA) (2017d). *Technologically Enhanced Naturally Occurring Radioactive Materials (TENORM)*. U.S. Environmental Protection Agency. Available at: www.epa.gov/radiation/technologically-enhanced-naturally-occurring-radioactive-materials-tenorm, last accessed: July 1, 2017.

Environmental Science Division of Argonne National Laboratory (EAD.ANL) (2017). *Is Uranium Radioactive?* Environmental Science Division of Argonne National Laboratory for the United States Department of Energy (DOE), Office of Environmental Management (EM). Available at: http://web.ead.anl.gov/uranium/faq/uproperties/faq5.cfm, last accessed: May 21, 2017.

International Energy Association (IEA) (2017). *About Nuclear Fission*. International Energy Association. Available at: www.iea.org/topics/nuclear, last accessed: April 12, 2017.

Los Alamos National Laboratory (LANL) (2017a). *Uranium. Periodic Table of Elements*. Los Alamos National Laboratory. Available at: http://periodic.lanl.gov/92.shtml, last accessed: May 15, 2017.

Los Alamos National Laboratory (LANL) (2017b). *Plutonium. Periodic Table of Elements*. Los Alamos National Laboratory. Available at: http://periodic.lanl.gov/94.shtml, last accessed: May 15, 2017.

Los Alamos National Laboratory (LANL) (2017c). *Thorium. Periodic Table of Elements*. Los Alamos National Laboratory. Available at: http://periodic.lanl.gov/90.shtml, last accessed: May 15, 2017.

McGinnis, E. (2017). *Nuclear Energy: Clean, Constant, and Cool*. U.S. Office of Nuclear Energy. Available at: www.energy.gov/articles/nuclear-energy-clean-constant-and-cool, last accessed: July 3, 2017.

Nuclear Energy Institute (NEI) (2017a). *Top Global Nuclear Generating Countries (2016) Billion kWh*. Nuclear Energy Institute. Available at: www.nei.org/Knowledge-Center/Nuclear-Statistics/World-Statistics/Top-10-Nuclear-Generating-Countries, last accessed: April 12, 2017.

Nuclear Energy Institute (NEI) (2017b). *The Advantages of Nuclear Energy*. Nuclear Energy Institute. Available at: www.nei.org/Why-Nuclear-Energy/Reliable-Affordable-Energy/Electricity-Supply, last accessed: July 1, 2017.

Nuclear Energy Institute (NEI) (2017c). *Water Desalination. Knowledge Center*. Nuclear Energy Institute. Available at: www.nei.org/Knowledge-Center/Other-Nuclear-Energy-Applications/Water-Desalination, last accessed: July 1, 2017.

Nuclear Energy Institute (NEI) (2018). *Voices for Nuclear Energy*. Nuclear Energy Institute. Available at: www.nei.org/Knowledge-Center/Other-Nuclear-Energy-Applications/Water-Desalination, last accessed: July 24, 2018.

National Park Service (NPS) (2017). *The Manhattan Project*. National Park Service. Available at: www.nps.gov/mapr/index.htm, last accessed: April 12, 2017.

Nuclear Regulatory Commission (NRC) (2014–2016). *Safety and Regulatory Issues of the Thorium Fuel Cycle (NUREG/CR-7176)*. Prepared by: Brian Ade, Andrew Worrall, Jeffrey Powers, Steve Bowman, George Flanagan, Jess Gehin, Oak Ridge National Laboratory, Managed by UT-Battelle, LLC, Oak Ridge, TN 37831-6170, M. Aissa, NRC Project Manager. Prepared for: Office of Nuclear Regulatory Research. U.S. Nuclear Regulatory Commission, Washington, DC 20555-0001, NRC Job Code V6299. Available at: www.nrc.gov/reading-rm/doc-collections/nuregs/contract/cr7176, last accessed: April 30, 2017.

Nuclear Regulatory Commission (NRC) (2017a). *Nuclear Reactors*. U.S. Nuclear Regulatory Commission. Available at: www.nrc.gov/reading-rm/basic-ref/students/what-is-nuclear-energy.html, last accessed: April 12, 2017.

Nuclear Regulatory Commission (NRC) (2017b). *Fertile Material*. U.S. Nuclear Regulatory Commission. Available at: www.nrc.gov/reading-rm/basic-ref/glossary/fertile-material.html, last accessed: July 1, 2017.

Nuclear Regulatory Commission (NRC) (2017c). *Backgrounder on Plutonium*. U.S. Nuclear Regulatory Commission. Available at: www.nrc.gov/reading-rm/doc-collections/fact-sheets/plutonium.html, last accessed: April 30, 2017.

Nuclear Regulatory Commission (NRC) (2017d). *Pressurized Water Reactors*. U.S. Nuclear Regulatory Commission. Available at: www.nrc.gov/reactors/pwrs.html, last accessed: April 30, 2017.

Nuclear Regulatory Commission (NRC) (2017e). *Backgrounder on Radioactive Waste*. U.S. Nuclear Regulatory Commission. Available at: www.nrc.gov/reading-rm/doc-collections/fact-sheets/radwaste.html, last accessed: April 30, 2017.

Office of Science and Technical Information (OSTI) (1971). *Breeder Reactors by Walter Mitchell, III, and Stanley E. Turner 1971*. The United States Atomic Energy Commission, Department of Energy, Office of Science and Technical Information. Available at: www.osti.gov/includes/opennet/includes/Understanding%20the%20Atom/Breeder%20Reactors.pdf, last accessed: April 30, 2017.

UN (2015). *World Population Prospects. The 2015 Revision. Key Findings and Advance Tables.* ESA/P/WP.241. Department of Economic and Social Affairs, Population Division, United Nations. Available at: https://esa.un.org/unpd/wpp/publications/files/key_findings_wpp_2015.pdf, last accessed: July 1, 2017.

U.S. Census Bureau (2017). *Total Midyear Population for the World: 1950–2050.* International Programs, U.S. Census Bureau. Available at: www.census.gov/population/international/data/worldpop/table_population.php, last accessed: July 1, 2017.

U.S. Geological Survey (USGS) (2014). *Minerals Yearbook. Thorium.* By Joseph Gambogi. Domestic survey data and tables were prepared by Annie Hwang, statistical assistant, and the world production table was prepared by Lisa D. Miller, international data coordinator. Available at: https://minerals.usgs.gov/minerals/pubs/commodity/thorium/myb1-2014-thori.pdf, last accessed: April 30, 2017.

U.S. Navy NHH (2017). *U.S.S. Nautilus.* Naval History and Heritage Command, U.S. Navy. Available at: www.history.navy.mil/browse-by-topic/ships/uss-nautilus.html, last accessed: April 12, 2017.

World Health Organization (WHO) (2011). *Safe Drinking Water from Desalination.* WHO/HSE/WSH/11.03. World Health Organization, Geneva, Switzerland. Available at: http://apps.who.int/iris/bitstream/10665/70621/1/WHO_HSE_WSH_11.03_eng.pdf, last accessed: July 1, 2017.

World Nuclear Association (WNA) (2017a). *How Uranium Ore is Made into Nuclear Fuel.* World Nuclear Association. Available at: www.world-nuclear.org/nuclear.../how-is-uranium-ore-made-into-nuclear-fuel.aspx, last accessed: April 12, 2017.

World Nuclear Association (WNA) (2017b). *Emerging Nuclear Energy Countries.* World Nuclear Association. Available at: www.world-nuclear.org/information-library/country-profiles/others/emerging-nuclear-energy-countries.aspx, last accessed: August 15, 2017.

World Nuclear Association (WNA) (2017c). *Nuclear Power in the World Today.* World Nuclear Association. Available at: www.world-nuclear.org/information-library/current-and-future-generation/nuclear-power-in-the-world-today.aspx, last accessed: June 1, 2017.

World Nuclear Association (WNA) (2017d). *Generation IV Nuclear Reactors.* World Nuclear Association. Available at: www.world-nuclear.org/information-library/nuclear-fuel-cycle/nuclear-power-reactors/generation-iv-nuclear-reactors.aspx, last accessed: May 21, 2017.

World Nuclear Association (WNA) (2017e). *Advanced Nuclear Power Reactors.* World Nuclear Association. Available at: www.world-nuclear.org/information-library/nuclear-fuel-cycle/nuclear-power-reactors/advanced-nuclear-power-reactors.aspx, last accessed: August 21, 2017.

World Nuclear Association (WNA) (2017f). *Nuclear Fuel Cycle. Nuclear Power Reactors.* World Nuclear Association. Available at: www.world-nuclear.org/information-library/nuclear-fuel-cycle/nuclear-power-reactors/nuclear-power-reactors.aspx, last accessed: June 21, 2017.

Part III

Renewable Energy Security

Energy security entails sustainable production, accessibility, and the use of reliable logistics systems and technologies. Chapters 7–10 will cover these areas, as well as the operational, regulatory, and technological factors that contribute to a nation's energy security.

7 Wind Power Security

First, there is the power of the Wind, constantly exerted over the globe. . . . Here is an almost incalculable power at our disposal, yet how trifling the use we make of it! It only serves to turn a few mills, blow a few vessels across the ocean, and a few trivial ends besides.

What a poor compliment do we pay to our indefatigable and energetic servant!

Henry Thoreau in 'Paradise (To Be) Regained', *Democratic Review* (November 1848). Collected in *A Yankee in Canada: With Anti-Slavery and Reform Papers* (1866), 188–9.

Wind Power: An Overview

"Wind power" is defined as the technology utilizing the wind as a means of producing mechanical energy, i.e., electricity. Wind turbines transform the wind's kinetic energy into mechanical energy, that may be used for a) particular tasks, e.g., windmills pump water and mill wheat, or b) a wind turbine may transform the mechanical energy into electrical power that can be used for domestic, industrial and public utilities (DOE, 2017a).

Wind is a Type of Solar Energy

Wind energy is closely associated with solar energy; in fact, it is a type of solar energy. As the sun's rays warm the earth's surface, winds are generated. The patterns of wind circulation, velocity, direction, force, and pressure are impacted by the planet's physiographical features including the geological, floral, and aquatic landscapes, for the following factors:

- winds are not propagated regularly in a homogeneous manner, velocity and direction, hence the atmospheric temperatures are irregular and uneven;
- geological irregularities, mountain ranges, geophysical eruptions and tectonic disturbances are present throughout the earth's surface area;
- at the same time the earth rotates, hence is constantly in motion.

Wind energy incorporates the methods and technologies that aim to transform wind energy and its attributes into a sustainable system of storing and utilizing its full potential.

Innovative technologies and the strategic utilization of renewable energy segments enhances a nation's energy mix. Especially for nations that are dependent on foreign imports and nations whose fossil fuel reserves will eventually be depleted, the ability to explore domestic energy segments will help it defend against supply disruptions and thus enhance its leveraging position. For nations to achieve their energy security goals and to eliminate increasing reliance upon foreign energy, exploring renewable systems can offer substitute energy sources as part of a nation's diversification strategy.

While the current renewables industry greatly relies on government subsidies, energy majors launch innovative research projects in an effort to contribute to the energy mix, increase energy capacity, and eventually create alternative energy production that is feasible from an economic and operational perspective.

History

Wind energy is one of the oldest energy modes used, both to power windmills and sail ships.

The world's most ancient windmill was discovered in Persia (modern-day Iran) and extends back to 1000 BC. It was designed with a water pumping mechanism and would also mill grains. A hub and spoke system was designed with a set of wedges that were tied alongside one another to create vertical oars that rotated around a "hub" or central point. Meticulously designed outside partitions made certain that the wind would be streamlined towards a particular direction and activate the windmill system. A similar mechanism was utilized for millennia until the industrial revolution introduced modern power methods. Nowadays, wind blades and generator systems are used to generate electricity.

In the maritime industry, seafarers since prehistoric times employed sail ships to utilize the wind forces and discover new lands. Ancient civilizations of the Greeks, the Egyptians, the Chinese, the Scandinavians and other nations on most continents knew well the capabilities of wind power: Sail ships have existed since at least 2,500 years ago, as archaeological evidence reveals.

- The sail ship of the Egyptian Queen Hatshepsut is dated to around 1493 BC.
- Sail ships are well described in many ancient Greek epic poems, such as, for example, Homer's Iliad, which depicts the fleet of the Greek sail ships navigating to besiege Troy in around 1300 BC. Also, in Homer's Odyssey, describing the adventures of Odysseus right after the Trojan war, Aeolos, the God and Keeper of the Winds, gave Odysseus a sealed sack that confined all winds, except for the favorable West Wind that would take him home.

In the late 1800s, the new term "wind turbines" was used to describe electricity-producing windmills. In 1888 Charles F. Brush employed the world's largest windmill to produce electrical power in Cleveland, Ohio. Later on, Brush Electric Co. was acquired by General Electric.

The year 1973 was a landmark in the use of wind energy and other alternative energy sources to mitigate the high petroleum prices caused by the Organization of Petroleum Exporting Countries (OPEC) oil embargo.

By 1990, the State of California hosted over 50% of the global wind capacity, i.e., exceeding 2,200 megawatts.

The Energy Policy Act of 1992 required the expansion of the energy mix and the implementation of optimum energy efficiency policies. It also ratified a production tax credit for electricity produced by wind power at the price of 1.5 cents per kilowatt hour.

The Energy Policy Act of 2005 reinforced the commercial and financial enticements for wind power and other renewable energy segments (EIA, 2008).

Commercial and Economic Development

By 2017 wind energy production has become more elaborate, with extensive offshore and onshore installations. The average wind turbine consists of 8,000 parts. Nations like China, the US, Germany, India, Spain and the UK heavily invest to increase their wind energy capacity (EIA, 2016–2017).

The US enjoys optimum results in terms of wind production: The US is number one globally in terms of wind power production, with the state of Iowa supplying over 31% of its electricity in 2015 (AWEA, 2016).

Energy penetration levels pertain to the ratio of electricity produced by a specific energy segment, and refers to the amount of power produced and ready for consumption. Denmark leads production and consumption with 40% of the global wind power penetration levels, Uruguay, Portugal and Ireland follow with at least 20% each, while Spain and Cyprus approach the 20% production, and Germany is at 16%. Large global economies such as Canada (6%), USA (5.5%), and China (4%) are heavily investing in wind power. According to the Global Wind Energy Council (GWEC) rolling five-year forecast, almost 60 GW of new wind installations were built in 2017, which will rise to an annual market of about 75 GW by 2021, to bring cumulative installed capacity of over 800 GW by the end of 2021 (GWEC, 2016).

According to GWEC, global development will be driven by the Asian continent, with China and India heavily investing in new installations. These countries are promising wind power players, as they combine a vast land area, high population and high economic growth that is spread throughout other Asian trade partners (GWEC, 2017).

Table 7.1 and Figure 7.1 demonstrate the rapid growth in wind energy installation within a decade, i.e., from 2007 to 2016.

The Principle of Mass Continuity

The principle of mass continuity says that over a given time and space, mass can never be generated or destroyed, hence an air oversupply cannot be accumulated in a particular

Table 7.1 Top nations'* wind power capacity (MW).

	Country	*2007*	*2016*
1	China	5,912	168,690
**2	European Union	56,614	153,730
3	United States	16,819	82,183
4	Germany	22,247	50,019
5	India	7,850	28,665
6	Spain	15,145	23,075
7	United Kingdom	2,389	14,542
8	France	2,477	12,065
9	Canada	1,846	11,898
10	Brazil	247	10,740
11	Italy	2,726	9,257

Notes:

* The table demonstrates the nations with the highest total cumulative installed electricity generation capacity from wind power, estimated in megawatts (MW).

** The EU values pertain to aggregate levels for all EU nations, yet individual EU member states like Germany, Spain, France and Italy are also included in the table as top producing nations.

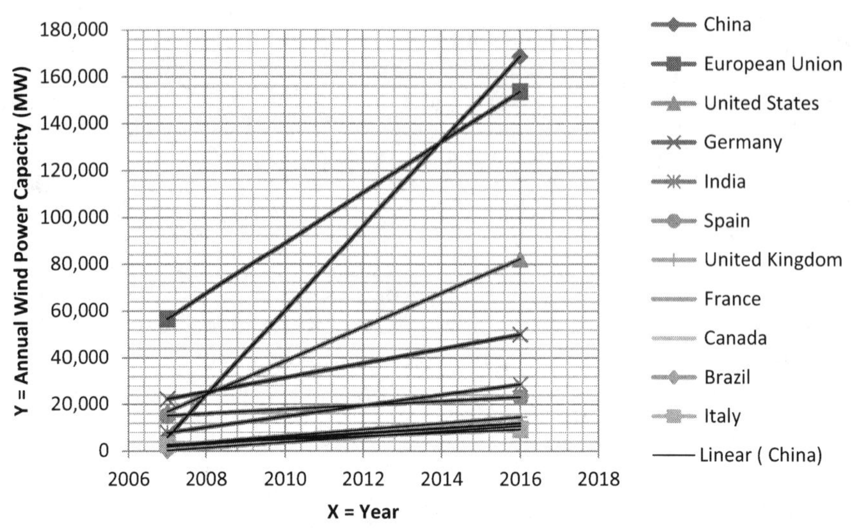

Figure 7.1 Top nations'* wind power capacity (MW).

Sources: Produced by the Author, based on data from: EIA (2007–2017a) and EIA (2007–2017b).

Note:

* The figure demonstrates the nations with the highest total cumulative installed electricity generation capacity from wind power, estimated in megawatts (MW).

space. As air is circulated, its pressure gets higher. As it soars, its temperature gets lower and condensation starts to surpass evaporation, therefore the undetectable steam condenses, creating clouds and subsequently rain. That is why there is often rainy and windy weather in close proximity to low pressure regions. The wind's velocity is specifically proportional to the pressure slope, suggesting that, as the pressure slope rises, so the wind's velocity likewise rises at that particular geographical location.

Wind Gradients: Divergence and Convergence Zones

The pressure gradient is defined as the atmospheric force variation among low and high pressure locations that are called isobars.

Compression fields are generated when the direction of the high and low pressure fields with air particles are converging, i.e., are coming close together. This is when we have a strong pressure gradient and strong winds are generated.

As the wind rotates inward to some degree, in the direction of the lowest pressure point, air converges, i.e., comes closer towards the center of the low pressure, closer to the earth's surface. Considering that the converging atmospheric flow has no place to move, it soars, and, as the air moves higher, the liquid vapor it contains reaches lower temperatures and gradually turns into rain. For this reason, low pressure locations are typically linked with atmospheric ambiance and challenging weather, i.e., windy, cloudy and rainy weather.

The opposite occurs when the highs and lows of air particles are diverging, i.e., spread apart while in motion, hence generating weaker winds.

Rarefaction fields are generated when there is a weak pressure gradient, i.e., weaker winds occur. Diverging causes air to move apart from the earth's surface, and causes air from higher atmospheric levels to come down in order to fill in the empty space generated at the lower levels. Consequently, air submerges, the velocity of the wind is mild, and the temperature rises, hence clouds will tend to evaporate. That is why good weather is typically linked with high pressure (NOAA, 2017).

Global Wind Energy

The world's greatest wind turbines produce sufficient electricity to supply hundreds of homes. Wind farms have dozens, and frequently hundreds of wind turbines strategically aligned in windy locations. Medium and small-sized wind turbines set up in an outdoor property can generate adequate electricity for an individual residence or business.

The world's greatest wind turbines have a height of 600 to 720 feet (180–220 meters), their blades have a length of over 250 feet (70 meters) and they can generate over 8 megawatts (MW) of electrical power, i.e., an amount sufficient to power thousands of homes, since one MW can power 1,000 homes. Modern-day wind farms consist of dozens of wind turbines, built onshore or offshore. Nations like China, the UK and Western Europe have subsidies and other financial incentives that encourage the development of wind farms.

Modern wind technologies entail wind turbines, where the wind force spins two or three propeller-like rotor blades. Wind turbines function as a reverse cooling fan mechanism. As opposed to employing electricity to generate wind as a cooling fan does, wind generators operate in a vertical axis and utilize wind to generate electricity. As the wind rotates the blades, it revolves a shaft, which is attached to a power generator and hence electricity is generated (DOE, 2017a).

Technologies and the Supply Chain

Wind is defined as air in movement, and yet, in fact, it is the solar power and fluctuations in temperature that contribute to the velocity and intensity of winds. As the solar rays increase the temperature in the earth's surface area, it causes hot air to rise and cool air to submerge the empty space in the areas closer to the surface. The modern wind energy technologies transform the wind's kinetic energy into mechanical energy, as the wind causes the turbine blades to rotate. Subsequently, the generator installed within the portal of the wind turbine construction transforms the captured energy into electricity.

The Supply Chain of Wind Energy

The supply chain of wind energy, as depicted in Figure 7.2, entails the production and use of raw materials such as metals, fiberglass, rubber and concrete, for the development of the infrastructure and superstructure of wind farms. The component manufacturing stage entails the assembly and installation of the blades, nacelle, towers and generator. It is worth noting that a single wind turbine requires the assembly of about 8,000 parts.

The study of each wind farm, both offshore and onshore, requires a unique windmill configuration, size of blades, type and number of units. The stages of land leasing, logistics and construction are conducted almost concurrently, and are always project-specific. Finally, the stages of operations, maintenance and repairs take place. Operations are ongoing, whereas maintenance and repairs are classified as planned vs dynamic maintenance systems: Planned maintenance systems signify the annual scheduled maintenance and repairs of the system, whereas dynamic maintenance systems pertain to systemic breakdown and the need for imminent, dynamic intervention.

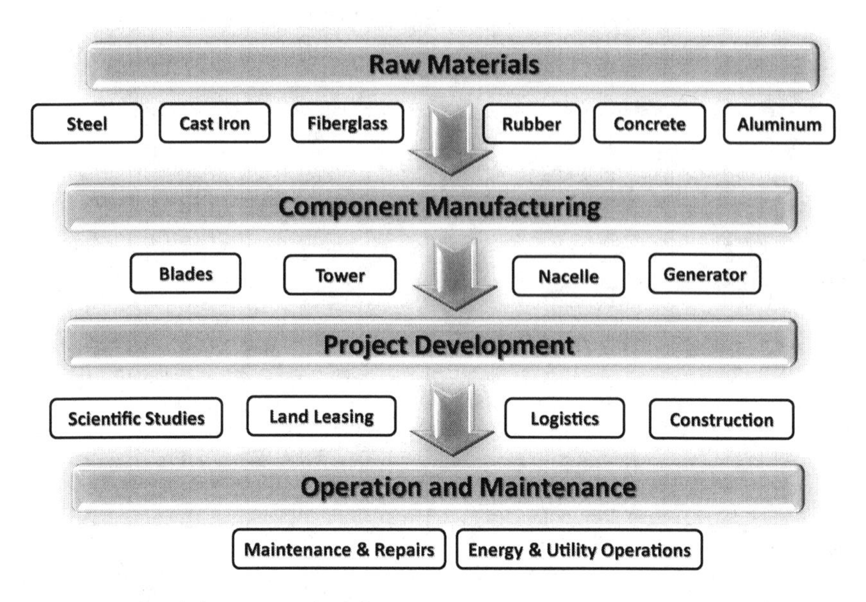

Figure 7.2 The wind energy supply chain.

Source: the Author, based on the US Bureau of Labor Statistics (BLS, 2017).

A wind farm network system links turbine towers with a voltage energy storage network, facilitating the transfer of energy upon demand to regional and some times to remotely located markets. Offshore wind systems require a sub-sea network management system. The installation of state-of-the-art systems may be an easy task for nations or regions that are recently building their wind energy network. Nevertheless, capacity issues and an aging network are common issues for locations that already have in place obsolete technologies with insufficient capacity and that require new transmission lines.

Hence, the building of a power system needs to take into consideration demographic information such as population density per region, and also government/state funding per region.

Furthermore, the network expansion over a long distance is another challenge: There is a direct correlation between the extension of energy transmitting cables and pertinent damages and losses. Any system deficiencies that are minimal over a short distance can become exacerbated as larger power levels are now transmitted via extensive networks spread over increased mileage.

Wind Turbine Types

Contemporary wind turbines belong to two primary categories:

1. the Horizontal-Axis Wind Turbine (HAWT) type, resembling the standard farm windmills, and;
2. the Vertical-Axis Wind Turbine (VAWT) style, i.e., the Darrieus model, that was given the name of its French inventor, and resembles a gigantic eggbeater.

The majority of sizeable wind farms are using horizontal-axis turbines. These possess two or three bladed wind turbines that are installed to function upwind, i.e., functioning when the blades are directly facing the wind.

Wind turbines may be installed onshore, i.e., on land, or offshore, i.e., in any water body like the ocean, coastlines or lakes.

Wind Turbine Sizes

Power turbines vary in dimensions and specifications from as small as 50 kilowatts to as sizeable as numerous megawatts.

Bigger wind turbines can benefit from economies of scale, hence can generate more energy at lower cost. Large-scale wind farms contain several turbine units arranged in a configuration in the landscape, hence feeding the electrical grid with sizeable energy volume. The largest-sized equipment generates adequate electricity to power 1,400 households.

Offshore units consist of wind turbines of greater size, that have enhanced power-production capacity and have limited transportation complications, as the systems are transported by ships. In contrast, the transportation of land-based wind systems can be quite complex, due to the weight and size limitations imposed by land infrastructure when transported on highways by trucks, or on rail by trains.

Wind turbines are found in a range of sizes and energy efficiency classifications. The turbines of a modest residence-sized wind equipment are sized at around 50 to 750 kilowatts.

Individual compact turbines of 100 kilowatts or less are typically employed for household utilities, satellite dishes, water pump and pressure tank systems. Furthermore, compact turbines may be utilized as part of hybrid or dual energy systems, to feed diesel generators, photovoltaic installations and power packs. While large-scale hybrid systems utilizing wind power are still in an experimental stage, smaller, stand-alone power system units or mini-grids are found in isolated rural areas, with limited access to the main grid.

Wind Turbine Size and Power Ratings

Wind turbine size and power ratings can determine the power production ratings. Wind resources are described as wind-power density classifications, which vary from class 1 (for minimum power density) to class 7 (to the optimum power density).

Table 7.2 Classes of wind power density at 10 m and 50 m(a).

*Wind power class**	*10 m (33 ft)*		*50 m (164 ft)*	
	Wind power density (W/m2)	*Speed(b) m/s (mph)*	*Wind power density (W/m2)*	*Speed(b) m/s (mph)*
1	100	4.4 (9.8)	200	5.6 (12.5)
2	150	5.1 (11.5)	300	6.4 (14.3)
3	200	5.6 (12.5)	400	7.0 (15.7)
4	250	6.0 (13.4)	500	7.5 (16.8)
5	300	6.4 (14.3)	600	8.0 (17.9)
6	400	7.0 (15.7)	800	8.8 (19.7)
7	1000	9.4 (21.1)	2000	11.9 (26.6)

(a) Vertical extrapolation of wind speed is founded on the 1/7 energy law.

(b) Mean wind speed is founded on Rayleigh speed distribution of corresponding mean wind energy density.

Source: RREDC.NREL (2017) Wind Energy. US National Renewable Energy Laboratory. Renewable Resource Data Center. Available at: http://rredc.nrel.gov/wind/pubs/atlas/tables/1-1T.html, last accessed: June 21, 2017.

Sufficient wind resources (starting from level 3 and over, with an average yearly wind velocity of 13 miles per hour and above are observed in several regions.

For a wind turbine to generate an optimum energy volume, its blades must be designed in a manner that captures wind streams from a large space. To achieve this, the industry increases the size of wind blades, wind structures and wind turbines. As a consequence, a significant amount of weight and structural pressure is placed on the construction.

Torque Efficiency

Torque, or "the moment of force" or "the twisting moment," is a way of measuring the amount of energy generated when a twisting force hits an object (e.g., the wind hits the blades), causing its rotation. The spinning point is referred to as the rotation axis. Torque is measured in Newton-meters. It is mathematically symbolized as

$$\mathbf{T = F * r * \sin \Theta}$$

Where:

- F = Force applied
- r = Moment arm, or the distance from rotation axis to the force pressure point
- Θ = the angle that force creates with the axis of the lever arm. The lever arm is the perpendicular distance from the rotation axis, to the force's action point.
- Torque is maximum when the Θ angle is 90°

Finally, it is worth noting that torque is a vector quantity, where clockwise torques are negative, whereas counterclockwise torques are positive.

Figure 7.3 depicts the torque efficiency and its impact on wind energy.

Turbine Configurations

- **Wind turbine rotor or blade**, transforming the wind's kinetic power into torque, also known as spinning shaft power. The blade design will determine its efficiency, i.e., the blade's ability to generate high amounts of energy. Hence, blades need to be large, yet with an efficient engineering configuration, and an optimum design. Blade design can be flat or curved. Flat, wide blades are the oldest types, easy to manufacture and use, and yet are the least energy-efficient: Their design generates a slower rotation, as the blades are pressing against the wind, hence constraining the energy output. Curved blades are characterized by an "aerofoil shape" that increases energy production as the wind moves more speedily on the curvy top area of the blade, compared with the flat bottom area. This aerodynamic design improves efficiency.
- A **drivetrain** is typically comprised of all the essential elements required for a turbine to generate electricity, i.e., a generator and a gearbox. The gearbox is designed to attach the low-speed base to the wind generator blades, and the high-speed base connected to the generator (DOE, 2017b).
- **The tower** is designed to reinforce the turbine's structure, drivetrain and rotor. It is typically produced from tubular stainless steel, or steel lattice and concrete. Based on the premise that height and wind speed are directly proportionate, there is a trend to build more elevated towers with wider turbine blades, in order to generate more power and produce more electricity (DOE, 2017c).

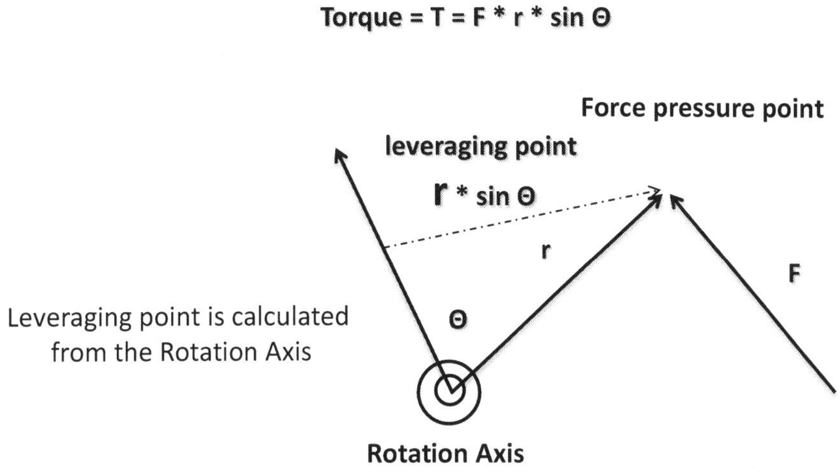

Torque = T = F * r * sin Θ
Force pressure point
leveraging point
r * sin Θ
r
F
Θ
Leveraging point is calculated from the Rotation Axis
Rotation Axis

Figure 7.3 Wind energy, torque.

Source: the Author.

Wind
Energy
Nacelle: includes the generator, gear box and shaft
Hub
Blade
Tower
Transformer
May be within tower

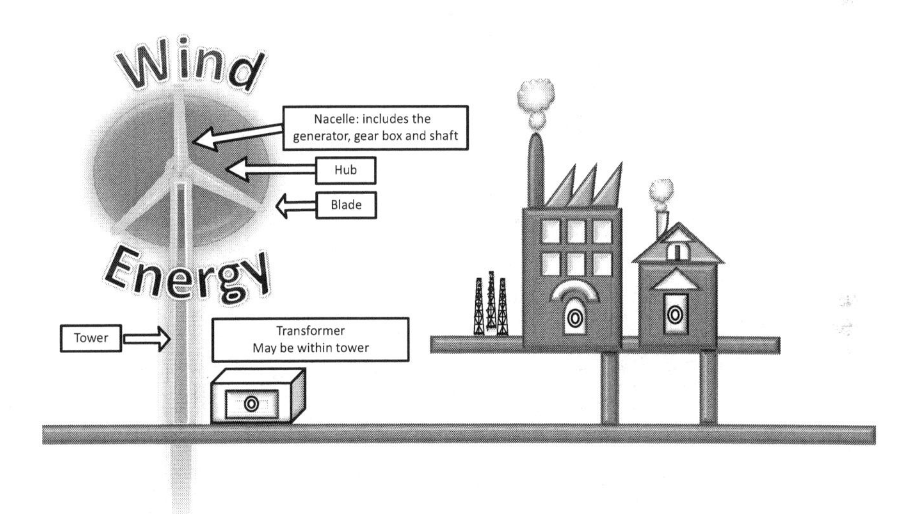

Figure 7.4 Wind energy: The tower.

Source: the Author.

- **The gear box** is a costly component with a bulky construction within the turbine. Science and technology innovators aim to develop "direct-drive" power generators that operate at reduced spinning rates, hence gear boxes will not be needed (ANL, 2017a).
- **Additional gear** includes interconnection devices, controllers, ground support gear, electrical wires, and so on (ANL, 2017b).

Pitch and Yaw Control

Horizontal-Axis Wind Turbine (HAWT)

The HAWT systems entail larger wind turbines that require rotor brakes, yaw control brakes and blade pitch control brakes, i.e., pitch is the blade angle adjustment and yaw is turbine rotation.

The design of a HAWT requires directional orientation, i.e., they must be facing the wind as a prerequisite for the blades to swivel.

The drawback of a Horizontal-Axis Wind Turbine (HAWT) system is its bulkier construction, and the inability to perform effectively in gales, whirlwinds, and other turbulent weather conditions.

Vertical-Axis Wind Turbine (VAWT)

By comparison, a VAWT system has a lean design with fewer parts. The rotor blades of a conventional VAWT system still capture wind energy regardless of their orientation, thus offering optimum, consistent productivity and increased reliability.

The drawback of a Vertical-Axis Wind Turbine (VAWT) is that it can't perform well under extreme weather conditions, due to the perpendicular location of the blades, i.e., the rotational axis is vertical with respect to the ground (DOE, 2017a).

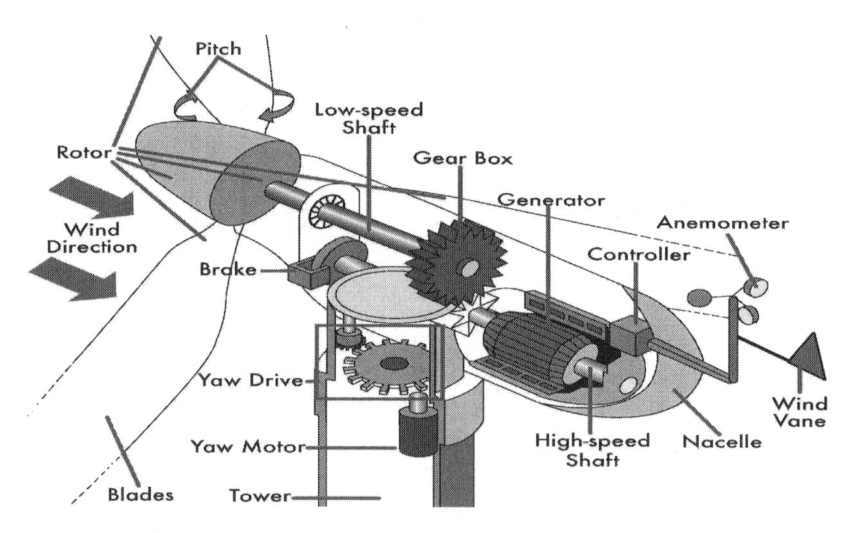

Figure 7.5 Horizontal wind turbine.

Source: ANL, 2017a.

Subsidy Classifications

Power subsidies and incentives are classified into five distinct program types:

1 **Direct expenditures to producers or consumers**. These are federal government plans that offer direct cash offerings that offer a monetary benefit to energy consumers or producers.
2 **Tax expenditures**. These are mainly conditions stipulated in the Internal Revenue Code (IRC, or Tax Code) – Title 26 of the United States Code – that decrease the tax liability of companies or persons who take particular measures that impact energy generation, circulation, sharing, usage, or storage.
3 **Research and development**. The federal government possesses a comprehensive system of financing energy research and development (R&D) pursuits targeted at a selection of objectives, for instance expanding US energy resources or increasing the performance of different energy technologies used for consumption, production, conversion, and ultimate consumer systems. While these programs do not directly impact present energy generation, consumption, and price levels, their effective implementation may impact the future energy market.
4 **National electricity plans funding government and remote utilities**. The national energy strategy introduces to the marketplace large volumes of electrical power, stating that "public bodies and cooperatives" will have priority in the sales transaction. The US Department of Agriculture's Rural Utilities Service (RUS), as an instrument of the US government, facilitates components of the electric energy sector by means of loans, loan warranties and low interest rates.
5 **Loans and loan guarantees**. The US government offers monetary support for specific energy systems either by ensuring the loan repayment or by financing energy industry entities directly. The US Department of Energy (DOE) offers monetary support for progressive clean energy systems with limited atmospheric or anthropogenic emissions in cases where these technologies are classified as high-risk, and thus are not entitled to private funding (EIA, 2015).

Regulations and Subsidies

Wind energy in several nations obtains monetary and alternative incentives such as production tax credit (PTC), including property tax exemptions, ruled acquisitions, and additional marketplaces for "green credits."

- The Energy Improvement and Extension Act of 2008 stipulates for wind energy certain credit expansions, such as for micro-turbines. Nations like Canada and Germany offer rewards for the building of wind turbines, including tax breaks or feed-in tariffs establishing prices over the average rates.
- Household wind generators that are linked to the grid might take advantage of grid electricity storage capacity, hence swapping acquired electrical power with regionally created energy when accessible. The excess energy generated by modest household turbines can, in certain areas, be supplied into the grid, for the energy company to buy and re-sell, hence providing an energy credit for the household entities to cover their energy expenses.

Benefits, Risks, and Recommendations

Benefits

Sizeable wind turbines produce electricity at a minimum cost and optimum efficiency compared with more compact turbines, for two reasons:

1 lengthier rotor blades can obtain energy from a greater rotor-swept area, i.e., wind cross-section.
2 More elevated wind towers are typically exposed to stronger winds, hence they have the ability to capture and produce more energy, with increased consistency and continuity of production.

Wind energy represents a clean renewable energy source that generates limited environmental emissions or damage. While the initial turbine installation cost represents the highest investment needed for power generation, there is minimum operational cost for the natural generation of wind. Mass construction and technological innovations contribute to a low-cost wind energy. Moreover, several nations provide subsidies and tax incentives to promote the advancement of wind energy (DOE, 2017d).

Environmental Benefits

Wind power is a natural, renewable energy source, hence of limited cost and endless power supplies. In contrast to traditional energy plants, wind farms generate green, clean, environmentally friendly electric power. The long-term impact on the atmosphere is much less challenging than that of non-renewable energy sources (ANL, 2017c; DOE, 2017d).

Risks

Environment

When wind turbines are installed near residential areas, the blades can be a source of noise pollution. Aesthetic pollution is also possible, when the wind turbines alter the scenery.

Offshore wind farms have less aesthetic and noise effects. Moreover, they offer continuous, hence more dependable and higher generating capacity compared with onshore facilities.

The blades' movement in land-based wind power can be harmful to birds, bats and their habitats, but not nearly as much as do cars, power lines, and high-rise buildings. Despite the fact that wind power facilities have diminished ecological impact as opposed to fossil fuel facilities, the majority of these issues have been remedied or reduced tremendously by means of technological evolution or by effective site selection.

Accessibility

Wind turbines are automated, unmanned systems that function remotely, frequently in isolated regions, resulting in high maintenance and repair costs.

Safety

The electro-mechanical equipment in HAWT structures is situated high up, thus requiring employees to work aloft. This increases the budget allocated in maintenance costs, including compensation for hardship and working aloft payments, as well as the need for additional Personal Protection Equipment (PPEs).

On the other hand, VAWTs are considered as a safer configuration, as the majority of these wind generator systems are situated at a surface level. This improves safety standards.

Investment

Although in the past decade or so the installation investment has decreased substantially, the technological innovation necessitates a larger primary investment compared with fossil-fueled generation devices. Approximately 80% of the wind turbine investment is in the equipment installation.

Wind harvesting includes a configuration of several wind turbines that are linked to the electric power grid. Onshore and offshore wind farms represent affordable power sources that can easily compete with other energy sources in terms of cost. However, the set up of offshore wind turbines entails higher installation, assembly and repairs and maintenance expenditure, compared with onshore turbines.

Supply Sustainability

The principal concern in selecting wind as an energy resource is that the frequency and intensity of the winds are irregular, non-sustainable, and hence cannot produce a pre-determined energy output. Furthermore, while wind-produced electrical power may be stored, wind energy in its primary form cannot be saved. In addition, superior wind farms are frequently situated in isolated regions, distant from big cities and regions where the buyers' markets are located.

Competition

Land usage is a significant input in the onshore wind energy industry, i.e., it is a key factor of production. Wind energy companies may have to compete with other energy segments for the utilization of this land and, on many occasions, the usage of land by other forms of energy production may generate higher output compared with wind energy. On the other hand, offshore wind turbines are not subject to competition, and modern wind farms onshore can always share land usage with agricultural activities.

Hence, a future challenge and opportunity for the wind energy industry is to be counted upon as a reliable energy supply source (ANL, 2017c; DOE, 2017d). However, modest onshore wind harvesting can supply the grid, as well as remote off-grid regions.

The output of wind energy is rather continual on an annual basis, yet varies greatly over daily, weekly or monthly time scales. Consequently, in order to provide a dependable energy supply, it must be utilized in conjunction with other energy sources. As the wind power activities expand in a particular location, there is a necessity to enhance the grid.

Recommendations

Improve designs with performance optimization – requirements: Higher cycle rates, higher loads, optimum dependability.

To address risks and challenges, energy experts need to implement energy management strategies including:

- the storage of surplus capacity;
- geographically dispersed turbines;
- dispatching ability of alternative energy sources (i.e., the ability to switch on and off the alternative supply);
- the ability to convey the captured capacity to nearby regions, but also import energy from these regions as needed;
- the utilization of Vehicle-to-grid (V2G) solutions, i.e., a technique where plug-in electric powered automobiles, for instance Battery Electric Vehicle (BEV), Battery-Only Electric Vehicle (BOEV), Full Electric Vehicles (FEV) and Plug-in Hybrid Electric Vehicles (PHEV), contact the power grid station to inquire into the sale or purchase of power.

The ability to cut down production when demand is low can, on most occasions, alleviate any energy supply issues.

Moreover, weather predicting enables the power supply chain to be prepared to adjust the power supply when fluctuations or disruptions are expected.

Capacity Factor

The capacity factor represents the ratio of a generator's yearly energy output in a specific site, divided by the power it may possibly have generated if its output was at 100% capacity. This is a valuable way of determining the generator's potential to produce close to its theoretical maximum, while at the same time optimizing the return on investment ratio.

Considering that wind input is not consistent, a wind tower's energy output can never reach its optimum capacity, i.e., the sum of the power generator's installed capacity multiplied by the aggregate hours per annum. In wind energy, optimum capacity factors are attained in advantageous locations, and as a consequence of design upgrades in the wind generator towers.

National power grids and transmission networks are designed in a manner that mitigates energy supply and demand fluctuations, such as plant production interruptions and blackouts, as well as increased energy demand, through the allocation of auxiliary power sources. However, in the wind energy segment the "production on demand" arrangement may not be applicable, as wind forces depend on weather conditions and cannot provide a consistent performance. On the other hand, efficient power generation plants may be able to deliver their nameplate capacity around 95% of the time.

Wind Energy Partnerships

From 1974 to 1987, the National Aeronautics and Space Administration (NASA) Glenn initiative led the US Wind Energy Program to further develop large wind horizontal-axis turbines, i.e., the prevailing technologies currently utilized to convert wind power into mechanical power. Utilizing funds from the National Science Foundation and the US Department of Energy, NASA built and activated the first experimental 100-kilowatt wind turbine technology in Cleveland, Ohio.

Among other noteworthy projects, NASA has invented 13 experimental wind turbines, which included four principal designs:

1. the MOD-0A (200 kilowatts);
2. the MOD-1 (2 megawatts, the first turbine in 1979 over 1 megawatt);
3. the MOD-2 (2.5 megawatts);
4. the MOD-5B (3.2 megawatt), which was the largest wind turbine operating in the world — with a rotor diameter of nearly 100 meters (330 feet) and a rated power of 3.2 megawatts (Office of Science, 2017).

Modern wind farms, both onshore and offshore, have developed in terms of technologies and practices. The future of wind energy seems to be appealing, as new technologies and large-scale wind farms are developed in onshore and offshore sites. The next stage would be to generate sustainable, reliable, and affordable energy production.

References

American Wind Energy Association (AWEA) (2016). *U.S. Number One in the World in Wind Energy Production. Wind Supplied Iowa with Over 31 Percent of its Electricity Last Year*. American Wind Energy Association, February 29, 2016. Available at: www.awea.org/MediaCenter/pressrelease.aspx?ItemNumber=8463, last accessed: June 12, 2017.

Argonne National Laboratory (ANL) (2017a). *Upper Greater Plains, Wind Energy. Horizontal Turbine Components*. The Western Area Power Administration and the US Fish and Wildlife Service. The Upper Great Plains Wind Energy Programmatic EIS.NREL. Public domain. Available at: www.plainswindeis.anl.gov/guide/basics/index.cfm, last accessed June 21, 2017.

Argonne National Laboratory (ANL) (2017b). *Wind Energy Development Programmatic EIS*. Bureau of Land Management. Available at: http://windeis.anl.gov/guide/basics, last accessed June 21, 2017.

Argonne National Laboratory (ANL) (2017c). *Wind Energy Basics. Wind Energy Guide. Wind Energy Development*. Programmatic EIS. US Department of Energy. Available at: www.windeis.anl.gov, last accessed: June 21, 2017.

Bureau of Labor Statistics (BLS) (2017). *The Wind Energy Supply Chain, Careers in Wind Energy, Diagram 1*. US Bureau of Labor Statistics. Available at: www.bls.gov/green/wind_energy, last accessed: June 21, 2017.

Department of Energy (DOE) (2017a). *How Wind Turbines Work*. US Department of Energy. Available at: https://energy.gov/eere/wind/how-do-wind-turbines-work, last accessed: June 21, 2017.

Department of Energy (DOE) (2017b). *Advanced Drivetrain Manufacturing*. DOE, Wind Energy Technologies Office. Available at: https://energy.gov/eere/wind/advanced-drivetrain-manufacturing, last accessed: June 21, 2017.

Department of Energy (DOE) (2017c). *The Inside of a Wind Turbine*. DOE, Wind Energy Technologies Office. Office of Energy Efficiency and Renewable Energy. Available at: https://energy.gov/eere/wind/inside-wind-turbine-0, last accessed: June 21, 2017.

Department of Energy (DOE) (2017d). *Advantages and Challenges of Wind Energy*. US Department of Energy. Available at: https://energy.gov/eere/wind/advantages-and-challenges-wind-energy, last accessed: June 21, 2017.

Energy Information Administration (EIA) (2008). *Energy in Brief: How Much Renewable Energy do We Use?* August, last revised: September 2008. US Energy Information Administration. Available at: www.eia.gov, last accessed: June 21, 2017.

Energy Information Administration (EIA) (2015). *Direct Federal Financial Interventions and Subsidies in Energy*. US Energy Information Administration. Available at: www.eia.gov/analysis/requests/subsidy, last accessed: June 21, 2017.

Energy Information Administration (EIA) (2007–2017a). *Annual Reports, International Electricity Generation Data*. US Energy Information Administration. Available at: Eia.doe.gov, last accessed: June 12, 2017.

Energy Information Administration (EIA) (2007–2017b). *International Energy Statistics*. US Energy Information Administration. Available at: www.eia.gov/beta/international, last accessed: June 12, 2017.

Energy Information Administration (EIA) (2016–2017). *Electric Power. Data for 2016 & 2017*. US Energy Information Administration. Available at: www.eia.gov/electricity/monthly, last accessed: June 12, 2017.

Global Wind Energy Council (GWEC) (2016). *Global Wind Report, 2016*. Global Wind Energy Council. Available at: http://gwec.net/publications/global-wind-report-2/global-wind-report-2016, last accessed: June 12, 2017.

Global Wind Energy Council (GWEC) (2017). *Wind in Numbers*. Global Wind Energy Council. Available at: http://gwec.net/global-figures/wind-in-numbers, last accessed: June 12, 2017.

NASA Glenn (2008). *NASA Glenn Responds to 1970s Energy Crisis*. NASA Glenn Research Center. April 10. Available at: www.nasa.gov/centers/glenn/about/history/70s_energy.html, last accessed: June 21, 2017.

National Oceanic and Atmospheric Administration (NOAA) (2017). *Origin of Wind*. National Oceanic and Atmospheric Administration, National Weather Service, US Dept of Commerce. Available at: www.srh.noaa.gov/jetstream/synoptic/wind.html, last accessed: June 21, 2017.

Office of Science (2017). *Sample Records for Mod-0 Wind Turbine*. Records by U.S. Office of Science, based on research by National Aeronautics and Space Administration (NASA), Department of Energy (DOE). Available at: www.science.gov/topicpages/m/mod-0+wind+turbine.html, last accessed: January 20, 2019.

8 Hydroelectricity and Ocean Energy Security

> We have many a monument of past ages . . . In them is exemplified the power of men, the greatness of nations, the love of art and religious devotion. But the monument at Niagara has something of its own, more in accord with our present thoughts and tendencies. It is a monument worthy of our scientific age, a true monument of enlightenment and of peace. It signifies the subjugation of natural forces to the service of man, the discontinuance of barbarous methods, the relieving of millions from want and suffering.
>
> Nikola Tesla, Speech at Niagara Falls Opening Ceremony, January 12, 1897 (Tesla Society, 2019).

Hydropower: An Overview

The oceans hold vast potential for sustainable, dependable, and cost-efficient energy production. Ocean energy has tremendous capabilities to create different types and large volumes of alternative energy, bearing in mind that over two thirds of Planet Earth consists of water. There are several ways of harnessing the ocean currents, ocean winds, and salinity gradients, (i.e., the use of pressure-retarded reverse osmosis technology and conversion systems), but also through tidal, wave, and ocean thermal energy (BOEM, 2017a; BOEM, 2017b).

A hydroelectric dam produces energy by gathering water from a river's flow, which is regulated by dam operations. Dams lead to water reservoirs that can also have multiple functions, such as drinking water tanks, recreation, or wildlife sanctuaries.

Hydropower pertains to energy generated from the flow of water and employed in the production of electricity. The power of water has been utilized for hundreds of years, where kinetic energy was converted into mechanical energy. Water mills were common in ancient civilizations as water power was used to mill wheat and grains. Water wheels were installed in waterways to collect circulating water in containers positioned around the steering wheel.

Hydropower History

Throughout the history of mankind there have been systematic efforts to harness significant accumulations of waterbodies. Large rivers like the Nile in Egypt or the Tigris and Euphrates in Mesopotamia and the Middle East needed to be controlled and their power utilized for agricultural purposes.

- The first flood control findings globally were found in Mesopotamia (meaning "land among two rivers" in ancient Greek). Ancient dam construction was found in Mesopotamia (i.e., modern-day Iraq and a small part of Iran) as a means of preventing frequent flooding incidents. It is worth noting that flooding and change of water course was encountered most frequently in the Tigris and Euphrates rivers. This is because these rivers contained higher levels of silt compared with any other river globally, including the Nile in Egypt.

- The most ancient systems of irrigation were designed in Mesopotamia and Egypt as a means of preventing floods, controlling surface water, and cultivating land. Irrigation systems greatly enabled these ancient civilizations to flourish, and prevented them from vanishing.
- The Jawa Dam in Jordan was built in 3000 BC, hence it is considered as the oldest dam in the world. It consists of a series of gravitational dams with a stone wall 30 feet high (9 meters), and 3.5. feet wide (over 1 meter). The meticulous architecture prevented water pressure from destroying the wall with rock-fill substances behind the upstream wall.
- The Sadd-el-Kafara Dam ("Dam of the Infidels") was built in 2500 BC in Ancient Egypt, situated almost 16 miles (25 kms) from Cairo. The dam was built to protect the region from flooding, and was 364 feet long (111 meters), 46 feet high (14 meters) with a crest of 187 feet wide (57 meters), and a base width of 321 feet (98 meters). It was actually ruined by a flood during its construction, as presumed by the erosion evidence found in its underdeveloped downstream side. The lack of a canal or ditch prevented the diversion of water and the absence of silt or other sediment blueprint confirms that the dam was ruined before its construction was completed.
- Ancient Yemen was well known for its water engineering marvels. The Great Dam of Marib was built in 800 BC and was considered as one of the emergency response miracles of ancient architecture. The dam's length of 2,100 feet long (650 meters) was two times greater than the Hoover Dam, and it had a height of 50 feet (15 meters). The Marib Dam's elaborate construction of mudbricks was designed to accumulate the water and control its distribution as desired. The region has always been prone to heavy rainfalls (southwest monsoon) each year, especially from June to September. Hence, the gates would close to prevent the region from overflow, yet, when needed, its sophisticated irrigation system would use the vast volume of water to nourish regional farms expanding through 25,000 acres (i.e., 10,000 hectares).
- In the 11th century AD European "tide mills" were built in England, France, and Spain. Compact tidal dams were constructed through the coastline in small channels and waterways. Tidal waters were captured by the means of a water wheel, thus causing stones to grind wheat and grains.
- It is worth noting that the modern technologies did not change much: A contemporary dam still captures water power by the means of steel turbines to generate electrical power.

1759: Daniel Joncaire created a compact ditch or modest channel over the Niagara Falls to develop energy for his sawmill. Although Joncaire was not an inventor, he is the first person to capture the Falls' power in order to generate any kind of energy.

1770s: During the mid 18th century Bernard Forest de Bélidor, a French hydraulic inventor and military engineer, wrote the revolutionary book *L'architecture hydraulique*. In these four volumes published from 1737–1753, STEM principles, with focus on integral calculus and use of horizontal and vertical axes, were used to resolve hydraulic construction challenges (EIA, 2017).

1805: The Porter brothers, i.e., Augustus and Peter, acquired the Niagara Falls region to supply hydroenergy for their grist mill, saw mill, and tannery constructed alongside Joncaire's old ditch.

1853: An ambitious plan to generate electricity was materialized through the Niagara Falls Hydraulic Power and Mining Company (Niagara Frontier, 2017).

1879: During the 19th century electricity was created by hydropower. In 1879 the world's first hydroelectric energy plant was constructed at Niagara Falls. These are three waterfalls crossing Ontario, Canada, and New York, USA (Online Niagara, 2017).

1881: Jacob F. Schoellkopf led Niagara's first hydroelectric generating station and generated sufficient energy to light up the Niagara Falls and the nearby village.

Nikola Tesla with George Westinghouse developed the pioneering Niagara Falls' hydroelectric power plant, the first in the world to create consistent generation and distribution of electricity. This was the ultimate triumph of Nikola Tesla's Alternating Electric Current (AC) that surpassed Direct Current Power (DC) introduced by Thomas Edison.

1882: The first DC plant used wave power to generate power for a paper mill in Appleton, Wisconsin, on the Fox River.

1887: The first hydroelectric plant was established in San Bernadino, California.

1889: The first AC plant in Williamette Falls Station, Oregon, was established.

1893: General Electric built the first three-phase AC system for commercial use. This innovative system enabled energy to be transported through lengthier ranges and was adopted in a commercial setting in the first AC hydropower facility at the Redlands Power Plant in California, implementing a three-phase electrical generator to provide a reliable sustainable power supply through an energy transmission length of seven miles. The plant generated a 250 kW power output by utilizing water from a close-by Mill Creek through the use of water wheels.

In the same year the Austin Dam, America's first dam particularly developed to produce hydropower, was established in Austin, Texas.

1920s: Engineer Dexter P. Cooper proposed an ambitious plan of utilizing tidal energy in Passamaquoddy and Cobscook Bays, and industrializing the Maine region in the U.S. To promote his work, Mrs. Cooper approached their neighbor, Mr. Franklin D. Roosevelt, who had recently moved to the White House. After deliberation, the plan was rejected due to the high investment needed. The Great Depression in 1929 halted this ambitious plan, yet its seeds were planted in the contemporary tidal energy projects.

1931: Building commenced on the Boulder Dam, a hydropower plant to be subsequently called Hoover Dam. Its construction on the Colorado River involved the hiring over 20,000 employees throughout the Great Depression.

1936: Operations commenced at the Hoover Dam, developing over 130,000 kilowatts of electricity.

2017: China, Brazil, Canada, and the U.S. were the major countries producing almost half of the world's electricity from hydropower (NHA, 2017).

Commercial and Economic Development

A hydroelectric plant pertains to the generation and distribution of electricity created by utilizing the gravitational pressure of falling or flowing water, such as waterfalls and streams, to create hydraulic energy and electricity. Hydroelectric energy is the globe's major renewable energy form suitable for electricity conversion. It is the most dependable alternative energy segment, capable of mitigating both planned energy demand along with sudden and unforeseen demand generated from disruptions and shortages. Hydropower offers 71% of alternative sources of electricity and 16.4% of the global electricity from aggregate energy resources, attaining 1,064 GW of annual production (USGS, 2017a, 2017b; World Energy, 2016).

Hydropower is a significant renewable power type, signifying over 16% of aggregate power production. Numerous possibilities are available for the global advancement of hydropower, and the most prevailing sources verify the existence of over 10,000 TWh per annum of unused hydropower capabilities globally. Several dams were designed for other functions and additional hydropower infrastructure was constructed at a subsequent stage. The U.S. alone contains over 80,000 dams. Among these merely 3% or 2,400 dams generate electricity, and the remaining serve for tourism, reservoirs for entertainment, navigation, agricultural wetlands, flood management, irrigation, and community water system purposes (DOE, 2017b; DOE, 2017c).

Hydroelectric Dam Power

A dam is a constructed barrier or buffer designed across a body of water. Its purpose is two-fold: First, to obstruct the water flow and, second, to create a water reservoir that not only prevents flooding but, in addition, supplies water for irrigation purposes, domestic, industrial, and community usage, fish farming and maritime operations. Electricity is created through hydropower and dams' construction. The use of dams may also be extended for water accumulation, preservation, or distribution, and allocation of water for different purposes as required. Other constructions include:

- **Floodgates** may be opened or closed to handle inbound and outbound water movement. This is achieved by the means of embankments such as flood dikes, levees, and flood banks designed to avert flooding.
- **Flood barriers** protect the dam and the nearby region from overflow or outpouring.
- **Dikes and levees** are frequently created to restore submerged acreage and structures across a coastline or embankment. Additionally, they help water-stream controlling or protection of designated areas.

Types of Ocean Energy

Ocean energy includes several classifications such as wave, tidal energy, current, thermal, and salinity gradients, i.e., difference in salinity levels. Among these energy types, the most popular types include a) wave power, b) tidal power, c) current power, d) Ocean Thermal Energy Conversion (OTEC).

1. **Wave power** produces power from waves on the ocean surface, whereby floating equipment or buoys are utilized.
2. **Tidal power** gathers energy from the flow of tidal waters (high and low tides): As the water enters and exits coastal regions it runs through turbines (EPA, 2017).
3. **Current power**, in which marine energy is obtained from harnessing of the kinetic energy of marine currents, such as the Gulf Stream. While technologies still need to evolve for this energy segment, there is tremendous future potential for power generation (BOEM, 2017b).

4 **Ocean Thermal Energy Conversion (OTEC)** entails the use of temperature differences in the water (California Energy Commission, 2017).

Figure 8.1 depicts a typical hydroelectric dam, Figure 8.2 shows the configuration of wave power, Figure 8.3 demonstrates the arrangement of tidal power and Figure 8.4 demonstrates the arrangement of OTEC.

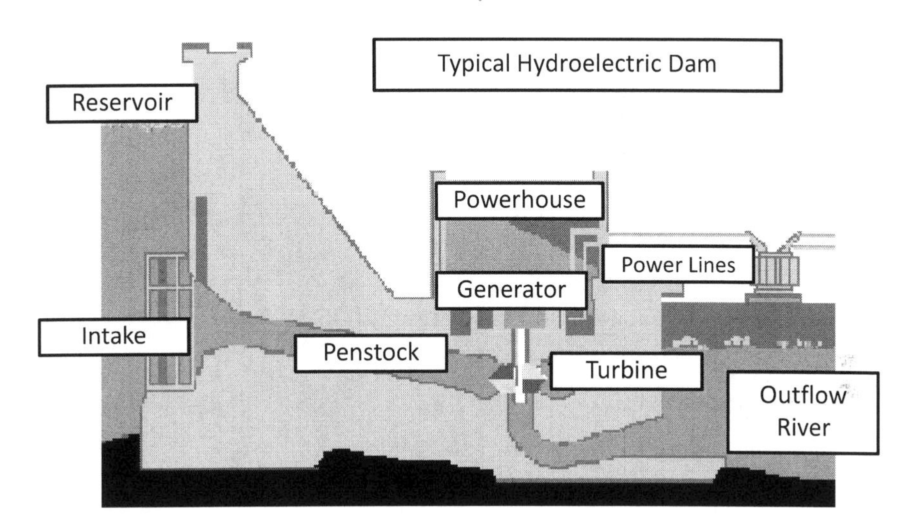

Figure 8.1 Hydroelectric dam.

Source: USGS, 2016. *Hydroelectric Power: How It Works*. The USGS Water Science School, U.S. Department of the Interior, U.S. Geological Survey. Available at: https://water.usgs.gov/edu/animations/hydrodam.gif, last accessed: December 15, 2016.

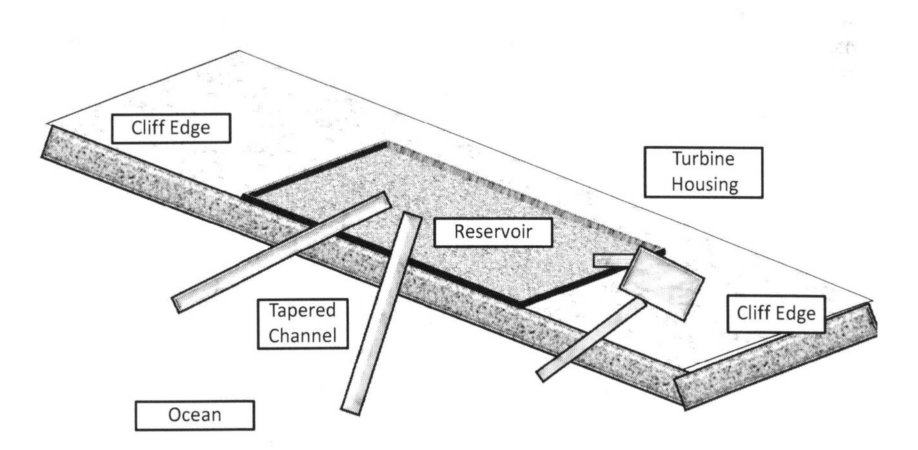

Figure 8.2 Wave power.

Source: the Author, adapted from: EIA, 2016a. *Hydropower Explained. Wave Power*. U.S. Energy Information Administration, adapted from the National Energy Education Development Project (NEED). Available at: www.eia.gov/energyexplained/index.cfm?page=hydropower_wave, last accessed: December 15, 2016.

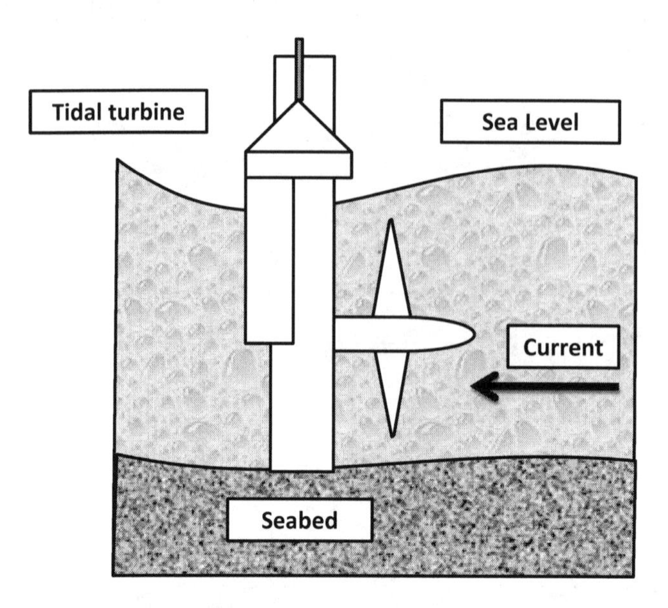

Figure 8.3 Tidal power.

Source: the Author, based on EIA, 2016b. *Hydro Power Explained: Tidal Power*. U.S. Energy Information Administration. Adapted from National Energy Education Development Project (public domain). Available at: www.eia.gov/energyexplained/index.cfm?page=hydropower_tidal, last accessed: December 15, 2016.

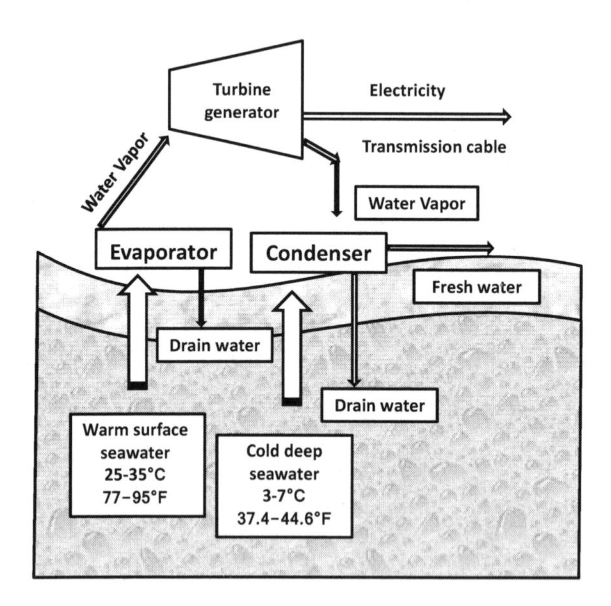

Figure 8.4 Ocean thermal power.

Source: the Author, based on EIA, 2016c. *Ocean Thermal Energy Conversion – Energy Explained. A Diagram of an Ocean Thermal Energy System*. U.S. Energy Information Administration. Available at: www.eia.gov/energy explained/index.cfm?page=hydropower_ocean_thermal_energy_conversion, last accessed: December 15, 2016.

Top Producing Countries

As seen in Figure 8.5 and Table 8.1, China is the world's greatest producer of hydroelectric energy, with other major nations being Brazil, Canada, and the United States (EIA, Beta International, 2017). Figure 8.6 and Table 8.2 show that the hydropower installed capacity regions are led by Asia – in particular East Asia – Europe, and North America. Figure 8.7 shows the hydroelectric energy consumption by region (2015), which is led by East Asia and followed by the Americas (North and South), Europe, and Eurasia.

About 66% of the commercialization capabilities are still to be developed. Unexploited hydroelectric resources are still plentiful in Latin America, Africa, India, and China. Generating electric power by the use of hydroelectric power has certain benefits in relation to alternative energy segments (USGS, 2017b).

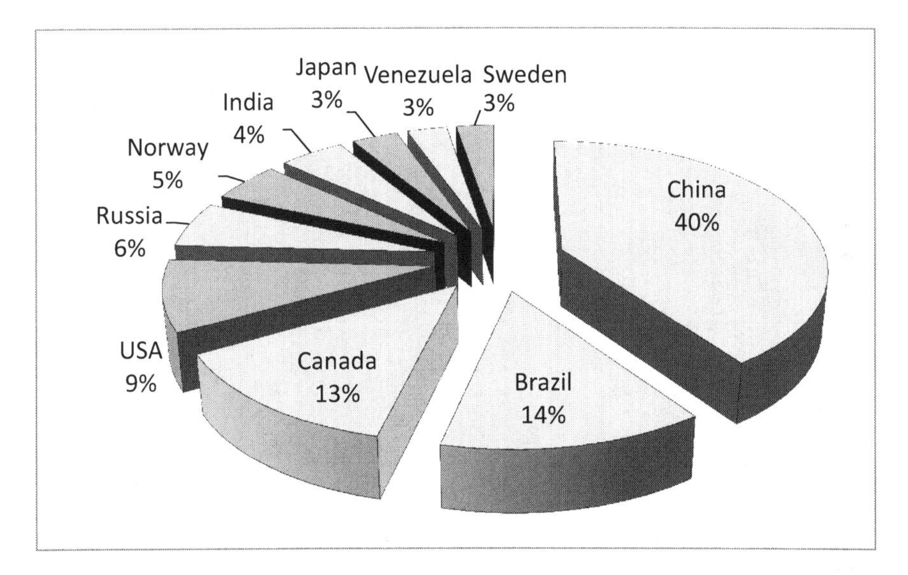

Figure 8.5 Top hydropower-producing countries (Mtoe).

Source: EIA, Beta International, 2017. *Beta International Energy*. U.S. Energy Information Administration. Available at: www.eia.gov/beta/international, last accessed: June 1, 2017.

Table 8.1 Top hydropower-producing countries.

Country	*Mtoe*
China	96.9
Brazil	32.9
Canada	32.3
USA	21.5
Russia	13.8
Norway	12
India	11.1
Japan	7.85
Venezuela	6.84
Sweden	6.36

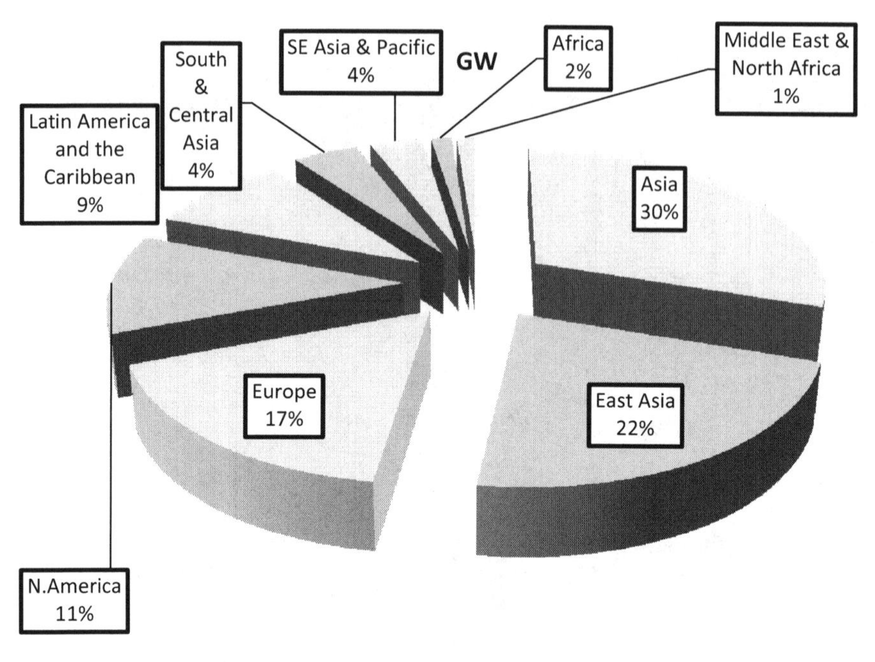

Figure 8.6 Hydropower installed capacity by region (Mtoe).

Source: EIA, Beta International, 2017. *Beta International Energy*. U.S. Energy Information Administration. Available at: www.eia.gov/beta/international, last accessed: June 1, 2017.

Table 8.2 Hydropower installed capacity by region (GW).

Region	*GW*
Asia	511
East Asia	381
Europe	293
N. America	193
Latin America and the Caribbean	159
South & Central Asia	72.3
SE Asia & Pacific	57.8
Africa	22.9
Middle East & North Africa	20.6
Total	1710.6

Technologies, Operations, and Processes

Hydropower Facilities: Impoundment, Diversion, and Pumped Storage

The three main categories of hydropower facilities are: Impoundment, diversion, and pumped storage.

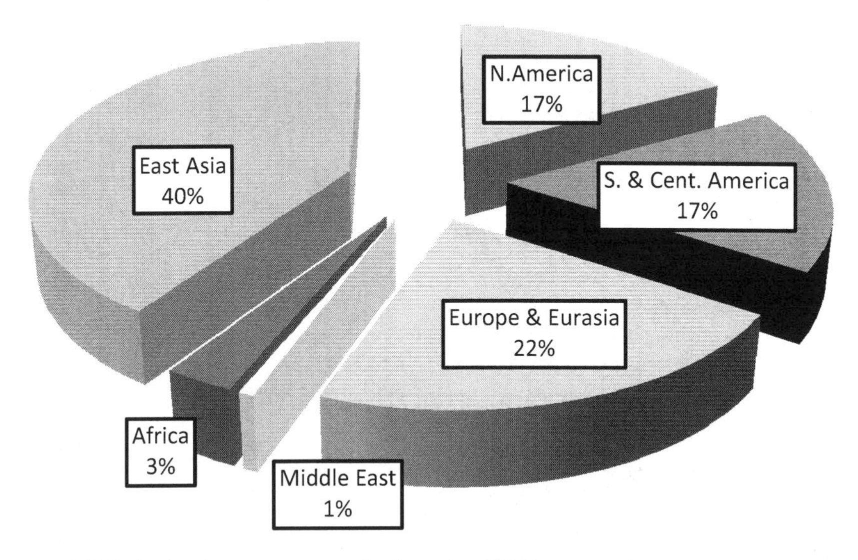

Figure 8.7 Hydroelectric energy consumption by region, 2015 (percentage).

Source: EIA, Beta International, 2017. *Beta International Energy*. U.S. Energy Information Administration. Available at: www.eia.gov/beta/international, last accessed: June 1, 2017.

1 Impoundment

An impoundment unit is a most typical hydroelectric power plant of a larger size, utilizing a dam to accumulate stream water in a water tank. As water is discharged from the tank it passes through and rotates blades. As the generator is activated, it generates electricity. Depending on the size and purpose of the impoundment, energy is generated to either supply a constant regional demand or to partially contribute as an alternative energy segment (DOE, 2017d).

2 Diversion

A run-of-the-river (ROR) hydroelectricity or diversion unit is a type of hydroelectric generation plant whereby little or no water storage is provided. It streams a river's part by way of a penstock or channel, without the typical utilization of a dam.

Hydropower may occasionally operate without the use of dams. This is accomplished through the diversion of a part of the water deriving from rapid-flowing streams, such as in the Niagara Falls. Diversion locations are typically found in the vicinity of waterfalls or mountains to ensure that the span of the diversion is large enough to obtain ample head for the turbine. Efficient diversion channels can be cost effective. In instances when the ingestion unit directly streams through the "penstock" pipeline, the penstock surges the water all the way down to the generator and produces head force with increasing vertical decline. This specific head pressure fundamentally catches the kinetic water energy and a power generator transforms the mechanical energy to electrical energy. The water leaves the turbine and comes back to the stream through the "tailrace."

Run-of-the-river micro-hydro energy units comprise:

- Small business solutions for a micro hydropower system.
- Although micro-hydro energy units are typically site-specific, their systems share certain common components, such as:
    - A penstock, which is a mechanism that directs or channels the flow of water. This is a conveyance-channel, pipe system or pressurized pipeline configuration used to control the water supply.
    - A turbine, water wheel or pumping system used to convert the force of streaming water into rotational energy.
    - Consequently, an alternator or generator is used to convert the rotational energy into electrical energy.
    - A regulator is used to manage the alternator or generator.
    - A wiring system is utilized to finally produce the electrical power.
    - Additionally, several systems employ an inverter to transform the DC of lower voltage electrical power into 120 or 240 Volts of AC electrical power (UNWTO, 2017).

3 Pumped Storage

Pumped storage is a different type of hydropower. Its operation resembles a battery that preserves the electrical power created by different energy resources such as wind energy, photo voltaic, or even nuclear energy, for future consumption. Energy is accumulated by means of moving water to a water tank positioned at a higher (upstream) level, from a subsequent reservoir at a decreased (downstream) level. During times of low demand for power, a diverted tank facility reserves power through pumping water from a lower-level tank to a higher-level tank. Also, during times of high demand for power, the water is returning to the lower-level tank and spins a turbine, thus creating electricity.

Hydroelectric Power Plants: Sizes

Hydroelectric power plants are constructed in different sizes ranging from pico, small and micro-hydro plants, as depicted in Table 8.3. These function to satisfy the power demand of anything from a power plant or small community to large hydroelectric units that are capital-intensive and technology-intensive, and designed to meet the demands of larger markets (DOE, 2017d; IREA, 2012). While different sources provide different classifications and generating capacities, the following table depicts the most prevailing classes.

A Typical Hydroelectric Power Plant

A typical hydroelectric power plant is a system with three parts:

1. an electric-powered unit where the electrical energy is created;
2. a dam that may be opened and shut as required to regulate the water flow; and
3. a water tank where water can be kept.

The water at the rear of a dam streams via an ingestion and forces its way through turbine blades, making them turn. A generator rotates by the spinning of a turbine to generate

Table 8.3 Types of hydropower sizes.

Hydropower sizes	*Generating capacity*	
Pico hydro	1 to 5 kW (kilowatts)	Reduced pollution from fossil fuels. Relative environmental pollution due to industrial, instalment, and distribution processes.
Micro hydro	6 to 100 kW (kilowatts)	
Small hydro	1 to 30 MWs (megawatts)	Can produce enough electricity for a farm building, small industrial unit, or small town.
Large hydro	31 megawatts (MW) or greater	

Source: the Author, based on data from DOE, 2017d. *Types of Hydropower Plants*. Office of Energy Efficiency & Renewable Energy, U.S. Department of Energy. Available at: https://energy.gov/eere/water/types-hydropower-plants, last accessed: June 1, 2017.

electricity. The volume of electricity that can be created is determined by the distance where the water falls and the volume of the water streaming via the plant. Hence, the electricity produced can be carried through long-distance electric-powered outlines to households, industrial facilities, and corporations. Hydroelectric energy supplies nearly 20% of the global electricity production.

Regions rich in wave power sources are: Northern Canada, West Scotland, South Africa, Alaska, NW USA, and Australia.

There are different types of hydropower units:

a the **large-scale hydroelectric facilities** utilizing the stream energy of a whole river behind the dams. Large-scale hydroelectric units are designed to generate electrical energy for national or large social utility undertakings. The majority of such units comprise of a dam and a tank where stream water is preserved. As the water is discharged from the tank, it whirls and spins the turbines thus generating electricity. This is an easy and reliable solution for electricity production when energy demand increases.

Dammed units may also be utilized as energy storage units. In times of increased energy needs such units function just like a conventional hydro-energy, where water is discharged from the top tank deposit flowing via turbines and rotates generators to create electrical power. Nevertheless, in times of reduced electricity demand, electrical power from the power grid rotates the generators in reverse, hence water is discharged from the bottom tank deposit to the top tank deposit. The water is placed in this location until the demand for energy requires electricity generation (DOE, 2013a).

b **small facilities** that reap the benefits of water streams in public water amenities or irrigation infrastructures (DOE, 2013b).

c **damless run-of-river units**, so-called diversions directing a portion of the water flow via an energy plant prior to the water entering the main river. Damless hydro energy does not require a dam to develop stress but instead takes advantage of the natural water stream to generate electrical power. These turbines consist of low head systems with limited ecological effects. The construction of damless hydro energy ensures that the turbine blades rotate slowly hence do not impede the fish migration, while enabling fish to break free. Facilitating technologies such as fish elevators and fish ladders are used at dams to assist fish and other fauna migrate, i.e., flow unhampered upstream, among streams and dams. One of the main benefits of damless hydro-energy plants is that the turbine motors are designed for site-specific purposes, thus ensuring maximum overall efficiency.

Hydro energy has the lowest maintenance and operational costs, as well as low energy charges to consumers. According to the Energy Information Administration, over 50% of electricity in the U.S. derives from renewable energy, among which 7% is created by hydropower.

Technologies for Wave Energy

Wave power is accumulated from the wind forces shaking sea waters. Certain global oceans have tremendous capabilities and output as powerful long-term wave energy hits the coastline.

Ocean energy is kinetic energy, generated from the fast-flowing deep sea waves that are mighty enough to move rotary engine blades. As the waves are elevated, they surge into a compartment pulling out the oxygen and upon a wind turbine that is connected to an energy alternator. As the wave level fluctuates inside the chamber, pressure moves turbines and becomes energy that can be transported or stored.

Wave-powered units are small-scale generators capable of providing energy to small- and medium-sized sites and structures, e.g., lighthouses, small industrial units, and so on (BOEM, 2017a).

An ongoing movement of sea currents transports vast volumes of sea water throughout global oceans. This velocity (kinetic energy) is converted into electricity (BOEM, 2017b).

Modern wave-energy transformation units can collect offshore and coastline power by the use of technologies such as the following.

1 **Point Absorber Systems**

These resemble floating buoys located on top of waves. They capture energy from widely fluctuating waves at a single area, thus transforming hydraulic or electro-kinetic energy to produce energy through an adjacent moored conversion system.

2 **Attenuators**

These are long buoyant constructions aligned to the waves' direction in a parallel manner. Wave fluctuations move to the flex tips adjacent to hydraulic pumps or other transformers. As wave energy is generated, umbilical cables transfer it ashore.

3 **Overtopping systems, Ocean Wave Energy Converters (OWECs)**

These systems include tanks that store forceful high-pressure waves. Discharged waters surge into the sea and, as they fall, they activate rotary engine blades. Internal blades used with buoyant structures generate additional energy point absorbers.

Technologies for Tidal Energy

Tidal energy is a form of ocean water power that transforms tidal forces into electrical power or alternative, similarly valuable, energy types. As high tides approach the coastline, water is captured in tanks and thus hydrokinetic energy is generated. Low tides release the water out of the tank and into the sea, just like a hydroelectric energy unit would do.

Tidal energy is harvested when constant fluctuations and flow velocity of the sea level occurs. It is brought on by the gravitational impact of the sun and the moon on Planet Earth. The vast forces of ocean waves and wind currents are subject to diffusion and movement (EIA, 2017).

There are three principal types of tidal technologies:

1 a dam with a barrage construction to gather tidal water flow in a reservoir;
2 subsea cables for distribution of energy to consumers;
3 tidal sluices and seaways.

The limited geographical suitability and cyclical harvesting availability are the main challenges of the system. However, if the selected site is situated near a powerful tide, production and energy efficiency rates are much improved.

High tides occur when the sun or the moon's orbit approaches a certain area of the ocean.

a **Gravitational energies** draw the water hence the sea levels soar in such areas, yet cause low tides in other areas when the sea levels decline. The gravitational impact of the moon upon the Earth is twice as strong compared with the impact of the sun upon the Earth, considering that the distance between the Earth and the moon is in greater proximity, compared with the distance between the Earth and the sun (EIA, 2017).

b **Tidal zones** are found in coastal regions where narrow water canals supercharge the water movement thus enhancing the density of tidal energy.

Intertidal zones, also known as littoral zones, are regions among tide marks, i.e., the seashore and foreshore. These are regions that appear as land masses above water level during low tides, whereas they submerge during high tides.

Tides create elaborate motifs throughout each sea region that frequently vary considerably from tidal motifs of neighboring waters (NOAA, 2017).

a **Diurnal tides**: These tidal cycles appear in areas with merely one high and one low tide each lunar day. Regions with these patterns are mainly the North Pacific Ocean (the regions of East Russia, Japan, and Northeast China). In the Americas there are two regions with diurnal tides: The Gulf of Mexico and Alaska.

b **Semidiurnal or semi-daily tides**: These tidal cycles take place in regions where the two high and two low tides of almost equal size occur each lunar day. This is the most common type of tides, found in every continent. Namely: In East Australia and New Zealand, most parts of the African coastline, most parts of western Europe including the UK and Scandinavia, the northern part of South America, and the U.S. East Coast.

c **Mixed semidiurnal tides**: In this pattern there is a difference in the height of the high and low tides. These tidal cycles are found on the U.S. West Coast, Chile, Middle East, Northern Mediterranean, and several regions in Australasia (NOAA, 2017a; NOAA, 2017b).

Tidal Streams, Barrages, and Tidal Lagoons

Scientists have discovered several methods of harvesting tidal energy as part of hydrokinetic energy in order to seize the shift of sizeable water masses from one region to another. Modern technologies offer a wide spectrum of applications in oceans, canals, rivers, and other aquatic settings, mainly based on turbines and pertinent engineering systems. There are three major methods to produce tidal energy: Tidal streams, barrages, and tidal lagoons.

Tidal Barrages

A barrage is a sizeable dam constructed through a water estuary, gulf, or part of a coastline, with turbines and gates constructed into the barrage wall.

This technology resembles a dam's capability to capture energy through altering the water levels as they move in and out of a river or channel. Tidal energy is generated through allowing the water pressure to circulate via the tidal turbines. When tides are high, the water streams via channels in the barrage through the gates. This movement spins the turbines or air flows via pipelines and consequently spins the turbines to generate electricity.

Sluice gates on the barrage are sliding gates designed to monitor and regulate water levels at the critical times of the tidal cycle. A two-way tidal energy technique produces electrical power from the incoming and outgoing tides.

During the high tides stage, the barrage gates close and inbound water is retained, so that the difference between inside and outside water levels exceeds 16 feet (5 meters) or so. Subsequently the water outflows and passes again via the turbines so that electricity is generated once more. A disadvantage of this system entails its ability to generate energy exclusively during the high and low tide times, and not at times that the water remains motionless. Another disadvantage pertains to potential environmental impact to the seabed.

Technologies for Current Energy

The U.S. leads the investment of innovative technologies for current power. This is a primary stage of prototypes' testing, and still a lot of research and experimental testing is needed to better understand the technological and commercial capabilities of the hydro-energy segment.

Among the innovative technologies available, the most prevailing configuration that is expected to be used in the next decade or so resembles the wind turbine technologies with horizontal axis turbines. The commercialization stage is still at a preliminary phase and includes various in-stream current and tidal turbine designs. The majority of innovative units are installed in near-shore ocean locations with tidal currents, or fixed flow within fresh-water environments such as streams and rivers.

Technologies for Ocean Thermal Energy Conversion (OTEC)

The Ocean Thermal Energy Conversion (OTEC) methodology pertains to production of energy via the utilization of elevated water temperature. This was a 1881 concept by Jacques D'Arsonval, a French engineer who observed the variations in ocean temperatures: The water surface is warmer due to the sun, whereas the temperature drops at the lower sea levels. Modern energy plants such as the Hawaii OTEC plant produce energy by using a variation among the cooler and warmer water temperatures, i.e., the sea bottom should be around 36°F (2.2°C) and the warmer temperatures at the top around 38°F (3.3°C) (DOE, 2013a).

Additional utilizations of this technique include water desalination and building temperature controlling (State of Hawaii, 2017).

Ocean Thermal Energy Conversion plants use sturdy pipelines that are immersed into the sea for over a mile to induce the cooler water's resurface on the top sea layers. This method has enormous capabilities to generate "billions of watts of electrical power," and it could become increasingly popular once the conversion process becomes more cost-efficient (DOE, 2013b).

Closed-cycle systems produce electricity by utilizing liquids of a low boiling point, e.g., ammonia, in order to revolve turbines. High-temperature water from the ocean's surface flows towards a heat exchanger, i.e., a technology that transfers heat from one liquid to another without permitting them to blend. The colder sea water flows via another heat exchanger and its vapor is condensed into a fluid that is subsequently reused via the system.

Open-cycle systems utilize warmer sea water from the ocean's surface to generate power. The sea water boils when it is pumped into a low-pressure tank. The expanding steam generates power as it moves through a power generator, whereas the remaining water gets desalinated so that clean fresh water is also generated.

Hybrid systems also exist that combine the features of closed- and open-cycle systems.

Benefits, Risks, and Recommendations

a **Environment**

- Minimal environmental pollution: Limited fluid, air-borne, or greenhouse contaminants.
- There is minimal visual impact, especially if submerged or offshore.
- Natural landscape may appear industrialized when wave power equipment is installed closer to the coastline or offshore units are in shallow waters.
- Decreased seashore and leisure activities. The capabilities of greater impact of marine installation sites may possibly decrease marine routes, boating, and amusement locations.
- Sound disturbances. The motion of attenuators or point absorbers or aquatic motion and absorption in the OWC and overtopping equipment may possibly create excessive persistent sound. However, this sound cannot be considerably higher than the original sound of the ocean waves.
- Turbines are detrimental to flora and fauna. Ecological infrastructure solutions like fish passages or fish ladders are necessary. Environmental damages may occur in rivers located in the vicinity of dams, impacting nature's flora and fauna. Several types of fish, like salmon, may be averted from moving upstream to spawn. Fish ladders and other technologies help fish rise up over dams and get into upstream breeding locations. However, the existence of wave power dams modifies their migration habits and damages fauna communities. Hydropower units may also trigger low dissolved oxygen levels within the water, which is damaging to river environments (NREL, 2010).

b **Economics**

- Substantial preliminary expenses. The increased system and machinery costs can prolong the repayment timeframe.
- High investment is needed to ensure durability and safety. Structures should be designed and preserved with the highest standards.
- High investment in infrastructure and superstructure including power facilities, dams, etc.
- As opposed to the cost-effective investments for fossil fuels, renewable energy is characterized by high expenditure levels. Namely, ocean energy costs surpass solar and wind energy costs by at least four times.
- Minimal running costs. Following installation there are low running costs and labor costs, with the exception of instances when machinery failure takes place.
- Hydropower is the most cost-effective method to produce electricity since the energy production is free once a dam has been constructed. Moreover, it is a green energy source that is easily obtainable, while it is continuously replenished by rainfall and snow.
- Design and development is the most critical stage as high investment is needed to construct an intricate system of dams, barricades, generators, and vessels.
- The subsequent process of managing, operating, and maintaining are low-cost.

- Tidal infrastructure can only be successful when built in specific sites, whereas it is unfit for other purposes such as for residential or modest production units.
- Limited operating and maintenance costs among the other renewable types.

c **Land**

- Geographic restrictions: A nation's geographic and oceanographic particularities will determine its ability to harness ocean energy. For example, ocean power is greater compared with that of a shallow sea, or even a river and a land-locked region. Hence, this energy type cannot be broadly utilized. Energy-dependent nations will still need to import energy or generate it from alternative sources.
- Careful site selection is needed, where constant robust energy flow is present.
- Extended subaqueous wires are needed resulting in elevated supply expenditures in order to transfer electrical energy from the immersed systems to the markets.
- Shoreline protection: By absorbing and utilizing tidal energy, a smaller amount of energy will be ramming upon the shore, that will actually assist in preventing harm to the coastline. Long-term degeneration of the cove habitat; shoreline deterioration.
- Accumulation of deposits and contaminants that enter or are generated in the interior of dams and barrages, yet are incapable of returning back into the sea.
- Destruction of flora and fauna. In particular, hazardous to fish along with other aquatic species who are trapped in the tidal system between the barrage and rotor blades.
- Optimum land utilization.
- Limited landscape alterations, limited aesthetic effects, since infrastructure is entirely built underwater.
- Water overflow and terrain deterioration are prevented by the construction of barrage systems.
- Optimum utilization of tidal region is possible, as leisure superstructures and activities are possible.
- Their fundamental limitation is the fact that only very specific locations are appropriate for large-scale tidal energy units.

d **Production efficiency**

- Renewable energy: Reliable, measurable inputs and production outputs.
- Machinery malfunction: Powerful seas, stormy weather and brine deterioration can harm the equipment, which may raise the development expense to enhance sturdiness and/or result in recurrent equipment failures. This particularly applies in instances with the amplified machinery intricacy.
- Impact on aquatic flora and fauna: Damage to marine species or disturbed aquatic life caused by the operations or installation of machinery.
- Small number of projects. There are comparatively limited industrial systems as opposed to other energy solutions, in particular solar and wind harvesting. This increases the impact of machinery breakdown to output.
- There are seasonal variations, with peak production being during winter time.

Recommendations

- Limited commercial benefits: Current technologies are incapable of fully capturing and utilizing ocean energy, hence there is limited interest in investing.
- Despite the tremendous future potential of harnessing the ocean power, the present supply chain consists of a few manufacturing, construction, installation, and operational key players.

- A large-scale investment strategy could potentially capture ocean energy and hence attain a more attractive cost–benefit analysis.
- This is a sustainable reliable type of energy, especially when compared with wind or solar power.
- A relatively new energy segment, with a few regional projects. Technology tests have mainly taken place in lab-based wave reservoirs.
- Its renewable nature makes ocean energy ideal as a substitute energy source that can power communities and public utilities during a crisis, extreme weather, etc.
- Strategic planning decision on constructing ocean energy plants. Ocean energy exists in every deep sea environment and is constantly buoying and moves regionally in adjacent areas.
- Conventional energy types are typically harvested at the source point of energy. However, in the case of ocean energy, it is difficult to identify the source point.
- There is a need to further test the facilities, production, and life-span and overall capabilities for sustainable performance in region-specific particularities, e.g., weather conditions, temperature, velocity, pressure, etc (State of Hawaii, 2011).

Case Studies

1 Three Gorges Dam, China

The 22,500MW Three Gorges Dam on China's Yangtze River is one of the greatest hydroelectric plants globally. This substantial undertaking contains a reservoir of over 373 miles (600 kilometers). Its construction required the displacement of over 1.2 million people and a large number of communities were intentionally submerged (i.e., 13 cities, 140 towns, and 1,350 villages). This large-scale enterprise produces at least ten times as much energy as the Hoover Dam.

The construction of this $29bn hydroelectricity plant commenced in 1993 and finished in 2012. The project consists of a dam of 181 meters (594 feet) in height and 2,335 meters (7,661 feet) in length, 50MW power generators and 32 turbine/generator units graded at 700MW each (Duddu, 2017).

2 The Grand Coulee Dam

The Grand Coulee Dam is America's greatest hydro plant, located on the Columbia River (N. Washington) and one of the greatest global dams. This gravity dam was constructed to develop hydroelectric power and produce irrigation water (Northwest Power and Conservation Council, 2008).

The dam is almost a mile long, i.e., it surpasses 5,223 feet (1,592 meters) and its height exceeds 550 feet (167.6 meters). Its foundation is constructed of solid granite and extends to 350 feet (106.7 meters). Over 70% of the electrical power produced in Washington State is created by hydroelectric facilities (DOE, 2015).

References

Bureau of Ocean Energy Management (BOEM) (2017a). *Ocean Wave Energy*. U.S. Bureau of Ocean Energy Management, Department of the Interior. Available at: www.boem.gov/Ocean-Wave-Energy, last accessed: June 1, 2017.

Bureau of Ocean Energy Management (BOEM) (2017b). *Ocean Current Energy*. U.S. Bureau of Ocean Energy Management (BOEM), Department of the Interior. Available at: www.boem.gov/Ocean-Current-Energy, last accessed: June 1, 2017.

California Energy Commission (2017). *Ocean Energy*. California Energy Commission. Available at: www.energy.ca.gov/oceanenergy, last accessed: June 1, 2017.

Department of Energy (DOE) (2013a). *Large-Scale Hydropower Basics*. Office of Energy Efficiency & Renewable Energy, U.S. Department of Energy. August 14. Available at: https://energy.gov/eere/energybasics/articles/large-scale-hydropower-basics, last accessed: June 1, 2017.

Department of Energy (DOE) (2013b). *Microhydropower System Components*. Office of Energy Efficiency & Renewable Energy, U.S. Department of Energy. August 15. Available at: https://energy.gov/eere/energybasics/articles/microhydropower-basics, last access: June 1, 2017.

Department of Energy (DOE) (2013c). *Ocean Thermal Energy Conversion Basics*. U.S. Department of Energy. August 16. Available at: https://energy.gov/eere/energybasics/articles/ocean-thermal-energy-conversion-basics, last accessed: June 1, 2017.

Department of Energy (DOE) (2015). *Top 10 Things You Didn't Know About Hydropower*. U.S. Department of Energy, April 27. Available at: http://energy.gov/articles/top-10-things-you-didnt-know-about-hydropower, last accessed: June 1, 2017.

Department of Energy (DOE) (2017a). *History of Hydropower*. Department of Energy. Available at: http://energy.gov/eere/water/history-hydropower, last accessed: June 1, 2017.

Department of Energy (DOE) (2017b). *Hydropower Technology Development*. U.S. Department of Energy. Available at: https://energy.gov/eere/water/hydropower-technology-development, last accessed: June 1, 2017.

Department of Energy (DOE) (2017c). *Types of Hydropower Plants*. U.S. Department of Energy. Available at: https://energy.gov/eere/water/types-hydropower-plants, last accessed: June 1, 2017.

Department of Energy (DOE) (2017d). *Types of Hydropower Plants*. Office of Energy Efficiency & Renewable Energy, U.S. Department of Energy. Available at: https://energy.gov/eere/water/types-hydropower-plants, last accessed: June 1, 2017.

Duddu, P. (2017). "The 10 biggest hydroelectric power plants in the world". *Power Technology*, October 27. Available at: www.power-technology.com/features/feature-the-10-biggest-hydro electric-power-plants-in-the-world, last accessed: June 1, 2017.

Energy Information Administration (EIA) (2017). *Hydropower Explained. Tidal Power*. U.S. Energy Information Administration. Available at: www.eia.gov/energyexplained/index.cfm?page=hydropower_tidal, last accessed: June 1, 2017.

Energy Information Administration (EIA) Beta International (2017). *Beta International Energy*. U.S. Energy Information Administration. Available at: www.eia.gov/beta/international, last accessed: June 1, 2017.

Energy Hawaii (2011). *Feasibility of Developing Wave Power as a Renewable Energy Resource for Hawaii*. Department of Business, Economic Development, and Tourism, Honolulu, HI. Available at: https://energy.hawaii.gov/wp-content/uploads/2011/10/Feasibility-of-Developing-Wave-Power-as-a-Renewable-Energy-Resource-for-Hawaii.pdf, last accessed: June 1, 2017.

Energy Quest (2017). *Wave Energy Supply Chain*. California Energy Commission. Available at: www.energyquest.ca.gov, last accessed: June 1, 2017.

Environmental Protection Agency (EPA) (2017). *A Student's Guide to Global Climate Change. Water Energy*. U.S. Environmental Protection Agency. Available at: https://archive.epa.gov/climate change/kids/solutions/technologies/water.html, last accessed: June 1, 2017.

International Hydropower Association (2017). *Types of Hydropower*. International Hydropower Association. Available at: www.hydropower.org/types-of-hydropower, last accessed: June 1, 2017.

International Renewable Energy Agency (IREA) (2012). *Renewable Energy Technologies: Cost Analysis Series* (Report). International Renewable Energy Agency. June 2012, p. 11. Available at: https://hub.globalccsinstitute.com/sites/default/files/publications/138178/hydropower.pdf, last accessed: January 14, 2017.

National Hydropower Association (NHA) (2017). *History of Hydro. Hydropower Milestones*. National Hydropower Association. Available at: www.hydro.org/tech-and-policy/history-of-hydro, last accessed: June 1, 2017. Based on the following sources (2009):

Energy Information Administration, *The Changing Structure of the U.S. Electric Power Industry 2000*, Appendix A: History of U.S. Electric Power Industry: 1882–1991, 2000.

The Foundation for Water and Energy Education, *Timeline of Hydroelectricity and the Northwest*, August 2008.

The National Energy Education Development Project, *Secondary Energy Infobook*, 2008.
U.S. Department of Interior, Bureau of Reclamation. *The History of Hydropower Development in the United States*, November 2008.
U.S. Department of Energy, Office of Energy Efficiency and Renewable Fuels, *History of Hydropower*, October 2004.
U.S. Library of Congress, *America's Story – Gilded Age (1878–1889)*, January 2009.

National Oceanic and Atmospheric Administration (NOAA) (2017a). *National Ocean Service*. National Oceanic and Atmospheric Administration, U.S. Department of Commerce. Available at: https://oceanservice.noaa.gov/education/kits/tides/media/supp_tide07b.htm, last accessed: June 1, 2017.

National Oceanic and Atmospheric Administration (NOAA) (2017b). *Tides and Water Levels: Types and Causes of Tidal Cycles – Diurnal, Semidiurnal, Mixed Semidiurnal; Continental Interference*. National Oceanic and Atmospheric Administration, U.S. Department of Commerce. Available at: http://oceanservice.noaa.gov/education/tutorial_tides/tides07_cycles.html, last accessed: June 1, 2017.

National Renewable Energy Laboratory (NREL) (2010). *Large-Scale Offshore Wind Power in the United States*, Sep 14. The National Renewable Energy Laboratory. Available at: www.nrel.gov/docs/fy10osti/40745.pdf, last accessed: June 1, 2017.

Niagara Frontier (2017). *The Historic Mill District of Niagara Falls, New York*. Niagara Frontier. Available at: www.niagarafrontier.com/milldistrict.html, last accessed: June 1, 2017.

Northwest Power and Conservation Council (2008). *Columbia River History Project*, October 31. Northwest Power and Conservation Council. Available at: www.nwcouncil.org/history/GrandCouleeHistory, last accessed: June 1, 2017.

Online Niagara (2017). *Niagara Falls Power*. Online Niagara. Available at: www.onlineniagara.com/niagara-falls-power, last accessed: June 1, 2017.

State of Hawaii (2011). *Feasibility of Developing Wave Power as a Renewable Energy Resource for Hawaii*. Department of Business, Economic Development, and Tourism. Available at: https://energy.hawaii.gov/wp-content/uploads/2011/10/Feasibility-of-Developing-Wave-Power-as-a-Renewable-Energy-Resource-for-Hawaii.pdf, last access: June 1, 2017.

State of Hawaii (2017). *Ocean Thermal Energy Conversion (OTEC)*. State of Hawaii. Available at: www.hawaii.gov/dbedt/ert/otec-nelha/otec.html, last accessed: June 1, 2017.

Tesla Society (2019). *Nikola Tesla's Speech at Niagara Falls Opening Ceremony, January 12, 1897*. Tesla Memorial Society of New York. Available at: www.teslasociety.com/exhibition.htm, last accessed: January 20, 2019.

United Nations World Tourism Organization (UNWTO) (2017). *Hotel Energy Solutions. Micro Hydropower System*. United Nations World Tourism Organization. Available at: http://cf.cdn.unwto.org/sites/all/files/docpdf/re10microhydropowersystemtaq20062011okfinal.pdf, last accessed: June 1, 2017.

United States Geological Survey (USGS) (2016). *Hydroelectric Power: How it Works*. The USGS Water Science School, U.S. Department of the Interior, U.S. Geological Survey. Available at: https://water.usgs.gov/edu/animations/hydrodam.gif, last accessed: December 15, 2016.

United States Geological Survey (USGS) (2017a). *Hydroelectric Power: How it Works*. USGS Water Science School, U.S. Geological Survey. Available at: water.usgs.gov/edu/hyhowworks.html, last accessed: June 1, 2017.

United States Geological Survey (USGS) (2017b). *Hydroelectric Power and Water. Basic Information*. The United States Geological Survey. Available at: https://water.usgs.gov/edu/wuhy.html, last accessed: June 1, 2017.

World Energy (2016). *Hydropower*. World Energy Organization. Available at: www.worldenergy.org/data/resources/resource/hydropower, last accessed: June 1, 2017.

9 Solar and Geothermal Energy

I think the future for solar energy is bright.

Ken Salazar, U.S. Secretary of the Interior (2009–2013)

Solar Energy: An Overview

Solar power is the most plentiful energy resource on our planet, which is exposed to 173,000 terawatts of solar power; that is over 10,000 times the global energy demand (Energy DOE, 2016).

Active and Passive Solar Systems

Solar systems are distinguished into active and passive:

a **Active solar systems** utilize technologies in order to transform solar power (light or heat) into energy. A collector and a fluid are used to absorb solar radiation. Fans or pumps circulate air or heat-absorbing liquids through collectors and then transfer the heated fluid directly to a room or a heat storage system. Active water heating systems usually have a tank for storing solar heated water (EIA, 2017a; EIA 2017b). Liquid technologies are frequently utilized when storage applications are involved, and match perfectly to radiant heating structures, boilers with hot water radiators, as well as for heat pumping and cooler absorption systems (Energy DOE, 2018a).

b **Passive solar space heating** is attained when the building is warmed up by natural radiation and convection as the solar rays hit the windows of specifically designed structures that are built to gather and dispense solar power (EIA, 2017c; EIA 2017d). Hence, passive solar heating can be as simple as letting the sun shine through windows to heat the inside of a building, thus combining energy efficiency and low power costs (EPA, 2017a; Energy DOE 2018b).

Building plans designed for passive solar heating typically contain windows facing towards the south, thus taking advantage of the solar rays as they reflect on the walls or floors when cold, and thus generating solar heat.

Photovoltaic (PV) vs Solar Thermal (CSP) Technologies

The two prevailing technologies are a) Photovoltaic (PV) solar panels, and b) Solar Thermal Technologies/Concentrating Solar Power (CSP).

Photovoltaic (PV) Solar Panels

Over 100 nations are using solar energy, with photovoltaics (PV) representing 97.9% of the market. Photovoltaic technologies utilize semi-conductor technologies to directly transform sunlight into electricity. Since the system collects energy from solar rays, it is operable during the day and in sunny climates. For maximum efficiency this system must be coupled with other power generators to ensure a constant supply of electricity. Also, solar panels reach their peak efficiency during the day because of limited energy storage capacity. Silicon is the prevailing material used to produce PV cells, while modern solar panels offer a conversion rate of up to 40% (Energy DOE, 2013).

Photovoltaic systems can be installed on the rooftop, offshore, i.e., floating at sea, on the wall, or even on vehicles as auxiliary power and air conditioning systems (NREL, 2018a).

Several multi-megawatt PV power plants have been built globally, while millions of buildings and other structures have installed rooftop photovoltaic systems. If PV panels were installed in 4% of the global deserts, the energy produced would be sufficient to satisfy the global needs for electricity on a daily basis (EIA, 2018a).

Solar Thermal Technologies/Concentrating Solar Power (CSP) Technologies

Solar Thermal Technologies are relatively new energy segments, and, despite their high efficiency rates, they currently represent only 2.1% of the market. Solar thermal systems are increasingly more appealing to the markets due to their large-scale energy production capabilities and their heat storage capacity. The distinct capability of CSP is its capacity to store heat energy in a most efficient and least costly manner. Heat is typically accumulated throughout the day and consequently converted into electric power over the night hence the system can operate 24/7.

There are three key classifications of solar thermal technologies, which are duly explained in the technology section:

1 linear concentrating systems, consisting of: a) parabolic trough systems, and b) Compact linear Fresnel reflectors;
2 solar power tower systems; and
3 solar dish/engine systems.

As the Earth spins on its axis, the sun seems to be moving around it. To ensure optimum capturing of energy, the solar thermal energy systems possess tracking devices to ensure that sunlight is always concentrated upon the receiver during each day and every season, despite the solar position changes (EIA, 2018a). On top of producing power, CSP technologies are succeeding in developing global markets that require solar fuels, process heat, and desalination (NREL, 2018b).

History of Solar Energy

Solar energy has impacted every planet in our solar system for millions of years. Humans have worshipped the sun since prehistoric times, and its thermal power has helped them survive, keep warm, take baths, and process food. Since at least **700 BC** humans have been using glasses and mirrors to distillate sunlight and thus light fires for heat, creating tools and weapons, hunting, food processing, and shaping metals. Archeological findings demonstrate that torches were used by at least **400 BC**, when flammable materials (such as sulfur or vegetable oil) were placed on one end of the stick. Torches were lit during the Ancient Olympic Games and the religious ceremonies in ancient temples of most continents (EERE DOE, 2016).

By the **3rd century BC**, when the Roman army attacked Syracuse, Archimedes, the Greek polymath (mathematician, inventor, physicist, engineer, and astronomer), invented the notorious "heat ray" to burn the Romans' wooden fleet: He ignited the fire by utilizing a number of small mirrors, which formed a sizeable parabolic reflector upon copper shields. Millenia later, photovoltaic technology would enable the optimum use of solar power.

History of Photovoltaics (PV)

In **1954** Bell Labs researchers (Chaplin, Pearson and Fuller) announced the invention of the first silicon solar cell, the forerunner of all solar energy electronics that could transform solar rays into electric power. Its energy efficiency of 4% improved over the years: Hoffman Electronics attained 9% in **1958** and 10% in **1959**. The modern PV modules typically have an optimum efficiency of about 15%.

The New York Times broadcasted the groundbreaking invention as "the beginning of a new era, leading eventually to the realization of one of mankind's most cherished dreams: The harnessing of the almost limitless energy of the sun for the uses of civilization" (Energy DOE, 2016).

The aeronautics industry adopted solar energy at an early stage: By the **late 1950s**, NASA used solar cells to support the power supply of American space satellites. The case study of this chapter demonstrates other remarkable solar projects by NASA.

The energy crisis in the **early 1970s** occurred in the advanced economies as oil and gasoline consumption grew due to automation, while supply was stagnant. The **1973** oil embargo by OPEC and the Suez Canal closure harmed the global supply chains by cutting energy supplies while prolonging the transportation time. Scientists were forced to explore alternatives to fossil fuel energy. The energy crisis inspired the use of solar modules to generate electric power for domestic and industrial purposes. High production costs combined with low energy efficiency hindered the demand for solar energy.

By the **late 1970s** photovoltaic modules (panels) were generating electric power in isolated off-grid sites where electric power lines were not installed (EIA, 2017a).

Solar Energy: Commercial and Economic Development

Photovoltaics are ranked third in terms of global capacity among renewable power segments, followed by hydro and wind energy. While the global demand for solar energy remains high, it is the global economy and national regulations that will determine the energy's future.

The global solar photovoltaics capacity totaled 232 GW in 2015, 315 GW in 2016 and 400 GW in 2017. The estimated global growth approximates to 30% each year. In 2017 the global capacity increased by 30% (29.3%), equivalent to 98.9 gigawatts (SolarPower Europe, 2018).

In 2017 the annual global production of solar energy represented about 1.8% of the electricity demand of the planet (IEA, 2018a).

As of 2018, the top solar-power-producing nations are led by China, the U.S., Japan, India, and Germany, as demonstrated in Figure 9.1.

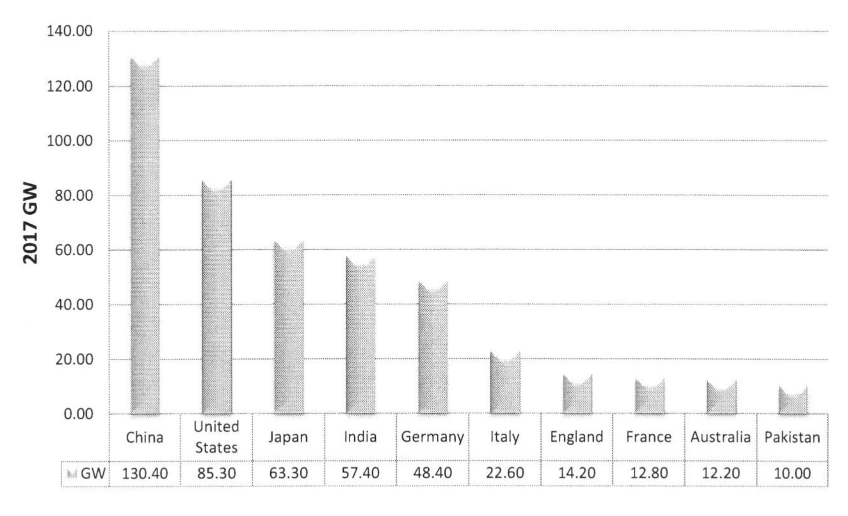

Figure 9.1 Top solar-power-producing countries (GW in 2017).

Source: IEA, 2018a, *Solar Energy; Solar Photovoltaics*. International Energy Agency. Available at: www.iea.org/topics/renewables/solar, last accessed: June 12, 2018.

China

China seems to be the frontrunner of renewable investments with its capacity expansion reflecting about 40% of the global aggregates. According to the nation's 13th five-year plan (2015–2020), fighting air pollution and attaining energy independence seem to be the nation's priorities. By 2022 China aims for a 30% renewable capacity growth, thus reaching an additional 1,150 GW. However, the two key impediments to China's growth are a) the increasing cost of renewable subsidies and b) establishing an integrated power grid (IEA, 2018b).

U.S.

The U.S. represents the second-largest economy for renewables, in terms of growth (Energy DOE, 2018c).

Solar PV has become cost-efficient to a great extent due to federal monetary incentives, such as the relaxed rules for credit availability, as well as the increased public awareness of the environmental rewards. In 2017 the Department of Energy declared that the national solar power segment had attained the 2020 utility-scale solar rate goal of U.S.$0.06/kWh, which is three years ahead of schedule and is moving toward the 2030 goal of $0.03/kWh (Energy DOE, 2018c).

India

Solar photovoltaics and wind energy combined exceed 90% of India's capacity growth since energy auctions have led to the lowest cost for both these renewable sources. India enjoys an impressive growth in terms of renewable energy capacity, which is anticipated to double by 2022 (IEA, 2017). The future of the market will be determined by

India's strategic efforts to reevaluate the financial feasibility of its utilities and address the challenges of power grid integration. At the same time, India contemplates the raising of anti-dumping duties; such an action could diminish the quantity of modules lest the country's manufacturers address the issue imminently.

European Union

The EU represents approximately one fourth of the global installation capacity, which may cover circa 4% of the region's need for electric power. The rapid growth of larger economies like India threaten to surpass Europe's growth in the near future.

Other growing markets include Mexico, the Netherlands, Brazil, South Korea, Spain, Egypt, etc. Large-scale solar projects are found in the U.S., China, India Germany, Japan, and Spain, while the greatest utility-scale plants are being built in Asia and Europe. The path to growth is more robust in nations with national legislation promoting the expansion of energy mix and energy independence.

Uses and Applications of Solar-Powered Technologies

The industrial production of solar cells and photovoltaic arrays has been increasing due to sustainable global demand. Greater installations such as large-scale arrays have the capacity to provide electric power to thousands of consumers. Multi-junction or tandem cells is a particular classification of solar cells employed in military operations. Other uses of solar energy include satellite systems, spacecraft applications, water pumping, power communications devices, providing electricity for individual companies or households, air conditioning, microwaves, etc. Additionally, solar energy is used to supply off-grid power for electric cars, recreational vessels and land vehicles, remote sensing, cathodic protection of pipelines, greenhouses, and roadside assistance telephone booths. Wrist watches and calculators represent another segment of the market (EIA, 2017d; EIA, 2018a).

Cost of Solar Power

As the manufacturing orders have increased and technological improvements were enabled by the use of new materials, the cost of PV has dropped dramatically. Installation costs have come down too with more experienced and trained installers (Energy DOE, 2018c).

Companies that have already covered initial investment must still cover high operating costs. As a result, certain projects will be terminated, while others may change scope or ownership. Higher module prices are anticipated to restrict supply and production.

There are two key factors that determine the choice of a solar power system: a) electrical or conversion efficiency, and b) cost efficiency. The power created by photovoltaic systems is determined by the installation location, weather conditions, the installed panels' orientation and slope, and the presence of sunlight during daytime. Due to the fact that PV panels reach their optimum efficiency levels when directed towards the sun, certain structures implement solar-tracking devices to boost power generation through revolving the panels across a single or double axis (EIA, 2017d).

Solar Energy: Technologies, Operations, and Processes

There are two main classifications of solar technologies: a) solar photovoltaic (PV) energy, and b) solar thermal energy, that uses Concentrating Solar Power (CSP) technologies.

Solar Photovoltaic (PV) Energy

Solar photovoltaic (PV) energy converts solar power into electric power. Photovoltaic technologies serve as photodiodes that produce current electricity by using solar cells to convert energy from the sun into a flow of electrons. Solar panels operate when protons or light particles absorb light photons (i.e., electromagnetic radiation) and release electrons, (i.e., electrically charged radiation) into an elevated energy status, thus serving as charge transporters for electric currents (NASA, 2002).

As seen in Figure 9.2, solar arrays are sheltered by a translucent protective glass that reflects almost 4% of the light. An anti-reflective coat (AR) is used to reduce this reflection and thus optimize the energy captured by the solar arrays.

Solar Thermal Energy: Concentrating Solar Thermal Power (CSP) Technologies

Solar thermal energy is based on Concentrating Solar Power (CSP) systems. These utilize mirrors to concentrate the sun's thermal energy to drive a conventional steam turbine to make electricity. The thermal energy concentrated in a CSP plant can be stored to produce electricity when needed, i.e., day or night. Concentrating Solar Power needs specific conditions to produce power, such as areas where direct sunlight is most intense (e.g., the U.S. Southwest) and contiguous parcels of dry, flat land.

There are three key classifications of solar thermal technologies, as reflected in Figure 9.3.

The above three types of solar thermal technologies, and any hybrid systems or variations, have three main structural modules in common: a) solar energy collectors; b) reflectors (mirrors) to attract the solar rays; and c) a receiver, to capture the solar rays and generate steam through the stimulation of a thermal-transfer liquid (EIA, 2018b). Hence, to generate steam, the thermal-transfer liquids move the heat via solar collectors

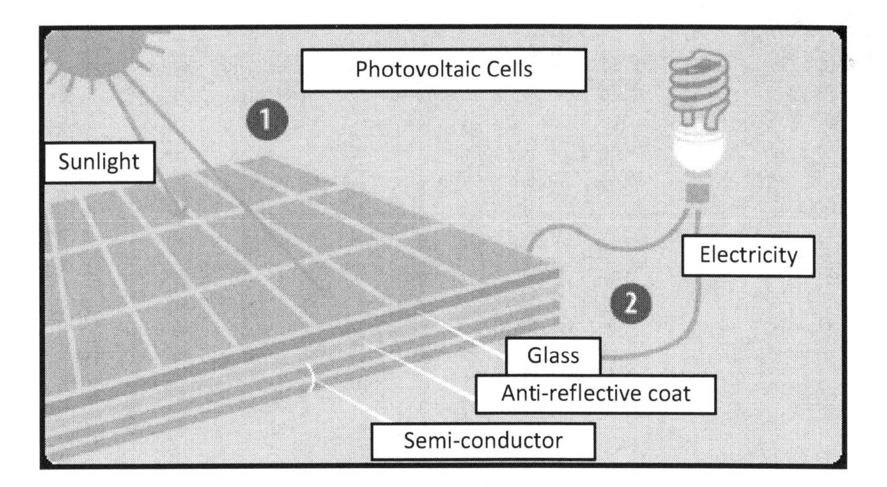

Figure 9.2 The key parts of a photovoltaic (PV) solar panel.

Source: EPA, 2017a. *A Student's Guide to Global Climate Change*. U.S. Environmental Protection Agency (EPA). Available at: https://archive.epa.gov/climatechange/kids/solutions/technologies/solar.html, last accessed: May 9, 2017.

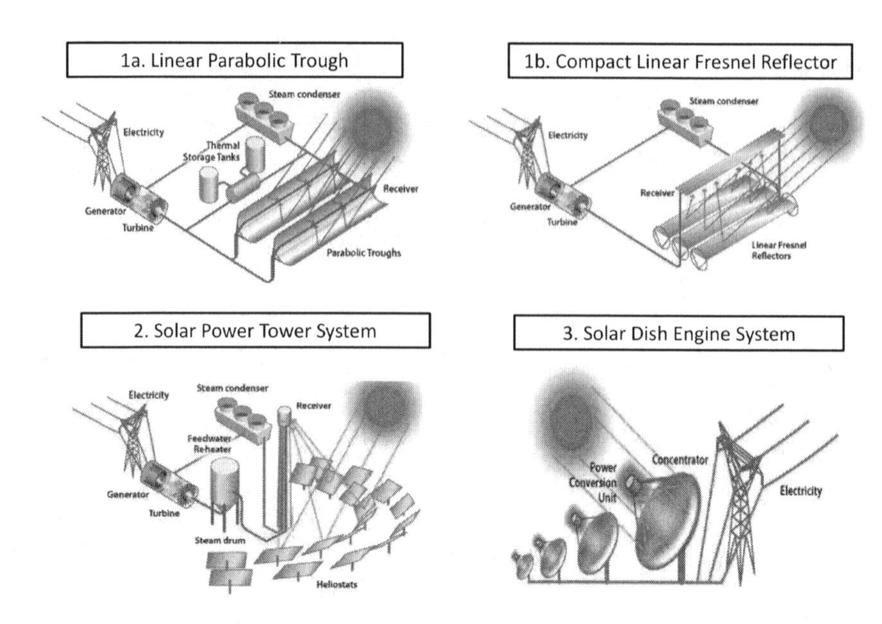

Figure 9.3 The key classifications of Thermal/Concentrating Solar Power (CSP) technologies.

Source: BLS, 2018. *Careers in Solar Power*. U.S. Bureau of Labor Statistics. Available at: www.bls.gov/green/solar_power, last accessed: June 6, 2018.

and a heat exchanger to the heat storage tanks within solar water heating systems (Energy DOE, 2018e). The steam power is transformed into mechanical energy (the sum of kinetic and potential power) in a turbine, and thus powers a generator to generate electricity.

Linear Concentrator Systems (LCS)

A linear concentrating collector power plant has a large number, or field, of collectors in parallel rows that are typically aligned in a north–south orientation to maximize solar energy collection. This configuration enables the mirrors to track the sun from east to west during the day and to concentrate sunlight continuously onto the receiver tubes.

There are two types of linear concentrator systems:

1 **Linear parabolic trough concentrator (LPTC)** systems entail the pioneering CSP methods that use parabolic troughs, i.e., elongated coiled reflectors that are positioned along the focal lines to serve as energy receivers.
2 **Compact linear Fresnel reflector (CLFR)** systems resemble LPTCs, yet the configuration of the reflectors varies. A receiver tube is placed over linear arrays of leveled reflectors. The solar rays' tracking is hereby optimized, as the reflectors direct the light towards the receiver.

Solar Power Tower Systems (SPTS)/Central Receiver Systems

A solar power tower system utilizes heliostats, i.e., a large park of leveled sun-trailing mirrors, in order to direct and concentrate solar rays towards a receiver installed on a steeple top.

The system's capacity to store thermal power enables the mechanism to generate power regardless of the weather conditions, i.e., also in cloudy days, or even during the night. Furthermore, it has the capacity to collect up to 1,500 times the solar rays.

Certain solar energy towers require the use of water as a fluid that transports the heat power. Cutting-edge configurations favor the use of molten salts, in particular nitrate, as a heat transfer fluid and for thermal storage (EIA, 2018b). The salt is heated in tubes and proceeds to the warm storage tank. This method is much preferred due to its impressive heat transfer and optimum power storage capabilities.

Solar Dish/Engine Systems (SDES)

Solar dish/engine systems use a mirrored dish similar to a very large satellite dish. To reduce costs, the mirrored dish is usually composed of many smaller flat mirrors formed into a dish-like, i.e., parabolic, shape. The parabolic surface directs and concentrates sunlight onto a thermal receiver, which absorbs and collects the heat and transfers it to a "Stirling" engine generator (NREL, 2018b). The Stirling engine transforms heat into mechanical power through fluid compression. As the fluid is heated into a turbine it expands, and thus it is converted into mechanical power and electricity.

Solar dish/engine systems always point straight at the sun and concentrate the solar energy at the focal point of the dish. A solar dish's concentration ratio is much higher than linear concentrating systems. The solar dish technology can be mounted at the focal point of the dish, making it well suited for remote locations, or the energy may be collected from a number of installations and converted into electricity at a central point (EIA, 2017b).

Optimizing via Hybrid Technologies: Concentrated Photovoltaic power and Concentrated Solar Power (CPV-CSP)

Hybrid technology comprises a significant scientific trend with promising potential in solar power technologies. Performance optimization is achieved via combining Concentrated Photovoltaic power and Concentrated Solar Power (CPV-CSP). The CPV-CSP hybrid technologies have rapidly evolved over the past few years, due to the prior testing of first- and second-generation technologies, namely CPV and CSP. Efficiency optimization is attained by executing two tasks concurrently. They use photovoltaic energy (CPV) at maximum capacity while also using the hot fluid storage capabilities offered in the Concentrated Solar Power system. Hence, a large capacity of electricity is produced plus storable thermal energy is obtained synchronously. This smart hybrid combination offers the best of both worlds, i.e., low cost with optimum efficiency.

Solar Energy: Benefits, Risks, and Recommendations

Environment and Safety

Benefits

- Solar is a green, clean, safe energy source with diminished harmful emissions or direct primary environmental impression (EIA, 2018a).
- Solar photovoltaics and thermal technologies are both regarded as safe, green methods that harness and utilize the sun's abundant renewable power.

- Smart module technologies optimize power performance of PV modules. Energy production is increased by continuously monitoring the input/output performance via maximum power point tracking (MPPT) systems.

Risks

- Secondary environmental risk may occur, since photovoltaic panels are produced using chemicals and toxic resources. Similarly, certain solar thermal systems utilize dangerous fluids as part of the heat-transporting process. Due to their dangerous goods (HazMat) nature, the process of utilizing and disposing is regulated in most parts of the world.
- The installation of certain large-sized panels may damage the ecological footprint.
    - Birds, small animals, and insects found near the light beam may be killed via overheating.
    - The deforestation and land-cleaning process, followed by the installation of an energy plant, may impact the regional flora and fauna.
    - The use of water may be needed in certain energy plants in order to cleanse solar power facilities (e.g., the collectors and concentrators) or for lowering the temperature of turbine generators.

Commercial, Technologies, and Economies

Benefits

- Solar energy is renewable, ample, and vast. It is available in most weather conditions and climates, even if it is cloudy or rainy.
- New generation technologies, especially hybrid systems, combine energy-efficiency and cost-efficiency.
- Solar installations can operate off-grid hence solar power can be produced in locations away from the electric power grid.
- In regions with large-scale production, solar energy companies enjoy a leveraged position and market share.
- The design, production, and installation stages are relatively simple, fast, and agile enough to fit different spaces, regions, and climates.
- Efficient, sturdy technologies ensure a long life span, i.e., 10–30 years.
- Economies of scale have been enabled due to technological advances and increased interest from governments and consumers. This leads to larger-scale manufacturing, increased reliability, and reduced costs (Energy DOE, 2016).
- Solar energy has gained government support leading to increasing R&D efforts. Testing has led to improved systems and hybrid solutions that attain increased demand and interest from buyers (Energy DOE, 2018i).

Risks

- Certain types of solar technologies may not offer consistent or sustainable energy, as their power production varies depending on weather, climate, time of the year, location, day or night, etc. Hence, limited energy-efficiency impacts cost-efficiency. New generation technologies, especially hybrid systems, resolve this challenge.

- Space-efficiencies are required. Despite the technological improvements, a large surface zone is needed to harvest and produce sufficient volumes of energy.
- Due to the great geographical disparity and the relatively concise supply chains, energy companies may be facing challenges as to the logistics and financial aspects of production.

Case Study: NASA, One of the First Solar Enthusiasts

NASA have been one of the first solar enthusiasts, as they have been consistently utilizing PV power on spacecraft since the 1950s. Solar energy has vast applications in aerospace including in communications, active cooling and heating, running the sensors, etc. *Vanguard I* was the second U.S. satellite in orbit (after *Explorer I*) and the oldest man-made object orbiting Earth to this day. During its 1958 launching, *Vanguard I* used small PV panels for radio power mission throughout its orbit of six billion miles (NASA, 2015).

Shortly after, *Explorer III, Vanguard II*, and *Sputnik III* were launched with PV-powered systems on board (Energy DOE, 2016; EIA 2018a).

In **1959** *Explorer VI* was launched with a PH array of 9,600 cells (EERE DOE, 2016).

Meanwhile, solar energy has been a part of NASA's space missions, from *Messenger*, the robotic spacecraft orbiting Mercury (2011–2015), to the *Juno* orbiting Jupiter (2016–2021), as shown in Figure 9.4.

Juno is humankind's most distant solar-powered space emissary, since planet Jupiter is located 500 million miles away from the sun, or approximately five times more distant from the sun compared with the Earth. Understandably, solar rays extending such a distance are rather dim, hence for *Juno*'s solar panels to gather and produce the required energy, they had to be installed over a large surface area.

Figure 9.4 NASA's *Juno* orbiting planet Jupiter.

Source: NASA, 2016b. *NASA Juno Mission Completes Latest Jupiter Flyby*. Editor: Tony Greicius for NASA, December 12 at 5:55 p.m. Available at: www.nasa.gov/feature/jpl/nasa-juno-mission-prepares-for-december-11-jupiter-flyby, last accessed: January 1, 2017.

To upsurge the energy output per kilogram, *Juno* and other modern spacecrafts use expensive premium-efficiency multi-junction solar arrays made of semiconductor materials such as gallium arsenide (GaAs) (EERE DOE, 2016). Hence, *Juno*'s three solar arrays have a length of 30 feet (9 meters) and are decked out with 18,698 singular solar cells made of silicon and gallium arsenide, each cell producing about 14 kilowatts of electricity (NASA, 2016b). NASA's *Juno* will continue its 53-day orbits around Jupiter in order to attain its pioneering space exploration objectives (NASA, 2018).

Case Study: Azuri Life Changing Technologies (LCT) Ltd, UK

Specialties: Solar Energy and Solar for Off-grid

Overview

Azuri Life Changing Technologies Ltd is a pioneering energy company established in 2012 and headquartered in Cambridge, England. Azuri delivers micro solar systems in sub-Saharan Africa, where 600 million people are living without electricity. For many countries in the region there are challenges to extending the grid. The standard problem with renewable energy is that users do not have access to the grid: They need to buy their own power station, which requires a capital investment that ranges too high in rural Africa. Azuri offers mobile technology, which means that solar can be used as a service and users can pay for electricity as they need it.

This has increased the need for finding an alternative energy solution. In sub-Saharan Africa, the region has a population that is growing faster than electrification rates, and solar presents a promising alternative for millions of people.

Azuri has the capacity to overcome the challenges of energy access, enabling technology deployment in emerging markets such as Kenya, Tanzania, Zambia, Uganda, and Nigeria.

Recognition

a Azuri was honored as a winner of a prestigious United Nations climate change award at a special ceremony at the 2015 United Nations Climate Change Conference in Paris. Azuri Technologies was named as one of three initiatives that are successfully mobilising Financing for Climate Friendly Investment.

b Azuri's CEO, Dr. Simon Bransfield-Garth, has been recognized as a Technology Pioneer by the World Economic Forum. Azuri joined a group of 23 of the most innovative technology start-ups from around the world who are poised to make a significant and positive impact on business and society. To be selected as a Technology Pioneer, a company must be involved in the development of life-changing technology that has been market proven and shows all the signs of being a long-standing market leader. Azuri was selected for its contribution to delivering affordable and clean energy to the world's off-grid population, improving global energy access through its PayGo solar solution.

c Azuri Technologies won the Sustainia Award 2012 for their Indigo solar solution tailored to the developing world.

Solving a Global Energy Security Problem

As discussed in Chapter 3 of this book, globally, over 1.2 billion people (that is, one out of seven people) do not have access to electricity in their homes. As a result, they are forced to burn expensive and polluting fuels for their basic lighting needs. Clean energy sources, like solar or wind, are often unaffordable in rural sub-Saharan Africa, where a typical farmer earns U.S.$2–3 a day and would struggle to afford a basic U.S.$70 solar system. On average, people in rural sub-Saharan Africa spend U.S.$2–4 a week on fuel and phone charging, which means their ability to save is limited.

> The most important thing is that my children can finally study at night.
>
> Teacher in Uganda

Technologies

Azuri has combined mobile technology and solar power to turn a development challenge into a business opportunity. Azuri solar home systems allow users to pay for solar power on a pay-as-you-go basis, while providing clean, safe, renewable power to families at a lower energy cost. All of this can be achieved without the need for any government subsidies or tariffs. Azuri is at the forefront of new technology innovation by providing a reliable service alternative to the grid for off-grid customers, delivering lighting, phone charging, radio, and TV access in affordable packages.

How Azuri has Changed Lives in sub-Saharan Africa

This contribution goes beyond energy security: It is science's ethical obligation to raise the communities with no access to the grid, and transform their lives in the sectors of health, education, learning, daily chores, work, and survival.

- 85% of Azuri customers make savings of up to 50% per week. This breaks down to:
  - 37% on school fees;
  - 28% on food;
  - 20% reinvested into their businesses.
- 97% of users claimed that children study more.
- Reduction of time spent collecting, cutting, and burning wood.

The benefits of solar energy are immediate for customers and Azuri's services have changed their lives. Using off-grid solar has improved users' health, safety, lifestyle, and financial position.

Azuri's customers are happy to find out that, since they have switched to solar, the air quality within their homes has improved dramatically. Azuri home solar systems don't produce any pollution or cause health issues.

The risk of fire is important to many of their customers as kerosene lamps can easily cause house fires. Often women would structure their day to be inside while the kerosene lamp was lit due to the risk of fire. This means that, for their children to study under the light of the kerosene lamps, someone would need to supervise them. Solar gives a choice to people on something as simple as how to structure their day.

Many women also reported that they can extend their day by up to three hours, as Azuri solar has made it easier for them to carry out their chores under the safety of a solar lamp.

Teachers and students in sub-Saharan Africa are among the most grateful clients of Azuri: Children can now study beyond sunset, thus improving their performance in school.

Case study sources: Courtesy of Azuri Life Changing Technologies (UK) Ltd.

- Interview.
- Company website: Azuri Life Changing Technologies (UK) Ltd. (2018a). Azuri Life Changing Technologies (UK) Ltd. Available at: www.azuri-technologies.com.

Geothermal Energy: An Overview

> Geothermal represents a clean, nearly inexhaustible baseload source of electricity, which makes it a viable renewable energy source.
>
> Timothy Unruh,
> EERE's Deputy Assistant Secretary for Renewable Power, DoE, USA

Geothermal energy pertains to heat power deriving from the earth's surface (in Ancient Greek, geo (Γεω) means earth and thermal (θερμότης) means heat). This is a renewable, sustainable, green type of energy deriving from the earth's surface but also generated underground from the earth's crust where extremely high temperatures of magma (i.e., molten rock) are found (Energy DOE, 2011; Energy DOE 2018g).

Figure 9.5 demonstrates the layers of Planet Earth and the temperatures in each layer.

This is an extremely efficient and sustainable energy segment that yields an output of 3–4 energy units for every single unit used as input. Hence, it is 300–400 times more efficient than an average energy segment that attains about 90% efficiency.

The majority of geothermal wells are drilled in shallow areas of about ten feet (3.048 meters) from the ground surface, where they retain an almost steady temperature. The vast majority of the planet's wells retain a temperature of 212°F (100°C) or above.

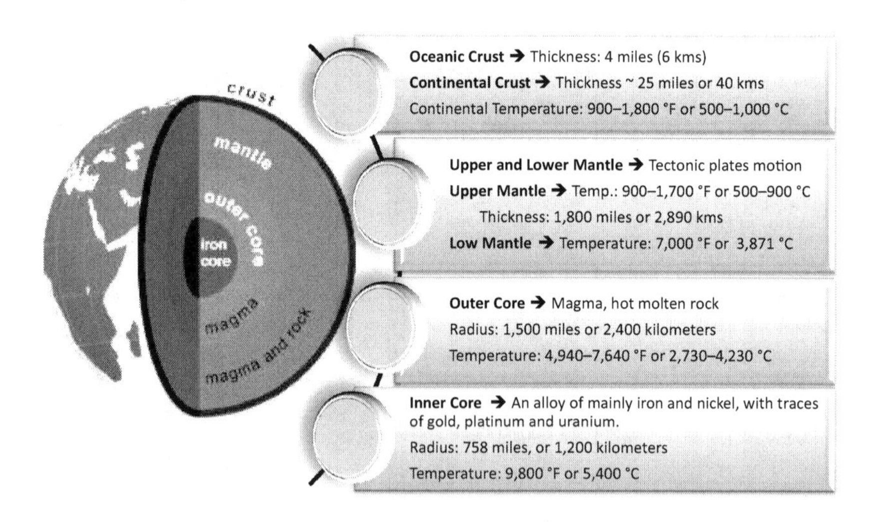

Figure 9.5 The Earth's interior.

Source: Comments by the Author. Image from EPA, 2017a. *A Student's Guide to Global Climate Change*. Environmental Protection Agency. Available at: https://archive.epa.gov/climatechange/kids/solutions/technologies/solar.html, last accessed: May 9, 2017.

This vigorous, environmentally friendly source can yield unlimited and uninterrupted quantities of power and only requires a small footprint to progress (Energy DOE, 2018g).

It can be used to produce electric power and generate heat, but also cool dehumidified air in buildings.

History of Geothermal Energy

Geothermal energy has been used since antiquity by ancient civilizations such as the Greeks and Romans, Egyptians, Indians, Native Americans, Chinese, Iranians, etc. (EPA, 2017b).

1904: The earliest dry steam geothermal energy plant was established in Tuscany, Italy, in a volcanically active area. More than a century later, the Larderello plant generates energy for over one million buildings.

1955: The Geysers is the world's greatest geothermal field owned and operated by Pacific Gas & Electric, California. In 1955 its first geothermal well was drilled, named Magma No. 1, signifying the commencement of a large-scale electrical power era. To date, the Geysers produces more electricity than any other geothermal field globally (State of California FTP Conservation, 2010).

1974: The first successful hot dry rock (HDR) project globally was attained at Fenton Hill, New Mexico. The project's success demonstrated that electricity could be produced from hot dry rock with minimum related micro-seismicity, of a magnitude < 1 on the Richter Scale. In 1978 an HDR geothermal energy plant was tested on-site, and commenced to produce electricity in 1980 (Energy DOE, 2016).

Geothermal Energy: Commercial and Economic Development

Geothermal wells are formed naturally in hydrothermal locations. Due to their immense depth, these wells are untraceable without the use of technologies. Traditionally, large-scale geothermal plants are situated in volcanically active areas, where they possess larger environmental footprints.

Among the nations possessing such large-scale geothermal activities are Iceland, New Zealand, Costa Rica, Kenya, El Salvador, and the Philippines. Meanwhile, the greater geothermal plants are situated in California (the Geysers). Almost half of the global installed capacity is situated on islands, such as the Philippines, Indonesia, Hawaii, New Zealand, etc.

There are approximately 1,500 potentially active volcanoes globally, in addition to the continuous belt of volcanoes on the ocean floor. About 500 of these have erupted in historical time (USGS, 2018). The majority of active geothermal areas (72%) are typically located near volcanoes, amidst adjacent tectonic plate borders (e.g., Iceland), whereas less than 20% are located in hot spots (e.g., Hawaii). The Ring of Fire, as demonstrated in **Figure 9.6**, is an area of significant geothermal and volcanic presence that encompasses the Pacific Ocean (EIA, 2017e).

It is estimated that by 2050 geothermal energy could reach about 3.5% of global electric power, reaching 200 GW of installed capacity and eliminating from the world 760 million tons of CO_2 emissions each year (World Bank, 2018; IEA, 2018b).

In 2015, 22 countries used geothermal power to produce over 76 billion kWh of electricity. In 2017, U.S. geothermal energy plants generated approximately 16 billion kilowatthours (kWh), followed by the Philippines, Indonesia, Turkey, and New Zealand.

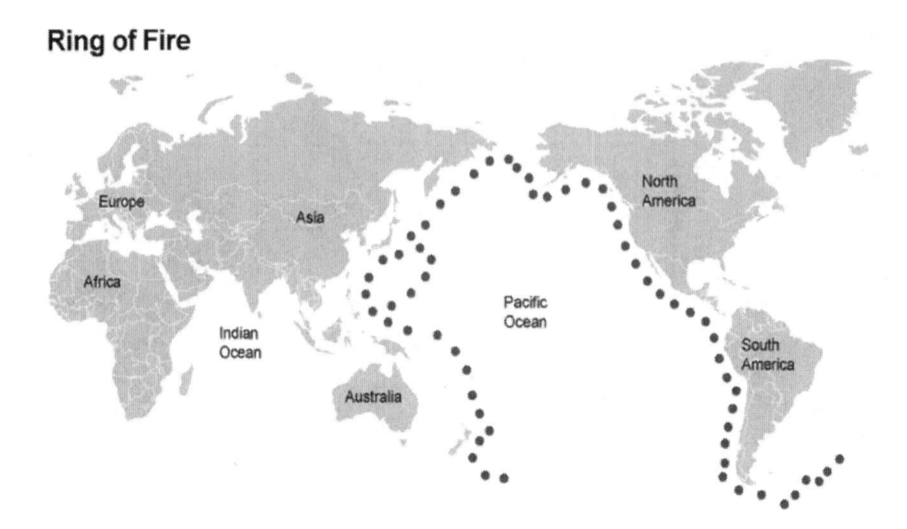

Figure 9.6 Thermal energy and the Ring of Fire along the edges of the Pacific Ocean.

Kenya was the seventh-greatest nation to produce electricity from geothermal energy, yet enjoyed the biggest share of aggregate electricity production from geothermal power, i.e., approximately 47% (EIA, 2017f). Other success stories include the Philippines, which relies on geothermal energy to produce about 60% of their electricity. Also, countries like Papua New Guinea have invested in geothermal power: The plant situated on Lihir island is capable of satisfying 75% of the country's electricity needs.

In 1995, global geothermal capacity was extended to 6,000 megawatts, and in 2017, i.e., 22 years later, the capacity has doubled to 12,894 megawatts. Although this is an energy resource with tremendous potential, geothermal only represents 0.5% of the electricity produced in the U.S. and is projected to increase to 0.6% by 2035 (Argonne National Lab, 2010). The transfer of innovative technologies used in unconventional oil and gas drilling to geothermal drilling has brought tremendous improvements and generated interest from governments and investors. Some of the most impressive technology transfer projects are discussed in the technology section that follows.

Geothermal Energy: Technologies, Operations, and Processes

Geothermal energy is extremely energy-efficient as it combines optimum land and water efficiency. Its power plants are very condensed compared with other renewable and non-renewable energy sources. For example, it occupies less land per GWh (404 m^2) compared with coal (3,642 m^2), wind (1,335 m^2), or solar photovoltaic arrays with a center station (3,237 m^2) (Geothermal Energy Association, 2012).

Throughout geologic time, volcanic upsurges appear to have benefited mankind. As volcanic elements are dissolved, they produce some of the most fruitful soils globally, leading to prosperity and civilization peaks of historic significance. The internal high temperatures related to recently formed volcanoes has been used to generate geothermal energy (USGS, 2018).

Geologists utilize several tools and technologies to discover geothermal reservoirs, while the most accurate method for discovering a geothermal reservoir entails drilling selected wells and measuring the temperature at great underground depths.

Geothermal power is found in three main geological forms:

1 volcanoes and fumaroles, i.e., openings in the Earth where volcanic gases (like carbon dioxide or sulphur dioxide) and steam are emitted (USGS, 2015)
2 hot springs
3 geysers (EIA, 2017f).

Many techniques have been developed to utilize geothermal energy, depending on the temperature and depth. Drilling is used to access hot water or steam reservoirs in great geological depths. As magma, or superheated molten rock, reaches the Earth's surface, it heats ground water confined in porous rock or fractured rock shells and fault zones (Energy DOE, 2011). Currently used systems do not enable the heat retrieval straight from magma, which is the most potent reserve of geothermal energy.

The two prevailing geothermal harvesting methods, as reflected in Figure 9.7, are direct use and deep geothermal.

Direct use of geothermal resource systems is the oldest and most prevailing method. It utilizes the hot water harvested from the Earth's depths to boil an operational liquid that vaporizes and then turns a turbine (Energy DOE, 2011).

It is utilized directly to industrial, manufacturing, agricultural, and other areas including greenhouses, controlling the buildings' temperature. Other heat-treatment processes include pasteurization of food, drying crops, dehydrating onions and garlics, temperature-control in fish farming, etc. In cold climates, e.g., Scandinavia, geothermal water is channeled under urban infrastructures to melt snow (NREL, 2004).

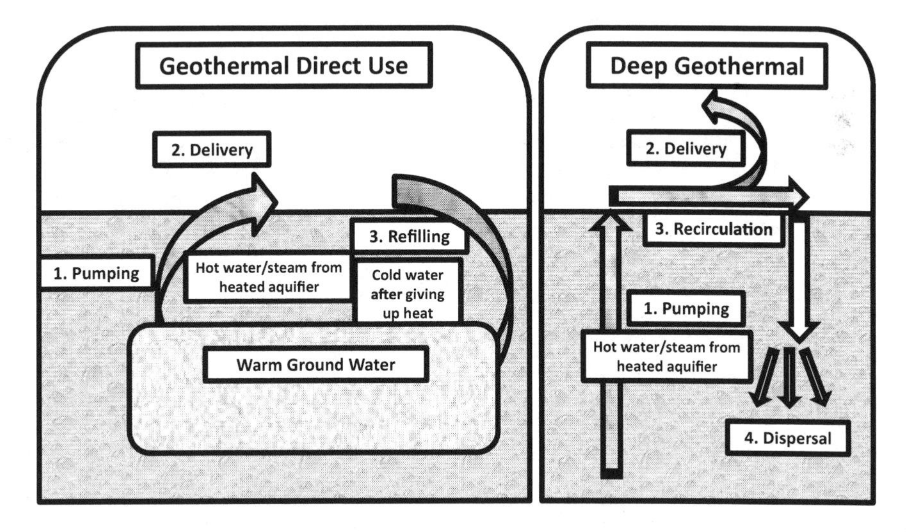

Figure 9.7 Geothermal direct use and deep geothermal energy.

Source: the Author, based on EPA, 2018. *Renewable Heating and Cooling. Geothermal Heating and Cooling Technologies*. Environmental Protection Agency. Available at: www.epa.gov/rhc/geothermal-heating-and-cooling-technologies, last accessed: March 12, 2018.

Ground Source Heat Pumps (GSHP), also known as GeoExchange or geothermal heat pump systems are central heating or cooling methods that move heat to or from the drilled well, where the steam from a basin is used to power a turbine/generator (EPA, 2017c).

This method utilizes the earth's constant temperature instead of the atmospheric temperature. The system consists of a heat pump and a heat exchanger, i.e., a piping configuration submerged near the Earth's surface to feed energy to the building. A dual-source heat pump integrates two heat pumps, i.e., an air-source and a geothermal heat pump. Also, certain models contain two-speed compressors and adjustable fans for convenience and cost-efficiency.

A New Plateau for Geothermal Energy

This is an era for geothermal energy to reach its full potential, as governments and industries increasingly understand and appreciate the potential of this significant energy segment.

Geothermal Reservoir Management Processes: Extending the Commercial Life of Critical Equipment

Ongoing optimization efforts seek to extend the commercial life of drilling technologies to the maximum. This way, production costs will decrease and demand for geothermal will increase.

Modern geothermal reservoir management processes have brought about notable improvements. One of them pertains to the reservoir life-span, which until recently was confined to 20–25 years of commercial life, i.e., just as long as an average well drilling project lasts. Improved reservoir management protocols include proactive maintenance, for example for the periodical recompletion and drilling of the producer/injector wells. As a result, the reservoir life-span can exceed the 25 years pertaining to the average project life and extend up to 100 years.

Drilling Equipment: High Powered Fiber Lasers and Polycrystalline Diamond Compact (PDC) Drag Bits

The quality of drilling equipment has improved over the years. Polycrystalline diamond compact (PDC) drag bits have been produced since the 1960s mainly for the oil and gas drilling industry. Although these were used in some geothermal ventures with mainly sedimentary reservoirs (e.g., Cerro Prieto in Mexico, etc.), their efficiency upon igneous harder rocks was not as sustainable.

Numerous research initiatives were funded by the U.S. Department of Energy and the oil and gas industry. Their research aimed to examine the interaction between the drilling bit with the rock, namely by testing how cutters instigate failure in hard rock.

It was observed that igneous and metamorphic rocks may comprise large quantities of abrasives like quartz (silica sand) or corundum, which may shorten the commercial life of the drilling equipment by imposing vibrations or quicker attrition that harms the drill bits. The test results showed that hard rocks are better drilled in tension rather than in compression. Hence, high-powered fiber lasers were developed to drill igneous, ultra-hard crystalline rocks.

Frictional Heating and Stress of the Cutters Impacts their Wear and Tear

Frictional heating and stress of the cutters impacts their wear and tear, and consequently costs and drilling efficiency. Research conducted evaluated the impact of various dynamics, including bit design, rotary speed, bit bounce, drilling fluid used, but also the interrelation between weight-on-bit (WOB) and drillable rock strength (Energy DOE, 2016).

Technology Transfer from Unconventional Oil and Gas to Geothermal Drilling

Unconventional drilling in extremely high temperatures deep into the Earth has been a challenge for geothermal energy that only advanced technologies could help overcome. For a number of decades investment for R&D had been limited as geothermal energy represented a small market segment. However, a renewed interest in geothermal has occurred over the past few years: Not only are governments and energy investors willing to invest in geothermal, but the technological solutions can be found and adopted from other energy segments. For example, as the oil and gas moguls address identical challenges in unconventional fracking operations, it appears that technology transfer is a feasible solution. A win/win scenario would be to see more energy partnerships, where pertinent non-renewable energy innovations are to be adopted by the renewable energy industry.

As a conclusion, the benefits of this energy segment are extremely promising and the industry anticipates increasing investments on technologies that will be able to bring geothermal to the next level of large-scale production.

Geothermal Energy: Benefits, Risks, and Recommendations

Environment and Safety

Benefits

- Geothermal power is an appealing renewable energy source. It offers unlimited sustainable power.
- It is environmentally friendly and carbon free, with limited noise pollution and limited impact on biodiversity.
- Its power plants can be situated in multiple-use lands that combine recreation, agricultural, and other activities.
- It is extremely energy efficient as it combines optimum land and water efficiency. Its power plants are very condensed compared with other renewable and non-renewable energy sources.
- GHG emissions (50g CO_2 eq/kWhe) are four times lower than solar PV, and 6 to 20 times less when compared with natural gas. Typically, geothermal power plants ingest less water over the lifetime energy output compared with most conventional generation technologies (Argonne National Lab, 2010).

Risks

- There is a debate as to whether geothermal energy contributes to global warming. Certain scientists argue that the process of extracting geothermal power from the Earth naturally causes high temperatures, which inevitably impact our planet. Other scientists posit that by converting this heat to energy we actually relieve the planet from excessive heat. As ongoing research projects aim to improve the technologies and our understanding of this fascinating energy source, we will be able to address the questions of this debate.

Commercial, Technologies, and Economies

Benefits

- Geothermal energy can meet the needs of a growing global population: The geothermal energy potential is unlimited. The global production can reach about 2,000 zettajoules (ZJ) of sustainable uninterrupted energy each year, i.e., 4,000 times the annual energy demand. This is a most positive potential, given the rapid increase of global population.
- Geothermal energy can achieve energy independence: In fact, all nations and regions can achieve energy independence, as geothermal energy is found anywhere in Earth's depths. High volcanic activity is not necessary to produce geothermal energy: Deep drilling in the Earth's depths in any global location will generate energy.
- Geothermal energy has tremendous strengths due to its high energy efficiency performance. Countries like the U.S., Philippines, New Zealand, etc., have proven the feasibility of increasing reliance on geothermal energy to feed the energy grid.
- The key benefits of these base load energy plants include energy-efficiency, cost-efficiency, and sustainability at optimum production levels. Geothermal energy plants generate electricity 24/7, unaffected by extreme weather or other factors.
- New technology transfer capabilities from unconventional oil and gas drilling (e.g., fracking) to geothermal will manage to reduce production costs and increase cost-efficiency.
- The quality of drills and internal hardware has improved over the years. Depending on the depth and geological morphology of the rock, the industry diversifies the use of drilling bits. For ultra-hard crystalline rock, high-power fiber lasers are used, whereas, for more conventional drilling, synthetic diamonds/polycrystalline diamond compact (PDC) drag bits are used. These developments have managed to lower the cost of drilling.

Risks

- There has been limited government support until recently: For several decades the geothermal industry lacked government support and industrial investments, thus its evolution was slower compared with other energy segments. Fortunately, over the past few years, the striking resemblance to unconventional oil and gas drilling has enabled technology transfer into this significant energy segment.
- Additional government support is needed to attain large-scale production and commercialization. Areas for improvement include tax credits, financial support to attain low production costs, and facilitating a faster process, e.g., of drilling permits.

Case Study on Partnerships: U.S. and New Zealand Agreement

An agreement between the U.S. Department of Energy (DOE) and New Zealand's Ministry of Business, Innovation and Employment (MBIE) was established on June 22, 2018, aiming to develop innovative cost-effective geothermal technologies and to boost their global production and marketability.

Presently the U.S. is the global frontrunner in installed geothermal capacity, exceeding 3.8 gigawatts (GW), while New Zealand is another global leader with almost 1.07 GW of installed capacity, which yields over 17% of the nation's electricity. New Zealand has an equally long tradition in implementing direct use geothermal systems to meet the nation's demand for cooling, heating, and industrial operations.

The projected collaboration fields comprise, among others, the cooperative development and advancement of modeling methods and instruments, and direct use programs and utilizations. Moreover, this partnership seeks to detect and resolve the key challenges that impede the growth of geothermal energy.

- One key topic is the association between geothermal production and induced seismicity, which often takes place during stages of geothermal energy harvesting and permeability enhancement. The new partnership will promote induced seismicity risk governance to mitigate the geological and safety hazards and risks. Furthermore, new research projects will seek to monitor and eliminate the side-effects of earthquakes as geothermal operations develop.
- Another topic entails the mineral recovery from geothermal brines. Geothermal liquids are frequently substantial cradles of rare earth elements and precious and semi-precious metals and minerals.
- Supercritical geothermal systems pertain to unconventional systems of ultra-high enthalpy (of high thermodynamic quantity). These wells are situated at great depths in the vicinity of or under the brittle–ductile transition (BDT) zone, i.e., by the layer where the reservoir liquid is in a supercritical state and where the temperature and pressure of clean water are over 752°F (400°C) and 221 bar respectively. These extreme conditions are expected to yield optimum energy-efficiency. Regions of supercritical geothermal systems in the U.S. include west coast regions, Hawaii, and Alaska; and in New Zealand the Taupo Volcanic Zone, specifically project Hades.

The agreement fortifies the partnership and renewable power objectives of the U.S. and New Zealand, and enhances their membership in the International Partnership for Geothermal Technology (IPGT), whose goal is to promote and enhance the growth of geothermal innovations by means of global partnerships (Energy DOE, 2018i).

References

Argonne National Lab (2010). *Life Cycle Analysis Results of Geothermal Systems in Comparison to Other Power Systems*. Argonne, Energy Systems Division, U.S. Department of Energy, ANL/ESD/10-5, August. Available at: www.evs.anl.gov/downloads/ANL_ESD_10-5.pdf, last accessed: May 9, 2017.

Azuri Technologies (2018a). *About Us*. Azuri Technologies. Available at: www.azuri-technologies.com/about-us, last accessed: April 12, 2018.

Azuri Technologies (2018b). *Azuri Paygo Solar Sales Top 100,000*. Azuri Technologies. Available at: www.azuri-technologies.com/news/azuri-paygo-solar-sales-top-100000, last accessed: April 12, 2018.

Azuri Technologies (2018c). *Azuri Recognitions*. Azuri Technologies. Available at: www.azuri-technologies.com/info-hub/recognition, last accessed: April 12, 2018.

Bureau of Labor Statistics (BLS) (2018). *Careers in Solar Power, U.S.* By James Hamilton, Office of Occupational Statistics and Employment Projections, Bureau of Labor Statistics. Available at: www.bls.gov/green/solar_power, last accessed: June 6, 2018.

Energy Efficiency and Renewable Energy, U.S. Department of Energy (EERE DOE) (2016). *The History of Solar*. Energy Efficiency and Renewable Energy, U.S. Department of Energy. Available at: www1.eere.energy.gov/solar/pdfs/solar_timeline.pdf, last accessed: May 9, 2017.

Energy Efficiency & Renewable Energy, Department of Energy (Energy DOE) (2006a). *A History of Geothermal Energy Research and Development in the United States. Reservoir Engineering 1976–2006*. Geothermal Technologies Program, U.S. Department of Energy, Energy Efficiency & Renewable Energy. Available at: www.energy.gov/sites/prod/files/2014/02/f7/geothermal_history_3_engineering.pdf, last accessed: June 6, 2018.

Energy Efficiency & Renewable Energy, Department of Energy (Energy DOE) (2006b). *Drilling 1976–2006. A History of Geothermal Energy Research and Development in the United States.* Geothermal Technologies Program, Energy Efficiency & Renewable Energy, U.S. Department of Energy. Eric Willis, James Bresee, Bennie Di Bona, John Salisbury, John Ted Mock, Allan Jelacic, Peter Goldman, Leland Roy Mink. Available at: www.energy.gov/sites/prod/files/2014/02/f7/geothermal_history_2_drilling_0.pdf, last accessed: February 18, 2017.

Energy Efficiency & Renewable Energy, Department of Energy (Energy DOE) (2011). *Geothermal Basics*. Energy Efficiency & Renewable Energy, U.S. Department of Energy. Available at: www.energy.gov/eere/geothermal/geothermal-basics, last accessed: June 6, 2018.

Energy Efficiency & Renewable Energy, Department of Energy (Energy DOE) (2013). *Solar Photovoltaic Technology Basics*. Energy Efficiency & Renewable Energy, U.S. Department of Energy, Aug 16. Available at: www.energy.gov/eere/solar/articles/solar-photovoltaic-technology-basics, last accessed: June 12, 2017.

Energy Efficiency & Renewable Energy, Department of Energy (Energy DOE) (2016). *Top 6 Things You Didn't Know About Solar Energy*. Energy Efficiency & Renewable Energy, U.S. Department of Energy, June 6. Available at: www.energy.gov/articles/top-6-things-you-didnt-know-about-solar-energy, last accessed: June 12, 2017.

Energy Efficiency & Renewable Energy, Department of Energy (Energy DOE) (2018a). *Active Solar Heating*. Energy Efficiency & Renewable Energy, U.S. Department of Energy. Available at: www.energy.gov/energysaver/home-heating-systems/active-solar-heating, last accessed: June 15, 2018.

Energy Efficiency & Renewable Energy, Department of Energy (Energy DOE) (2018b). *Energy Efficient Home Design*. Energy Efficiency & Renewable Energy, U.S. Department of Energy, January 18. Available at: www.energy.gov/energysaver/energy-efficient-home-design, last accessed: June 15, 2018.

Energy Efficiency & Renewable Energy, Department of Energy (Energy DOE) (2018c). *Solar Energy. 2018 Portfolio*. Technologies Office, U.S. Department of Energy. Available at: www.energy.gov/sites/prod/files/2018/02/f48/2018%20SETO%20Portfolio%20Book.pdf, last accessed: June 6, 2018.

Energy Efficiency & Renewable Energy, Department of Energy (Energy DOE) (2018d). *SunShot 2030*. Energy Efficiency & Renewable Energy, U.S. Department of Energy. Available at: www.energy.gov/eere/solar/sunshot-2030, last accessed: June 6, 2018.

Energy Efficiency & Renewable Energy, Department of Energy (Energy DOE) (2018e). *Heat Transfer Fluids for Solar Water Heating Systems*. Energy Efficiency & Renewable Energy, U.S. Department of Energy, January 18. Available at: www.energy.gov/energysaver/solar-water-heaters/heat-transfer-fluids-solar-water-heating-systems, last accessed: June 6, 2018.

Energy Efficiency & Renewable Energy, Department of Energy (Energy DOE) (2018f). *Science & Innovation. Energy Sources: Solar*. Energy Efficiency & Renewable Energy, U.S. Department of Energy. Available at: www.energy.gov/science-innovation/energy-sources/renewable-energy/solar, last accessed: June 6, 2018.

Energy Efficiency & Renewable Energy, Department of Energy (Energy DOE) (2018g). *Energy 101: Geothermal Energy*. Energy Efficiency and Renewable Energy, U.S. Department of Energy. Available at: www.energy.gov/eere/videos/energy-101-geothermal-energy, last accessed: June 6, 2018.

Energy Efficiency & Renewable Energy, Department of Energy (Energy DOE) (2018h). *Renewable Energy: Geothermal*. Energy Efficiency and Renewable Energy, U.S. Department of Energy. Available at: www.energy.gov/science-innovation/energy-sources/renewable-energy/geothermal, last accessed: June 6, 2018.

Energy Efficiency & Renewable Energy, Department of Energy (Energy DOE) (2018i). *Energy Department to Collaborate with New Zealand on Geothermal Energy Advancement*. Energy Efficiency & Renewable Energy, U.S. Department of Energy, Jun 22. Available at: www.energy.gov/eere/articles/energy-department-collaborate-new-zealand-geotherm, last accessed: June 25, 2018.

Energy Information Administration (EIA) (2017a). *Solar Thermal Collectors, Basics. Solar Explained*, last updated: December 15. Energy Information Administration. Available at: www.eia.gov/energyexplained/print.php?page=solar_thermal_collectors, last accessed: June 6, 2018.

Energy Information Administration (EIA) (2017b). *Solar Energy Explained*, last updated: December 15. Energy Information Administration. Available at: https://www.eia.gov/energyexplained/index.cfm?page=solar_home, last accessed: June 6, 2018.

Energy Information Administration (EIA) (2017c). *Solar Photovoltaics. Energy Explained.* Energy Information Administration. Available at: www.eia.gov/energyexplained/index.cfm?page=solar_photovoltaics, last accessed: June 6, 2018.

Energy Information Administration (EIA) (2017d). *More Than Half of Utility-Scale Solar Photovoltaic Systems Track the Sun Through the Day*. Energy Information Administration, April 24. Available at: www.eia.gov/todayinenergy/detail.php?id=30912, last accessed: June 6, 2018.

Energy Information Administration (EIA) (2017e). *Where Geothermal Energy is Found. Geothermal Explained.* Energy Information Administration, Last updated: November 28. Available at: www.eia.gov/energyexplained/index.php?page=geothermal_where, last accessed: June 6, 2018.

Energy Information Administration (EIA) (2017f). *Use of Geothermal Energy. Geothermal Explained.* Energy Information Administration. Available at: www.eia.gov/energyexplained/index.php?page=geothermal_use, last accessed: June 6, 2018.

Energy Information Administration (EIA) (2018a). *Solar Energy and the Environment.* Energy Information Administration. Available at: www.eia.gov/energyexplained/index.cfm?page=solar_environment, last accessed: June 6, 2018.

Energy Information Administration (EIA) (2018b). *Photovoltaics and Electricity. Basics. Solar Energy Explained.* Energy Information Administration, last updated: April 30. Available at: www.eia.gov/energyexplained/print.php?page=solar_thermal_power_plants, last accessed: June 6, 2018.

Environmental Protection Agency (EPA) (2017a). *A Student's Guide to Global Climate Change.* Environmental Protection Agency. Available at: https://archive.epa.gov/climatechange/kids/solutions/technologies/solar.html, last accessed: May 9, 2017.

Environmental Protection Agency (EPA) (2017b). *Geothermal Energy: A Student's Guide to Global Climate Change.* Environmental Protection Agency, May 9. Available at: https://archive.epa.gov/climatechange/kids/solutions/technologies/geothermal.html, last accessed: June 6, 2018.

Environmental Protection Agency (EPA) (2017c). *Renewable Heating and Cooling. Geothermal Heating and Cooling Technologies.* Available at: www.epa.gov/rhc/geothermal-heating-and-cooling-technologies, last accessed: May 9, 2017.

Geothermal Energy Association (2012). *Geothermal Basics Q&A*. Geothermal Energy Association. Available at: www.geo-energy.org/reports/PREPRINT-DRAFT_GeoBasicsQ&A_forWeb_Aug2012.pdf, last accessed: May 9, 2017.

International Energy Agency (IEA) (2017). "Solar PV grew faster than any other fuel in 2016, opening a new era for solar power". International Energy Agency, News, Oct 4. Available at: www.iea.org/newsroom/news/2017/october/solar-pv-grew-faster-than-any-other-fuel-in-2016-opening-a-new-era-for-solar-pow.html, last accessed: May 9, 2017.

International Energy Agency (IEA) (2018a). *Solar Energy; Solar Photovoltaics*. International Energy Agency. Available at: www.iea.org/topics/renewables/solar, last accessed: June 12, 2018.

International Energy Agency (IEA) (2018b). *Energy and Climate Change. World Energy Outlook Special Report.* International Energy Agency. Available at: www.iea.org/publications/freepublications/publication/WEO2015SpecialReportonEnergyandClimateChange.pdf, last accessed: April 25, 2018.

NASA (2002). *How do Photovoltaics Work?* By Gil Knier for NASA. Available at: https://science.nasa.gov/science-news/science-at-nasa/2002/solarcells, last accessed: January 1, 2017.

NASA (2015). *Vanguard Satellite, 1958.* NASA, March 17. Available at: www.nasa.gov/content/vanguard-satellite-1958, last accessed: January 1, 2017.

NASA (2016a). "NASA's Juno Spacecraft Breaks Solar Power Distance Record". News, January 13. Available at: www.jpl.nasa.gov/news/news.php?feature=4818, last accessed: January 1, 2017.

NASA (2016b). *NASA Juno Mission Completes Latest Jupiter Flyby*. Editor: Tony Greicius for NASA, December 12. Available at: www.nasa.gov/feature/jpl/nasa-juno-mission-prepares-for-december-11-jupiter-flyby, last accessed: January 1, 2017.

NASA (2018). *NASA Re-plans Juno's Jupiter Mission*. NASA, June 6. Available at: www.nasa.gov/feature/nasa-re-plans-juno-s-jupiter-mission, last accessed: June 12, 2018.

National Renewable Energy Laboratory (NREL) (2004). *Geothermal Technologies Program. Energy Efficiency and Renewable Energy*. National Renewable Energy Laboratory, U.S. Department of Energy, DOE/GO-102004-1957, August. Available at: www.nrel.gov/docs/fy04osti/36316.pdf, last accessed: January 1, 2017.

National Renewable Energy Laboratory (NREL) (2018a). *Solar Photovoltaic Technology Basics*. National Renewable Energy Laboratory, U.S. Department of Energy. Available at: www.nrel.gov/workingwithus/re-photovoltaics.html, last accessed: June 12, 2018.

National Renewable Energy Laboratory (NREL) (2018b). *Concentrating Solar Power*. National Renewable Energy Laboratory, U.S. Department of Energy. Available at: www.nrel.gov/csp, last accessed: June 12, 2018.

Solar Power Europe (2018). *European Solar Market Grows 28% in 2017*. Solar Power Europe, 9 February. Available at: www.solarpowereurope.org/fileadmin/user_upload/documents/Media/090218_press_release_European_Solar_Market_Grows_28__in_2017.pdf, last accessed: June 12, 2018.

State of California FTP Conservation (2010). *A Geysers Album*. Publication No. TR 49, State of California. Elena M. Miller, State Oil and Gas Supervisor, by Susan F. Hodgson, graphic design by Jim Spriggs and Alex G. Paman. Published by the California Department of Conservation, Division of Oil, Gas, and Geothermal Resources, Sacramento, California. Second Edition. Available at: ftp://ftp.consrv.ca.gov/pub/oil/publications/tr49.pdf, last accessed: June 12, 2018.

U.S. Geological Survey (USGS) (2015). *EarthWorld: Fumarole*. U.S. Geological Survey, U.S. Department of the Interior, October 5. Available at: www.usgs.gov/news/earthword-fumarole, last accessed: June 12, 2018.

U.S. Geological Survey (USGS) (2018). *What are Some Benefits of Volcanic Eruptions?* U.S. Geological Survey, U.S. Department of the Interior. Available at: www.usgs.gov/faqs/what-are-some-benefits-volcanic-eruptions?qt-news_science_products=0#qt-news_science_products, last accessed: June 12, 2018.

World Bank (2018). *Geothermal Energy is on a Hot Path*. The World Bank, May 3. Available at: www.worldbank.org/en/news/feature/2018/05/03/geothermal-energy-development-investment, last accessed: June 12, 2018.

10 Biofuel and Biomass Security

> The fact that fat oils from vegetable sources for engine fuels can be used may seem insignificant today, but such oils may become, in the course of time, as important as petroleum and the coal tar products of the present time.
>
> Rudolph Diesel, 1912

Introduction and Historical Overview

Biofuel: Definition and Overview

Biomass is one of the oldest sources of energy, as early humans burned wood and plants to create heat and light. Modern-day biofuels are produced by the use of wood, agricultural products, municipal solid waste, animal waste, etc. Pursuant to chemical treatment, biomass is used to generate electricity and also as an additive in diesel engines. Ethanol and biodiesel are the most common biofuels.

In essence, the biofuel industry is humankind's attempt to recycle all the humble unwanted materials on Earth. Hence, there is a noble quality about the biomass industry as it generates energy out of garbage and makes usefulness out of waste.

The new generation of biofuel energy also encompasses aquatic flora, especially algae. Recent scientific breakthroughs by companies like ExxonMobil have yielded algae with very high calorific value, suggesting a very promising future for algae-based biofuel.

The benefits of new-generation biofuel are immense, ranging from rocket fuel for light-weight aircraft to commercial aviation and land transport using renewable fuel for regular commercial operations.

Altair Paramount is an innovative company acquired by World Energy (MA), which produces renewable jet and diesel fuel. In 2017 the U.S. Navy officially inaugurated the "Great Green Fleet," a deployment of battleships driven by Altair's renewable fuel. Commercial airlines like United Airlines and Southwest Airlines, as well as transportation corporations like FedEx, have elevated the U.S. transportation fuel market in a new plateau of dedicated biofuel voyages.

Formation of Biofuels vs Fossil Fuels

Both renewable and non-renewable energy sources contribute to the mitigation of a nation's goals on energy, security, environmental, commercial, and so on.

A basic difference between fossil fuels and biofuels pertains to their geological versus biological process and formation method:

a Fossil fuels are non-renewable sources, produced by geological processes, i.e., the anaerobic decomposition of prehistoric flora and fauna buried beneath heavy layers of sediment.
b Biofuels are renewable sources, produced by biological and thermochemical processes of renewable hydrocarbon fixation. The substances used to produce biofuel include i) biogenic substances from flora and fauna, including algae, animal waste, wood, leaves, grass, paper, food, etc., as well as ii) waste from synthetic substances such as petroleum products, plastic, etc.

Waste Types Used to Generate Biofuel

The following types of waste are used to generate biofuel in waste-to-energy plants:

1 Municipal Solid Waste (MSW) or landfill garbage;
2 agricultural waste;
3 in items 1 and 2, biogas is generated when anaerobic bacteria, decomposed organic waste, or agricultural waste is converted into energy;
4 animal waste is used to create biogas with the assistance of anaerobic microorganisms. Biogas is generated in farming facilities such as digesters, manure ponds (i.e., lagoons), and other waste reception areas where animal waste is stored.

Biofuel Products

The two most frequently used biofuel categories are ethanol and biodiesel.

Ethanol is generated out of sugar and starch. It combusts with oxygen in order to form CO, CO_2, aldehydes, and water. Ethanol and other fuels are generated by the gasification method, which transforms biomass into synthetic gas (syngas), a compound of carbon monoxide and hydrogen. Biomass is exposed to high temperatures and depleted from oxygen.

To produce ethanol from starchy biomass like corn, two distinct types of fermentation are used: Dry-milling and wet-milling, with each process generating distinct products.

1 **The dry-milling method** pulverizes corn into flour and subsequently the fermentation method converts it into ethanol with co-harvests of distillers' grains and carbon dioxide. In the U.S., ethanol is generated by starch-based yield (typically corn) by the means of dry- or wet-mill processing. Dry mills represent approximately 90% of ethanol plants as they entail reduced capital costs and operating costs. However, this method produces less ethanol per corn bushel, thus driving ethanol biofuel producers towards identifying alternative production methods.
2 **The wet-milling method** distinguishes the protein, fiber, and starch found in yields such as corn before these compounds are treated. Although this method entails reduced capital costs and operating costs, it is becoming more popular due to its capacity to produce several biofuel products and thus meet the demand of various markets.

Biodiesel is generated when methanol or other alcohols are mixed with animal fat, vegetable oil or other used saturated oils. Its primary utilization is as an additive to eliminate emissions of particulate matters, hydrocarbons, and carbon monoxide. It can replace diesel fuel that derives from fossil fuel. Biodiesel fuels can be used in regular diesel engines without making any changes to the engines. Biodiesel can also be stored and transported using diesel fuel tanks and equipment.

Biodiesel is a fuel and a fuel additive, due to its methyl ester component. When used as an additive to diesel oil from fossil fuel, it is available in various grades. It is most often blended with petroleum diesel in ratios of 2% (B2), 5% (B5), or 20% (B20), with B20 being the most popular mixture. The number or ratio (%) signifies the percentage of biofuel existing as an additive. Biodiesel can also be used as pure biodiesel (B100).

Bio-oils process used to convert biomass into bio-oil is called fast or flash pyrolysis. The process involves the heating of compact solid fuels in the absence of air at temperatures between 842–932°F (450 and 500°C) for less than 2 seconds, and consequently condensing the resulting vapors within 2 seconds. The bio-oils produced are suitable for use in boilers or in turbines designed to burn heavy oils for the generation of electricity.

Fuel ethers or bioethers are produced from sugar beet, wheat, or from glycerol, a waste deriving from the production of biodiesel. They are developed through the etherification of iso-olefins with bioethanol. They are energy-efficient cleaner burning fuels with high octane. They seem to gain market share as additives to non-renewable energy sources, hence are anticipated to substitute petro-ether.

Methanol and biomethanol. The methanol or methyl alcohol currently used is a flammable volatile material generated from natural gas. It is generated as a result of fossil fuel exploration and processing. Assuming that the non-renewable substance is the first generation methanol, it is appropriate to purport that biomethanol is the second generation methanol produced from feedstock, municipal solids, and glycerine. Innovative projects in the car industry experiment with this new generation biofuel.

Methane, the primary component of natural gas, is a component of landfill gas, i.e., biogas that forms when garbage, agricultural waste, and human waste decompose in landfills or in special containers called digesters. It is generated from several methods, such as gasification, anaerobic digestion and pyrolysis (EIA, 2018a).

While during the pyrolysis and gasification methods methane is generated by increased temperatures and pressures, the anaerobic digestion method is simpler as it decomposes the yield through the oxygen deprivation. Acidic bacteria transform the smaller particles that have been formed in fatty acids. Subsequently, methanogenic microorganisms release a gas mixture containing methane.

Bioalcohols vs Bioethers

Bioalcohol is oxygenated fuel, i.e., fuel ether or alcohol that has been fermented from starches, sugars, or cellulosic biomass. Bioalcohol is obtained from organic processes, yet both biologically developed and chemically developed alcohols appear similar. Methanol, ethanol, propanol, and butanol are the most desirable fuels among the aliphatic alcohols, since they are appropriate for use in internal combustion engines and are suitable for both a biological and chemical synthesis. There is no chemical difference between biologically produced and chemically produced alcohols. Although they have lower energy density, they provide energy efficiency and high octane rating with limited greenhouse gas emissions.

The History of Biomass

Since prehistory, wood waste and lumber commodities have been used for food preparation, for heating, and for illumination. Timber was the primary power supply at a global level until the Industrial Revolution. Timber remains a significant fuel in several nations around the world, especially in developing nations (EIA, 2018b).

In **1897**, before petroleum diesel fuel became popular, Rudolf Diesel, the inventor of the diesel engine, experimented with using vegetable oil (biodiesel) and organic sources as fuel, some of which make today's biodiesel. He firmly believed in the future of biofuel, prior to the wide use of the conventional diesel from fossil fuels.

In **1908**, Henry Ford designed the Model T. As a flexible fuel vehicle, it could run on ethanol, gasoline, or a combination of the two.

By the **1920s** gasoline became the prevailing motor fuel. Standard Oil was the first company to implement ethanol additives into gasoline as an octane booster and to reduce engine pinging, i.e., knocking in spark-ignition of internal combustion engines.

From **1941 to 1945** there was an increase of ethanol production for fuel usage, due to an immense increase in fuel demand. It is worth noting that the demand for ethanol was generated for usage unrelated to war, i.e., for industrial and private usage.

In the decades following World War II, namely from **1945 to 1978**, there was significant reduction in the demand for ethanol, due to limited use for war fuel. The low price of conventional fuel contributed to this.

Consequently, from the **late 1940s until the late 1970s**, commercial fuel ethanol was scarce within the United States.

In **1997** the top U.S. car manufacturers commenced the mass production of flexible-fueled automobile models capable of running on dual fuel, i.e., ethanol (E-85) or gasoline. Notwithstanding the vehicles' capacity to use E-85, limited gas stations offered ethanol, therefore the majority of cars would only use gasoline (EIA, 2017a).

Commercial and Economic Development

Bioenergy is one of the stronger segments of renewable energy, representing 77% of the global production, followed by hydro energy (15%) and all the other categories (8%). Figure 10.1 and Table 10.1 depict the U.S. renewables consumption from 2013 to 2017 in trillion British thermal units (Btus).

Based on these values, Figure 10.2 demonstrates the percentage of U.S. renewables consumption for the year 2017. Biomass (44%) is the prevailing renewable energy, followed by hydroelectric power (25%), wind (22%), solar (7%), and geothermal (2%). A breakdown of biomass demonstrates that wood (19%) is the prevailing biomass,

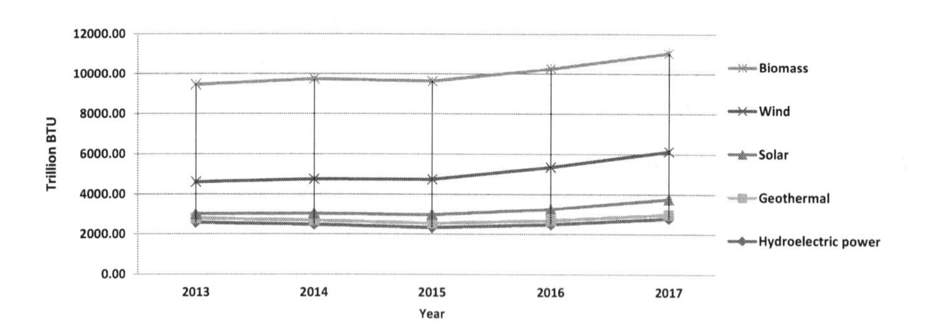

Figure 10.1 U.S. renewables consumption (trillion Btu, 2013–2017).

Source: EIA, 2018d. *Total Energy, Annual Data*. Energy Information Administration (EIA). Available at: www.eia.gov/totalenergy/data/annual/index.php, last accessed: June 24, 2018.

Table 10.1 U.S. Renewables consumption (trillion Btu, 2013–2017).

Renewables type	*2013*	*2014*	*2015*	*2016*	*2017*
Hydroelectric power	2,562.38	2,467.00	2,321.20	2,472.44	2,770.01
Geothermal	214.01	215.00	211.84	209.60	210.99
Solar	224.52	337.00	425.73	568.66	774.27
Wind	1,601.36	1,728.00	1,777.31	2,095.60	2,374.27
Biomass	4,850.19	4,992.00	4,897.91	4,913.26	4,913.25
Total	**9,452.49**	**9,738.00**	**9,633.95**	**10,259.55**	**11,015.80**

followed by waste (4%) and subsequently by various other biofuels. In the U.S. about 87% of bio-waste is converted into biofuels (EIA, 2017b).

Top Producing and Consuming Countries

The world leaders in biofuel development and use are the United States, Brazil, and Germany. As depicted in Figures 10.3, 10.4, and 10.5, the U.S. leads by far the global production and consumption of biofuels, ethanol, and biodiesel.

Figure 10.6 consists of four sub-figures and demonstrates the ratio of global energy resources and the share of bioenergy in the global primary energy mix (2014). Namely, Figure 10.6(a) shows the comparison among non-renewable energy (87%) versus the renewable energy (13%). Figure 10.6(b) shows the non-renewable energy segments, consisting of oil (29%), gas (24%), coal (29%), and nuclear energy (7%). As seen in Figure 10.6(c), bioenergy represents the vast majority of renewables with 77% of the total production, followed by hydro energy (15%) and other renewables (8%). Finally, Figure 10.6(d) breaks down the three main bioenergy segments, i.e., woody biomass (87%), agricultural crops and by-products (9%) and municipal/industrial waste (4%) (IEA Bioenergy, 2013).

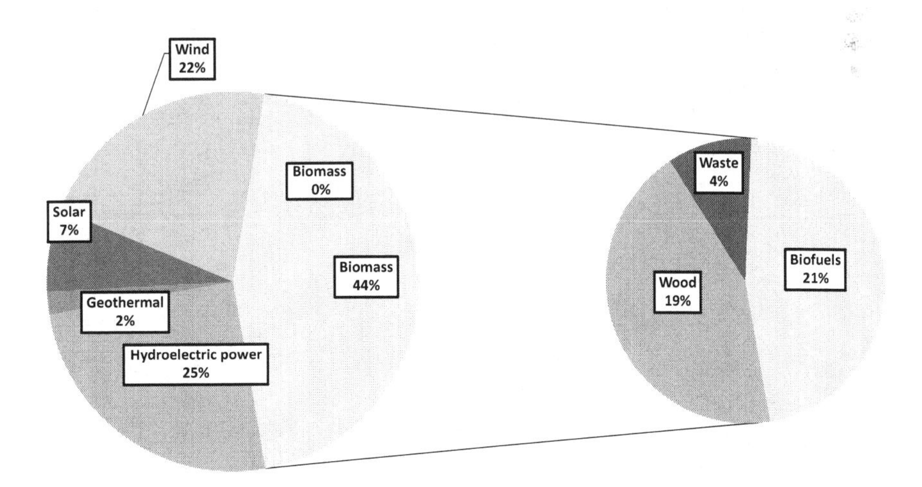

Figure 10.2 U.S. renewables consumption 2017 (%).

Source: EIA, 2018. *Total Energy, Annual Data*. U.S. Energy Information Administration (EIA). Available at: www.eia.gov/totalenergy/data/annual/index.php, last accessed: June 24, 2018.

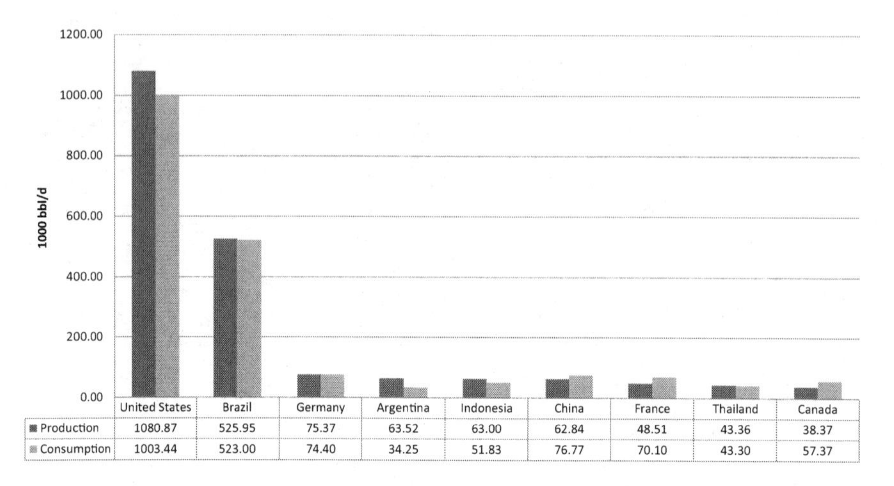

	United States	Brazil	Germany	Argentina	Indonesia	China	France	Thailand	Canada
Production	1080.87	525.95	75.37	63.52	63.00	62.84	48.51	43.36	38.37
Consumption	1003.44	523.00	74.40	34.25	51.83	76.77	70.10	43.30	57.37

Figure 10.3 Production and consumption of biofuels, 2016 (1000 bbl/d).

Source: EIA, 2017c. *International Energy Statistics, Total Biofuels Consumption*. U.S. Energy Information Administration. Available at:

www.eia.gov/beta/international/data/browser/#/?pa=000000000008&c=ruvvvvvfvtvnvv1urvvvvfvvvvvvfvvvou20evvvvvvvvvnvvuvo&ct=0&tl_id=79-A&vs=INTL.79-2-AFG-TBPD.A&vo=0&v=H&start=2000&end=2014, last accessed: June 24, 2017.

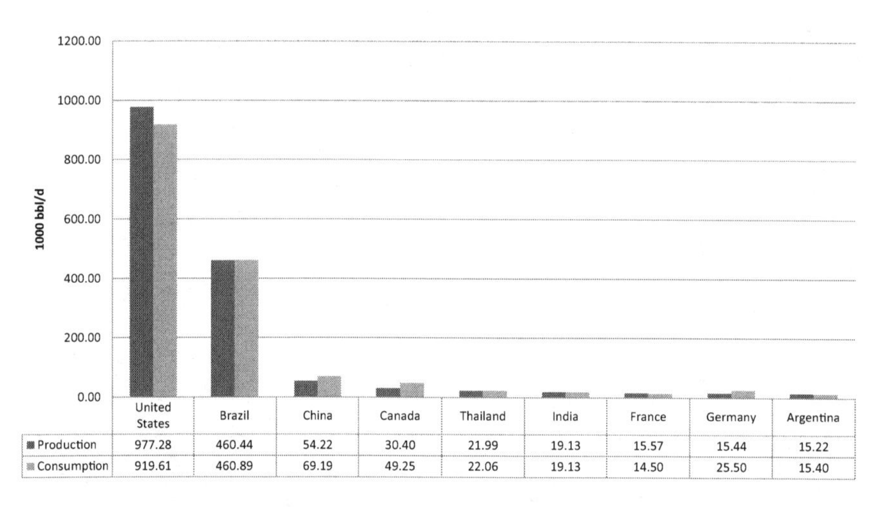

	United States	Brazil	China	Canada	Thailand	India	France	Germany	Argentina
Production	977.28	460.44	54.22	30.40	21.99	19.13	15.57	15.44	15.22
Consumption	919.61	460.89	69.19	49.25	22.06	19.13	14.50	25.50	15.40

Figure 10.4 Production and consumption of bio-ethanol, 2016 (1000 bbl/d).

Source: EIA, 2017c. *International Energy Statistics, Total Biofuels Consumption*. U.S. Energy Information Administration. Available at:

www.eia.gov/beta/international/data/browser/#/?pa=000000000008&c=ruvvvvvfvtvnvv1urvvvvfvvvvvvfvvvou20evvvvvvvvvnvvuvo&ct=0&tl_id=79-A&vs=INTL.79-2-AFG-TBPD.A&vo=0&v=H&start=2000&end=2014, last accessed: June 24, 2017.

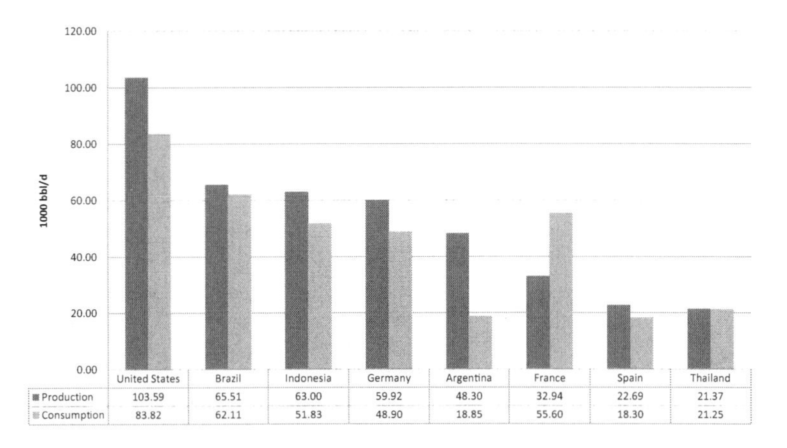

Figure 10.5 Production and consumption of biodiesel, 2016 (1000 bbl/d).

Source: EIA, 2017c. *International Energy Statistics, Total Biofuels Consumption*. U.S. Energy Information Administration. Available at:

www.eia.gov/beta/international/data/browser/#/?pa=000000000008&c=ruvvvvvfvtvnvv1urvvvvfvvvvvvfvvvou20evvvvvvvvvnvvuvo&ct=0&tl_id=79-A&vs=INTL.79-2-AFG-TBPD.A&vo=0&v=H&start=2000&end=2014, last accessed: June 24, 2017.

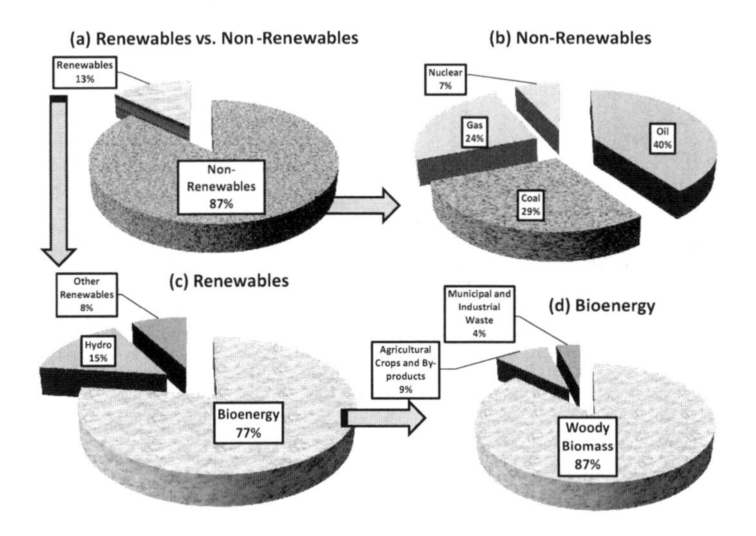

Figure 10.6 Global energy resources: Share of bioenergy in the global primary energy mix (2014).

Sources:

IEA Bioenergy, 2009. *A Sustainable and Reliable Energy Source*. Main Report. Paris: International Energy Agency. Available at: www.ieabioenergy.com/wp-content/uploads/2013/10/MAIN-REPORT-Bioenergy-a-sustainable-and-reliable-energy-source.-A-review-of-status-and-prospects.pdf, last accessed: June 24, 2018.

IEA Bioenergy, 2013. *A Sustainable and Reliable Energy Source. A Review of Status and Prospects*. International Energy Agency. Available at: www.ieabioenergy.com/wp-content/uploads/2013/10/MAIN-REPORT-Bioenergy-a-sustainable-and-reliable-energy-source.-A-review-of-status-and-prospects.pdf, last accessed: January 20, 2019.

Waste entails municipal solid waste, agricultural material/waste, animal waste, etc.

Algae and the use of aquatic flora is another promising biomass sector, with very promising technological and scientific innovations that have the potential to change the biofuel market.

Figure 10.7 depicts global biofuels, 80% of which produce biodiesel and the remaining 20% other products. Figure 10.8 demonstrates biomass and the making of ethanol: Corn (52%) is the prevailing feedstock used, followed by sugar cane (25%), molasses (8%), wheat (3%) and various others (12%).

Energy security and the need for national energy independence requires governments to have several energy alternatives. Although bioenergy leads the global renewable energy production, and new scientific breakthroughs suggest the tremendous future opportunities for growth, there are certain commercial and economic restrictions related to cost-efficiency: Growth is limited by the current regulatory framework in many countries. There is a need to revisit and modify the policies on subsidies and tax credit arrangements in order to support this promising energy segment.

Transport Fuels and Flexible Fuel Vehicles

Many biofuels, such as ethanol and biodiesel, are mainly used as transport fuels and are being used as additives to boost the octanes, protect the engine, improve the environmental performance, etc.

To ensure vehicle-specific energy efficiency, it is imperative to examine the potential interaction between the main fuel and biodiesel additive combination. Impact factors in the

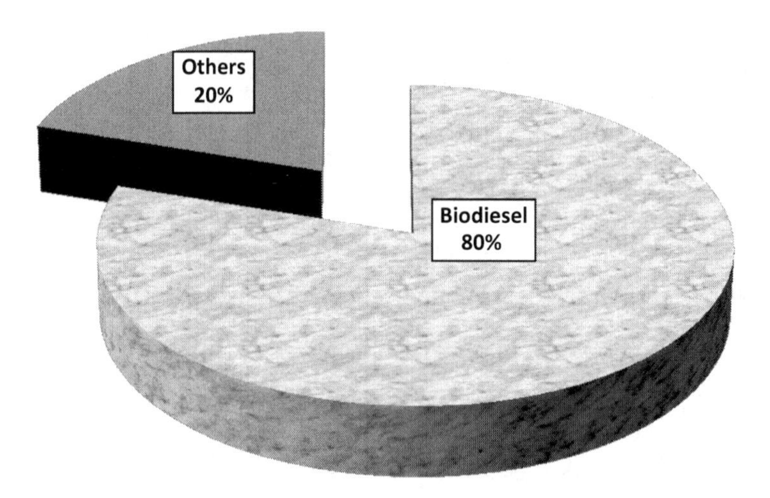

Figure 10.7 Global biofuels.

Sources:

IEA Bioenergy, 2009. *A Sustainable and Reliable Energy Source*. Main Report. Paris: International Energy Agency (IEA). Available at: www.ieabioenergy.com/wp-content/uploads/2013/10/MAIN-REPORT-Bioenergy-a-sustainable-and-reliable-energy-source.-A-review-of-status-and-prospects.pdf, last accessed: June 24, 2018.

IEA Bioenergy, 2013. *A Sustainable and Reliable Energy Source. A Review of Status and Prospects*. Paris: International Energy Agency. Available at: www.ieabioenergy.com/wp-content/uploads/2013/10/MAIN-REPORT-Bioenergy-a-sustainable-and-reliable-energy-source.-A-review-of-status-and-prospects.pdf, last accessed: June 24, 2018.

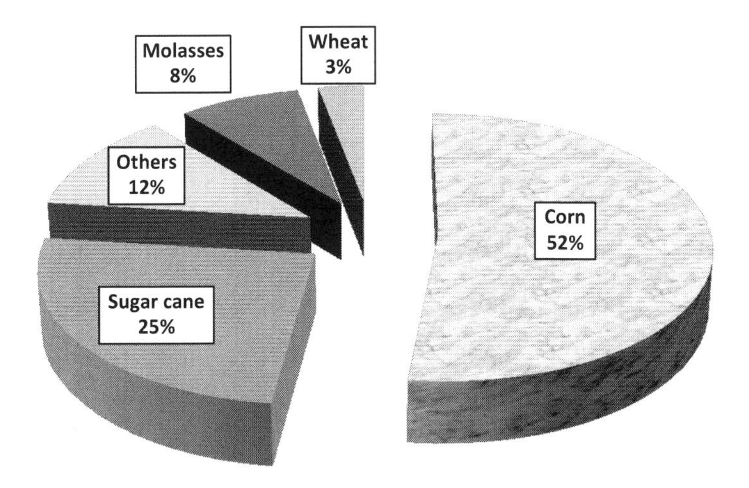

Figure 10.8 Biomass and the making of ethanol.

Source: EIA, 2017c. *International Energy Statistics, Total Biofuels Consumption*. U.S. Energy Information Administration. Available at: www.eia.gov/beta/international/data/browser/#/?pa=000000000008&c=ruvvvvvfvtv nvv1urvvvvfvvvvvvfvvvou20evvvvvvvvvnvvuvo&ct=0&tl_id=79-A&vs=INTL.79-2-AFG-TBPD.A&vo=0&v=H &start=2000&end=2014, last accessed: June 24, 2017.

decision-making process include the biofuel's blend, quality, feedstock type, plant processing, as well as the chemical quality and specifications of the conventional diesel.

Different additives and mixtures are suitable in different temperatures and climate conditions. For example, winter temperatures may impact the biofuel behavior and subsequently the engine performance, with short and long-term complications. For example, both biodiesel and No. 2 diesel contain compounds that may crystalize in cold climates.

Flexible Fuel Vehicles (FFVs) are using gasoline or gasoline–ethanol mixtures of approximately 85% ethanol (E85). Newer generation fuel typically contains increased ethanol content (DOE, 2018).

The Four Generations of Biofuels

Table 10.2 demonstrates the four generations of biofuels, with the respective information about the feedstock used as input, process, and products developed therefrom.

The Renewable Fuel Standard (RFS) program stipulates for the annual purchase of specific quantities of alternative fuels as per the following classifications (EPA, 2017a):

- **First-generation or conventional biofuel** uses conventional technology to treat vegetable oils, agricultural crops, and other arable farming products. It includes all types produced from complex carbohydrate feedstocks, which are derivatives of starch, sugar cane, vegetable oil, corn, grains, and so on. What makes the feedstock in this category a "first-generation fuel" is the fact that they include human foodstuffs, thus contributing to increased food prices, and diminished availability of land, fertilizers, and other agricultural resources. Considering they are conveniently produced by means of conventional engineering methods, they are also known as conventional biofuels.

Table 10.2 The four generations of biofuels.

1st Generation	*2nd Generation*	*3rd Generation*	*4th Generation*
Food vs fuel debate	*Non-food crops resolve the food vs fuel debate*		
Feedstock Sugar, starch, oil seeds Corn, cane, maze, wheat	*Feedstock* *Usage of seawater & waste water Non-food crops Inedible waste product Citrus peels Switchgrass, Sawdust	*Feedstock* *Fast growing plants **No need to occupy arable land Lignin trees (transgenic materials) Algae Botryococcus Braunii Chlorella Vulgaris Macroalgae	*Feedstock* *Boosts calorific value ** Optimum production ***High yield Algae Wheat and barley
Process Fermentation Hydrolysis Distillation	*Process* Mainly biochemical and thermochemical Enzyme treatment Fermentation Reactive distillation, transesterification	*Process* Mainly biochemical and thermochemical Advanced enzyme treatment Fermentation Distillation	*Process* Metabolic engineering Genetic manipulation to enhance biofuel production Mainly pyrolysis, gasification, solar-to-fuel.
Products Bioethanol Bio-oil, biodiesel Vegetable oil Bioalcohol, bioether Solid biofuels	*Products* Cellulosic ethanol & other alcohols Biohydrogen Biomethane Fischer-tropsch diesel Myco-diesel	*Products* Cellulosic ethanol & other alcohols. Syngas, biodiesel, biobutanol, biohydrogen, biomethane	*Products* Cellulosic ethanol & other alcohols. Syngas, biodiesel, biobutanol, biohydrogen, biomethane

Source: the Author.

This category is considered to have a negative Net Energy Gain (NEG), which according to energy economics pertains to the ratio between the energy consumed to harvest and the thermal efficiency of the energy accumulated from that harvest.

- **Second-generation biofuels, or advanced biofuels**, represent fuels generated from different types of biomass through biochemical and/or thermochemical processes, while encompassing all categories of organic carbon renewables.

  - **Biomass-based diesel** denotes renewable diesel power developed by alternative feedstocks. These include biodiesel treated with petroleum, or non-ester renewable diesel, including soybean oil, canola oil, animal fats, etc.
  - **Cellulosic biofuels** represent fuels such as biogas and cellulosic ethanol. Biofuels in this category can be manufactured from diverse types of organic carbons, encompassing both plant and animal materials, that are not appropriate for food consumption. These are generated by cellulose, hemicellulose or nonfood alternative feedstocks, including wood chips, corn, etc. (DOE, 2017a).

Advanced technologies are required for the conversion processes of the non-edible feedstocks as they require intricate efforts during the production process. For this reason second-generation fuels are known as "advanced biofuels."

This category increases in popularity as the feedstock is non-edible, hence does not deprive the food supply. They favor biodiversity and their effects on land use and the soil, water, noise, and air quality are non-intrusive. These fuels are environmentally friendly and can be used as additives and blending agents. Nevertheless, this energy segment is still under development in order to improve their thermal efficiency.

Second-Generation (Advanced) Biofuels: LC Biomass

The second-generation biofuels are produced by lignocellulosic biomass (i.e., generated from lignin, cellulose, and hemicellulose) and other plant biomass types (e.g., agricultural waste or timber waste).

Unconventional technology and more complex protocols are used in the second generation, as the process of collecting, storing, and treating the biomass types throughout the final generation of fuel is more complex and currently more costly.

Cellulosic Ethanol

The second-generation technologies develop modified versions created by cellulose and hemicellulose. It is developed by the transesterification method of Fatty Acid Methyl Esters (FAMEs). Biofuels are produced by lignocellulosic biomass, agricultural waste, or timber waste.

The cellulose components of biomass are dissolved into sugars, and this is how ethanol is generated. Recommended action includes:

- Investment, and STEM disciplines (i.e., Science, Technology, Engineering, and Mathematics) are needed in order to transform lignocellulosic feedstocks into a liquid form of energy that is compatible with transportation fuel vehicles.
- There is a need to further improve the oil yields of biomass used.
- Alternative solutions are required for water and land configurations.

The Net Energy Gain (NEG) for this category is marginal between break-even and positive. Nevertheless, the energy industry has also added a third generation of biofuels.

The 3rd and 4th Generation of Biofuels

The efficiency of the 3rd and 4th generations of biofuels will be determined by the scientific breakthroughs that will generate a biofuel that is sustainable while also encompassing the Triple-E notion: Energy efficiency, economy efficiency, and environmental integrity.

There is a plethora of ongoing technological and scientific projects that aim to meet the Triple-E prerequisites for the new era of biofuels. Currently, the industry considers the following biofuels and methods as the most promising candidates for the transition into renewable energy.

Aquatic Biofuels: Algae for Bio-crude and Ethanol Production

Algae is the principal representative of the third and fourth generation of biofuels. Algae, also known as microscopic algae or micro-algae, is an oil-rich aquatic flora that employs the solar light with water and carbon dioxide in order to produce biomass. The photosynthesis process is faster and more productive compared with terrestrial biomass.

The need for energy independence, combined with global market uncertainties, have generated great interest in algae culture, i.e., the farming of algae, and the potential of producing algae biofuels of optimum thermal efficiency.

Due to its high calorific value, algae attracts the attention of energy moguls, researchers, and investors. This organic compound is considered as the most promising among other biofuels with the potential to truly impact the renewable energy market.

Algae has the potential to replace hydrogen generated from fossil fuels, and eventually develop algal oil, jet fuel, ethanol, biodiesel (green diesel), bio-gasoline and other products.

Benefits of algae:

- It possesses high energy yield due to its oil-rich composition, with a natural oil content of over 50%.
- Algae can be cultivated and harvested in aquatic locations away from agricultural or other public utility or recreational areas. Hence, if harvested in an appropriate aquatic environment, the production of algae fuel will not negatively impact the food chain, and will not compromise the terrestrial soil or water use.
- Based on the principle that two thirds of Planet Earth consists of water, aquatic biofuels have ample space available for production and harvesting. This provides an excellent opportunity for versatile highly innovative harvesting configurations that can be region-specific and climate-specific. Some of the proposed harvesting configurations suggest floating screens and floating bags in the ocean. Literally, the ocean is the limit as to the algae ponds and algae wastewater treatment plants of the next generation.

Lignocellulosic Compounds

The future growth and efficiency of biofuels depends on the government and industry's ability to shift towards the utilization of second-generation biofuels, also known as lignocellulosic compounds. These will increase the energy efficiency of waste, and will further utilize unwanted materials such as flora and fauna compounds from the farming, timber, and other residues that are inappropriate for conventional usage.

CELLULOSIC ETHANOL COMMERCIALIZATION

A substantial improvement in second-generation biofuels entails the pre-treatment of the biomass and the application of by-products. As advanced biofuels such as cellulosic forms evolve in terms of technologies, processes, and market segment growth, traditional nutrients will no longer serve as biomass to supply the first-generation biofuels, but will feed the global food supply chain.

- Investment and technology is required to transform lignocellulosic feedstocks into a liquid form of energy, as required to run in the existing vehicle and industrial engines, which is compatible with transportation fuel vehicles' specifications.
- There is a need to further improve the oil yields of biomass used.
- Alternative solutions are required for land and water configurations.

Green Diesel or Hydrotreated Diesel

Green diesel is developed via the hydro-cracking process of organic oil-rich biomass. Hydro-cracking technologies utilize high-pressure, high-temperature methods. Catalystic treatment dissolves large compounds into smaller hydrocarbons for the fuel to be suitable for combustion engines.

As with other biofuels, diesel engines do not need modifications to utilize green diesel. Due to the capital-intensive technology-intensive processes, this fuel is not cost-efficient at the moment. However, several leading energy and engineering corporations such as ConocoPhillips, Valero, Man B&W and Wartsila are investing in the future evolution of green diesel.

Conclusions

As stipulated by the Renewable Fuel Standard (RFS) and according to the 2007 amendment of the Energy Independence and Security Act, the shift from the first-generation biofuels will serve both the traditional industries in agriculture and forestation, but will also eliminate financial and commercial losses in energy sources of low calorific value. Biotechnology and other STEM disciplines (Science, Technology, Engineering, and Mathematics) will boost the development and advancement of cellulosic biofuels of the second and third generation.

The energy industry needs to meet the increasing energy demands for the future, while taking into consideration the patterns of rapid population growth, and consequent global trade and transport growth. While recognizing the pivotal role of the fossil fuels, governments and industries alike seek for a long-term energy solution in an effort to mitigate the future needs when fossil fuels are scarce. The eventual transition between non-renewable and renewable energy sources must be smooth. The energy sources of the future must be sustainable, energy-efficient and engine-efficient, with higher octane ratings, increased thermal efficiency, cost efficiency, and environmentally friendly.

Technologies and Supply Chain Processes

According to the U.S. Department of Energy, biofuels have been in a "demonstration and pilot" phase. A great benefit they have is the ability to use similar equipment, cargo handling systems, and the infrastructure and superstructure of fossil fuel energy supply chains (DOE, 2014).

Converting Biomass into Biofuel

Biomass-based renewables are transformed into energy in a manner that resembles natural energy conversion processes that appear in plants. Figure 10.9 depicts the biomass categories, energy conversion methods, and energy products used as heat, electricity, and fuels.

Depending on the biomass feedstock available and the desired end biofuel products, the following conversion methods are applied.

Photosynthesis, Artificial Photosynthesis, Biomimetic Photosynthesis

Photosynthesis is a natural method of photochemical transformation, accumulation, and storage of solar energy with a commendable effectiveness and efficiency. Trees, plants, and other microorganisms store light energy as nutrient molecules intended for the plant, and transform it into chemical energy (glucose). In the process, chlorophyll, which is the green pigment in the chloroplast, is the catalyst that enables natural sunlight to transform CO_2 and oxygen into O_2 and glucose (DOE, 2014).

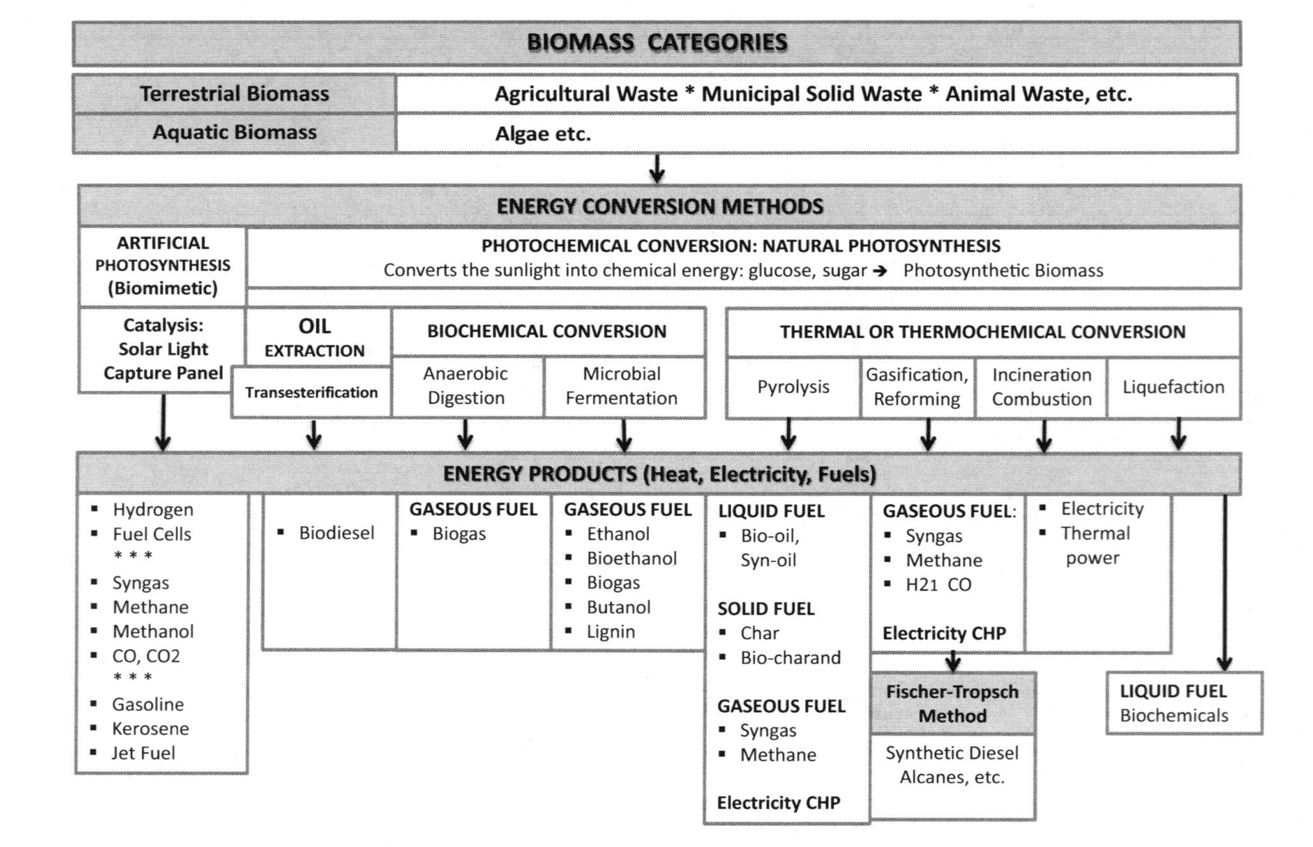

Figure 10.9 Biomass categories * conversion methods * energy products.

Source: the Author.

In this natural photochemical method, trees, plants, algae, photosynthetic bacteria, and other micro-organisms use solar-powered energy to yield biomass.

Thermal or Thermochemical Conversion

The thermochemical conversion process typically generates bioethanol and syngas, which is a compound of hydrogen and carbon monoxide. This is achieved by exposing the biomass to gasification, with high temperatures and chemical treatment. A catalyst serves as chemical facilitator, thus converting the synthetic gas to ethanol and other liquescent fuels.

The thermo-chemical conversion methods include: 1) combustion, 2) liquefaction, 3) pyrolysis, and 4) gasification.

COMBUSTION

Biomass combustion transforms biomass feedstock into energy through its transportation in high volumes of oxygen. The combustion process allows biomass to emit the carbon dioxide that was previously absorbed from the sunlight. Direct combustion is the most established method for transforming biomass into energy. As the biomass energy is released, it is transformed into transportation fuels such as ethanol and biodiesel, methane gas, etc. (EIA, 2018a).

LIQUEFACTION

Hydrothermal **Liquefaction** (HTL) is the thermochemical conversion also known as thermal depolymerization process where high pressures and moderate temperatures (392–662°F, i.e., 200°C to 350°C) are used to convert wet **biomass** macromolecules. Hydrolysis breaks the solid bio polymeric compounds into small molecules, thus producing bio-oil and other liquid components.

PYROLYSIS

Pyrolysis is a thermochemical breaking down of large biomass molecules into smaller ones at high temperatures (exceeding 400°F or 200°C) in the absence of air or halogens. The higher the pyrolysis temperatures, the higher the yield of biofuel. A thermal decomposition of the chemical particles transforms materials like cellulose, hemicellulose, lignin, etc., into flammable gases and carbons. The feedstock quality and chemical structure as well as the pyrolysis variables (i.e., operating temperature, pressure, oxygen content, and other conditions) will determine the output, i.e., the products generated, level of energy yield, greenhouse emissions, etc.

Typically, biomass pyrolysis generates three types of fuel, in a liquid, solid, and gaseous form, including liquid fuel (oxygenated oil or bio-oil), solid fuel (char or bio-charand) and gaseous fuel (syngas or synthetic gas, methane, carbon monoxide, carbon dioxide, etc.) (USDA, 2017).

GASIFICATION

The **thermochemical or biomass gasification method** converts solid biomass energy into a combustible gas with a high calorific value through a fumigation process. In this method, gaseous products are generated as an outcome of the oxidation and thermal cracking process.

Gasification is a procedure that transforms carbonaceous materials, such as organic components, into syngas, carbon monoxide, carbon dioxide, and hydrogen. The biomass quality as well as the gasification and fumigation protocols to be followed will determine the overall product quality and chemical properties.

Biochemical Conversion Methods

The most common biochemical conversion methods are microbial fermentation and anaerobic digestion (also known as biomethanation). These methods involve the use of bacteria, enzymes, and other microorganisms to break down biomass into liquid or gaseous fuels, such as biogas and bioethanol.

As a first step, a pre-treatment method is needed for the biochemical process in order to emit the cellulose and hemicellulose sugar from starch feedstock. A hydrolysis method is required in order for carbohydrates to break down into sugars.

MICROBIAL FERMENTATION

Fermentation is an irreversible process of converting carbon-based feedstocks like sugar cane and corn into ethanol, butanol, and other hydrocarbons. This large-scale production is achieved via the microbial (yeast) fermentation of sugars, distillation, dehumidification, and chemical modification, as needed. Cellulolysis is the saccharification of cellulose whereby saccharification pertains to the use of enzymes in order to convert feedstocks into sugars. Hence, the fermentation process requires the hydrolysis of saccharides, i.e., of feedstocks like cellulose, starch, and other carbohydrates (NCBI, 2018).

ANAEROBIC DIGESTION (AD) OR BIOMETHANATION

This is an intricate biological process where microorganisms break down biodegradable material while depleted of oxygen. Biogas is one of the byproducts that can either be combusted to produce electricity and heat, or be transformed into renewable natural gas and transportation fuels.

Anaerobic digester methods were created to optimize the mixing of additives with conventional fuels, reduce the fuel volume reduction, while improving its calorific value, etc.

Anaerobic digestion is a biological method during which organic compounds disintegrate in an oxygen-depleted atmosphere. During the digestion process, various microbes are present in organic compounds including industrial and edible waste material, animal manure, etc. Science and technology assist the method's monitoring and optimization (EPA, 2016).

The chemical reactions taking place throughout anaerobic digestion include hydrolysis, fermentation, and methanogenesis or biomethanation, i.e., the biological production of carbon dioxide and methane via microorganisms known as methanogens (EPA, 2016).

Benefits, Risks, and Recommendations

One of the great security benefits of biofuels pertain to the reduction of imported energy, and its contribution towards energy independence.

Biofuels

Benefits

Environment and safety:

- Green energy with moderate atmospheric emissions.
- Viable projects will be energy-efficient, cost-effective, and environmentally friendly.
- Second- and third-generation biofuels resolve the environmental and food-related debates, thus offering new opportunities for public involvement and support in future plans.
- Economic growth as alternative fuel: Empowerment and energy independence of regions deprived of fossil fuels or other energy resources.
- Second- and third-generation biofuels impose an environmental solution: As forests now have diminished capability to absorb the emissions from other fuel forms, the future biofuel forms may heal the environment.

Commerce, technology, and economics:

- The biofuel energy supply, just like any other energy segment, inevitably affects the energy prices, in particular those of fossil fuels.
- Can be used as rocket fuel for lightweight aircrafts.

Risks and Challenges

Environment and safety:

- Biodiversity and ecosystem challenges in the first generation of biofuels.
- Biofuels generate atmospheric emissions, although their greenhouse outflow is lower compared with fossil fuels. Depending on the biomass quality and chemical/organic composition, biofuels typically emit: Carbon particulate matter, carbon monoxide, nitrous oxides, and so on.
- The food versus fuel debate in first-generation biofuels. Biofuel production, especially in agricultural biomass, occupies arable land that could otherwise produce food. This increases food cost due to food scarcity.
- Energy-related land development leads to risks of deforestation, geological erosion, water treatment, edge tillage, etc.
- So, there is moderate environmental impact, but biofuels can eliminate greenhouse gases.

Commerce, technology, and economics:

- Increased cost due to technology-intensive treatment.
- Not yet ready for large-scale production. Low calorific value, e.g., ethanol's calorific value is 34% of the conventional fossil fuel diesel.
- There is a need to ameliorate biofuel yields by the use of biotechnology, cellulose processing, and other methods.
- Presently not sustainable financially without subsidies.
- The industry still needs to improve the energy efficiency vs cost optimization, in order for the quality and quantity of biofuel to become comparable with the prevailing energy segments, such as the fossil fuels.

Biodiesel

Benefits

Environment and safety:

- Biodiesel is a greener, clean-burning, carbon-neutral fuel. Certain plants, such as palm oil trees and soybeans, actually absorb CO_2.
- Biodegradable, when combusted it emits less pollutants than fossil fuel diesel.
- Non-toxic, free from sulfur and lead.
- Less corrosive than other fuels.

Commerce, technology, and economics:

- Biodiesel can be used in diesel engines without engine modifications, as its chemical composition is identical to fossil diesel.
- It is being safely blended with conventional diesel, and offers optimum performance.
- An effective solvent: It cleanses residues placed by fossil-fuel generated diesel and dissolves deposits in piping systems and tanks.

Risks and Challenges

- First-generation biodiesel entailed land clearing and burning. However, second Generation biodiesel is increasingly diversifying from conventional land treatment methods, and are implementing methods and technologies that are less intrusive, with minimum impact on the ecosystem, the food supply, or other anthropocentric factor.

Biobutanol

Benefits

Environment and safety:

- Environmentally safer with less CO_2 and GHG emissions.

Commerce, technology, and economics:

- High energy content, i.e., 20% higher than ethanol.

Biomethane

Benefits

Environment:

- Low carbon with limited impact on air quality.
- Limited noise pollution.
- Limited waste to landfill.

Commerce, technology, and economics:

- Its high calorific value makes the fuel suitable for electricity, cooking, and heating.
- Lower cost reduction of biomethane vs diesel.
- Prolongs engine life.

Risks and Challenges

- It is combustible and imposes fire hazards, just like most other fuels.
- It is considered a greenhouse gas (GHG), hence is subject to environmental regulations.

Bioethanol

Benefits

Environment and safety:

- It is environmentally friendly.
- Certain feedstocks such as corn and sugarcane actually absorb carbon emissions and other greenhouse gases (GHGs).
- It is biodegradable and non-toxic, with less environmental pollution impact.

Commerce, technology, and economics:

- Bioethanol is a flexible additive, available in various grades, i.e., E10, E15, E85, etc.
- It has a high calorific value: It serves as an octane booster both in its pure form and as an additive.
- It does not require engine modifications.
- It enhances the engines' compression ratio, and thus boosts energy efficiency. It enhances the primary fuel's calorific value.
- Bioethanol is cost-effective due to the subsidies available in the U.S., Europe, and other regions. It is a rather common fuel in Western Europe.
- Second-generation ethanol is more fuel efficient; consumes less power compared with the first-generation fuels that undergo the fermentation process.

Risks and Challenges

Environment and safety:

- Despite its greener performance, it still causes moderate pollution.
- Ethanol and ethanol-gasoline mixtures have greater evaporative releases from storage and transportation tanks and distributing gear, resulting in the generation of ground-level ozone and smog.
- Additional processing is required to eliminate the gasoline evaporative emissions prior to mixing with ethanol (EIA, 2018c).
- It contains about 33% of gasoline's energy content, hence has lower calorific value compared with fossil fuels.
- It is highly flammable, just like gasoline.

Commerce, technology, and economics:

- In cold climates, bioethanol has limited volatility hence vehicles have engine starting problems. As a remedy, the fuel mix should have increased gasoline intake, while the ethanol blend can be reduced to E70 (i.e., less than E85). Additionally an engine heater system may be installed.

Algae

Benefits

Environment and safety:

- It is environmentally friendly.
- Algae fuel is more eco-friendly compared with other fuels.
- It does not occupy arable lands and does not deprive water from the food industry as it is harvested in aquatic environments. It requires saline and waste water.
- It is biodegradable, non-toxic, and comparatively innocuous to the ecosystem if spilled.

Commerce, technology, and economics:

- Third- and fourth-generation algae offer fuel with a high calorific value.
- Price factors become more competitive as long as the tax credits are renewed.

Risks and Challenges

Algae production costs per unit mass (i.e., capital and operating expenses) are higher compared with other renewable energy segments. However, the recent scientific and technological breakthrough yields higher calorific values, making algae a most promising fuel for the future.

Case Studies

Case Study 1: ExxonMobil and the Future of Advanced (Aquatic) Biofuels: Algae

ExxonMobil is dedicated to sustainable investment in research on advanced biofuels. The development of algae-based biodiesel is a significant project among the corporation's several investments in innovative technologies with the transformative potential to enhance the production of energy, eliminate emissions, and optimize operational efficiencies (ExxonMobil, 2018).

ExxonMobil has signed an agreement with California-based Synthetic Genomics Inc. (SGI) as part of its ongoing research into advanced algae biofuels. The research develops algae that demonstrate significantly enhanced photosynthetic efficiency and energy production via selection and genetic engineering of higher-performance algae strains. The agreement builds on recent ground-breaking scientific discoveries of biological conduits regulating lipid generation and growth in cutting-edge algal strains (*Renewable Energy World*, 2017).

While algae using CO_2 to generate fat is not novel, the amount of fat generated by the algae in this project is truly remarkable. The SGI scientists collaborated with their colleagues at ExxonMobil as nutritionists of sorts, modifying the segment of the algae genome accountable for the absorption of nitrogen, a vital nutrient.

This has resulted in a revolutionary enhancement of algae's mass as fat by approximately 40%, i.e., over double the fat content of natural algae.

- Success in producing algae-based biodiesel at a commercial stage will offer numerous tangible benefits.
- The fuel releases fewer greenhouse gases compared with most conventional energy sources, and that will strengthen the continuing transition to greener fuel resources (ExxonMobil, 2017a).
- Fat algae are a significant achievement as they hold more lipids that can be transformed into energy-rich, yet clean and sustainable, biodiesel.
- Furthermore, unlike other biofuel feedstocks, algae generation at an industrial scale would not impact food production (ExxonMobil, 2017b).

Case Study 2: The "Food vs Fuel" Debate

Over 60 nations have biofuel mandates, due to environmental, technical, and economic challenges. The food vs fuel argument imposes a criticism on the biofuel energy industry, for producing energy by using arable land and feedstocks that could be used as food (e.g., crops like corn and sugar). Environmental, socioeconomic, and other energy segment groups have criticized the first generation of biofuels by introducing the ethical dilemma of sacrificing food resources and depriving social segments from nutrition.

Critics passionately raise the "food vs fuel" debate, as well as environmental and socioeconomic factors – all of which are associated with most, if not all, energy sources.

Upon examining the extensive media coverage on the "food vs fuel" debate, some critics purport that all biofuel feedstocks belong to the same category and thus are equally impactful on the environment, the arable land, and the food argument. Other critics focus on "the tremendous impacts of corn yield." The following debate points aim to shed light on several misconceptions, while motivating all energy industry segments (not just the biofuel professionals) to work together towards common goals related to the protection of the ecosystem.

1 The Biomass Cookie-Cutter Argument

- There are several feedstock segments used in the production of biofuels, most of which are not edible. The biofuel industry's scientific and technological advancements have created environmentally friendly biofuels. Cellulosic feedstocks from second generation and beyond significantly alleviate the food vs fuel challenge. They use fewer food-related feedstocks such as sugar or corn, but also less arable land and fertilizers. For example, cellulosic ethanol is typically used from grass, sawdust, fast-growing trees, and other non-edible material (DOE, 2017b).
- There are several biofuel segments requiring different production processes, and with diverse environmental, land-use, and water-use requirements.
- Corn production represents a small segment of biofuel production. The section above (pages 334 and 347) duly explains that the use of agricultural crops has been less than 9% of the global production of biofuels. Namely, these feedstock segments include: Woody biomass (87%), agricultural crops and by-products (9%) and municipal waste (4%).

- The biofuel production has moved away from the first-generation biofuels and has worked on alleviating any environmental or "food vs fuel" challenges. The second, third, and fourth generations of biofuels are greener, use non-arable lands, and cultivate yield of high calorific value, with limited impact on the food chain. For example, algae is a rapidly growing and most promising sector that is unrelated to the use of arable land and food.

2 Global Poverty, Food Insecurity and The Choice of Feedstock Argument

Critics of biofuels associate food insecurity in poverty-stricken regions with biofuel production and the use of sugar, corn, wheat, maize, oilseeds, etc. However, upon examining the types of feedstock used in less-developed countries (LDCs), it appears that they mainly use a) inedible oil-rich plants like jatropha and pongamia, and b) low-water usage crops like tropical sugar beet, sweet sorghum, and cassava.

3 Global Poverty and The High Cost of Food Argument

Critics of biofuels associate the high prices of food with biofuel production. In fact, the price of food increases due to the higher price of energy, pesticides, and fertilizers.

4 Socioeconomic Impact

Social elimination and inequality within rural areas; domestic migration associated with energy-producing areas.

5 Environmental Impact and The Use of Arable Land Argument

It is worth noting that most types of biofuels do not use arable land. Most biofuel yields require alkaline sodium-rich soils, which are unfitting for the production of food yield.

Regulations

2005: The Energy Policy Act of 2005 was responsible for regulations that ensured gasoline sold in the United States contained a minimum volume of renewable fuel, called the Renewable Fuels Standard (RFS). The regulations aim to sustain, or even increase, the use of renewable fuel, mainly ethanol made from corn.

2007: The Energy Independence and Security Act of 2007 expanded the Renewable Fuels Standard to require that 36 billion gallons of ethanol and other fuels be blended into gasoline, diesel, and jet fuel by 2022. The Act institutes the annual percentage values for the Renewable Fuels Standard program for advanced biofuel, cellulosic biofuel, biomass-based diesel, and total renewable fuel. The rules apply to all diesel and gasoline produced or imported each calendar year (EPA, 2013).

The Renewable Fuels Standard was established by the U.S. Environmental Protection Agency in collaboration with the U.S. Department of Energy and Department of Agriculture. It is a federal program established under the auspices of the U.S. Energy Policy Act of 2005 (EPACT), as a modification of the U.S. Clean Air Act (CAA). Its scope broadened when the Energy Independence and Security Act of 2007 (EISA) further revised the CAA (EPA, 2017b). The RFS was developed to lower carbon emissions, but also to ensure that alternative energy is used as part of the national strategy for energy independence and energy security.

The program also demands a specific amount of renewable energy to substitute or decrease the volume of petroleum-based transport fuel, jet fuel, or heating oil sold in the U.S. on an annual basis. It is an excellent tool to classify diverse biofuels according to biomass origin, as well as the technologies and other scientific tools used.

It is worth noting that certain biofuels may be marginally considered, e.g., as first- or second-generation fuels, and thus fit into more than one RFS category. A case-specific, company-specific investigation is necessary for government and industry professionals to better evaluate in which classification each fuel better fits. Hence, this dual classification is a reminder that science and technology constantly evolve, and a single scientific breakthrough has the power to transform the energy industry.

In 2018, EPA announced the final Renewable Volume Obligation (RVO) numbers under the Renewable Fuels Standard, as follows:

1. The volume requirements for total renewable fuel is set for 19.29 billion gallons including 15 billion gallons for conventional biofuel.
2. Advanced biofuel is set for 4.29 billion gallons, including 288 million gallons of cellulosic biofuel.
3. The biomass-based biodiesel amount remains at 2.1 billion gallons (EPA, 2018).

As a general conclusion, it is worth noting that the biofuel industry has demonstrated great improvements from the second generation and beyond, making it one of the most promising alternative fuel industries for decades to come.

References

Department of Energy (DOE) (2014). *Photosynthesis and Biomass Growth.* National Renewable Energy Laboratory, U.S. Department of Energy (DOE). Available at: https://energy.gov/sites/prod/files/2014/06/f16/biomass_photosynthesis.pdf, last accessed: March 25, 2018.

Department of Energy (DOE) (2017a). *Alternative Fuels Data Center, Renewable Fuel Standard.* Alternative Fuels Data Center (AFDC), U.S. Department of Energy. Available at: www.afdc.energy.gov/laws/RFS, last accessed: April 30, 2017.

Department of Energy (DOE) (2017b). *Starch- and Sugar-Based Ethanol Production.* Alternative Fuels Data Center (AFDC), U.S. Department of Energy. Available at: www.afdc.energy.gov/fuels/ethanol_production.html, last accessed: April 30, 2017.

Department of Energy (DOE) (2017c). *Bioenergy, Biofuel Basics.* U.S. Department of Energy. Available at: www.energy.gov/eere/bioenergy/biofuels-basics, last accessed: March 25, 2018.

Department of Energy (DOE) (2018). *Flex-Fuel Vehicles. The Official U.S. Government Source for Fuel Economy Information.* Office of Energy Efficiency and Renewable Energy, U.S. Department of Energy and U.S. Environmental Protection Agency (EPA). Available at: www.fueleconomy.gov/feg/flextech.shtml, last accessed: January 25, 2018.

Energy Information Administration (EIA) (2017a). *Renewable & Alternative Fuels.* U.S. Energy Information Administration. May 3. Available at: www.eia.gov/renewable, last accessed: June 1, 2017.

Energy Information Administration (EIA) (2017b). *Biomass Explained. Waste-to-Energy (Municipal Solid Waste).* U.S. Energy Information Administration, January 24. Available at: www.eia.gov/energyexplained/?page=biomass_waste_to_energy, last accessed: January 25, 2018.

Energy Information Administration (EIA) (2017c). *International Energy Statistics, Total Biofuels Consumption.* U.S. Energy Information Administration. Available at: www.eia.gov/beta/international/data/browser/#/?pa=000000000008&c=ruvvvvvfvtvnvv1urvvvvfvvvvvvfvvvou20evvvvvvvvvnvvuvo&ct=0&tl_id=79-A&vs=INTL.79-2-AFG-TBPD.A&vo=0&v=H&start=2000&end=2014, last accessed: June 24, 2017.

Energy Information Administration (EIA) (2018a). *Biomass Explained. Biomass—Renewable Energy from Plants and Animals*. U.S. Energy Information Administration, adapted from The National Energy Education Project (public domain), update: June 21. Available at: www.eia.gov/energyexplained/index.cfm?page=biomass_home, last accessed: June 21, 2018.

Energy Information Administration (EIA) (2018b). *Biomass Explained: Wood and Wood Waste*. U.S. Energy Information Administration, update: June 20. Available at: www.eia.gov/energyexplained/index.cfm?page=biomass_wood, last accessed: June 21, 2018.

Energy Information Administration (EIA) (2018c). *Biofuels: Ethanol and Biodiesel Explained. Ethanol and the Environment*. U.S. Energy Information Administration. March 23. Available at: www.eia.gov/energyexplained/index.cfm?page=biofuel_ethanol_environment, last accessed: June 24, 2018.

Energy Information Administration (EIA) (2018d). *Total Energy, Annual Data*. Energy Information Administration. Available online: www.eia.gov/totalenergy/data/annual/index.php, last accessed: June 24, 2018.

Environmental Protection Agency (EPA) (2013). *Renewable Fuel Standards for Renewable Fuel Standard Program (RFS2) Final Rulemaking*. U.S. Environmental Protection Agency, August 15. Available at: www.epa.gov/renewable-fuel-standard-program/2013-renewable-fuel-standards-renewable-fuel-standard-program-rfs2, last accessed: April 30, 2017.

Environmental Protection Agency (EPA) (2016). *Anaerobic Digestion and its Applications*. U.S. Environmental Protection Agency. EPA/600/R-15/304 | October 2015 | Land Remediation and Pollution Control Division. Prepared by Costa, Allison, Ely, Charlotte, Pennington, Melissa, Rock, Steve, Staniec, Carol and Turgeon, Jason. National Risk Management Research Laboratory, Cincinnati, OH 45268. Available at: www.epa.gov/sites/production/files/2016-07/documents/ad_and_applications-final_0.pdf, last accessed: September 12, 2017.

Environmental Protection Agency (EPA) (2017a). *Renewable Fuel Annual Standards*. U.S. Environmental Protection Agency. Available at: www.epa.gov/renewable-fuel-standard-program/renewable-fuel-annual-standards, last accessed: January 25, 2018.

Environmental Protection Agency (EPA) (2017b). *Overview for Renewable Fuel Standard*. U.S. Environmental Protection Agency. Available at: www.epa.gov/renewable-fuel-standard-program/overview-renewable-fuel-standard, last accessed: April 30, 2017.

Environmental Protection Agency (EPA) (2018). *EPA Finalizes RFS Volumes for 2018 and Biomass Based Diesel Volumes for 2019*. U.S. Environmental Protection Agency. Available at: www.epa.gov/newsreleases/epa-finalizes-rfs-volumes-2018-and-biomass-based-diesel-volumes-2019, last accessed: January 1, 2019.

ExxonMobil (2017a). "The fat, fit, fantastic fit machine". *EnergyFactor*, ExxonMobil, June 19. Available at: https://energyfactor.exxonmobil.com/science-technology/fat-fit-algae-biofuel, last accessed: June 3, 2017.

ExxonMobil (2017b). "Ka-pow! Algae debuts a fat, fit new look". *EnergyFactor*, ExxonMobil, June 22. Available at: https://energyfactor.exxonmobil.com/science-technology/algae-comic, last accessed: June 3, 2017.

ExxonMobil (2017c). *Algae Energy Farmer* [Image]. ExxonMobil. Available at: http://corporate.exxonmobil.com/en/company/multimedia/energy-lives-here/energy-farmer, last accessed: June 3, 2017.

ExxonMobil (2018). *Advanced Biofuels and Algae Research*. ExxonMobil. Available at: http://corporate.exxonmobil.com/en/energy/research-and-development/advanced-biofuels/advanced-biofuels-overview, last accessed: June 24, 2018.

International Energy Agency (IEA), Bioenergy (2009). *A Sustainable and Reliable Energy Source*. Main Report. Paris: International Energy Agency (IEA). Available at: www.ieabioenergy.com/wp-content/uploads/2013/10/MAIN-REPORT-Bioenergy-a-sustainable-and-reliable-energy-source.-A-review-of-status-and-prospects.pdf, last accessed: June 24, 2018.

International Energy Agency (IEA), Bioenergy (2013). *A Sustainable and Reliable Energy Source. A Review of Status and Prospects*. Paris: International Energy Agency. Available at: www.ieabioenergy.com/wp-content/uploads/2013/10/MAIN-REPORT-Bioenergy-a-sustainable-and-reliable-energy-source.-A-review-of-status-and-prospects.pdf, last accessed: June 24, 2018.

IEA, 2013, IEA, 2006, and IPCC, 2007. http://www.ieabioenergy.com/wp-content/uploads/2013/10/MAIN-REPORT-Bioenergy-a-sustainable-and-reliable-energy-source.-A-review-of-status-and-prospects.pdf . Last access: Jun *24* 2017

National Center for Biotechnology Information (NCBI) (2018). *Compounds, Methane. Compound Summary for CID 297.* U.S. National Library of Medicine, National Center for Biotechnology Information. Available at: https://pubchem.ncbi.nlm.nih.gov/compound/methane#section=Top, last accessed: June 24, 2018.

Renewable Energy World (2017). "ExxonMobil extends research agreement for advanced algae biofuels". *Renewable Energy World*, January 27. Available at: www.renewableenergyworld.com/articles/2017/01/exxonmobil-extends-research-agreement-for-advanced-algae-biofuels.html, last accessed: June 3, 2017.

U.S. Department of Agriculture (USDA) (2017). *Biomass Pyrolysis Research.* U.S. Department of Agriculture, Agricultural Research Service. Available at: www.ars.usda.gov/northeast-area/wyndmoor-pa/eastern-regional-research-center/docs/biomass-pyrolysis-research-1/what-is-pyrolysis, last accessed: September 12, 2017.

11 Cyber Security in the Era of Big Data

> The threat we see now is coming at us from all sides.
>
> This is a wider range of threat actors.
>
> A blended threat of nation-states using criminal hackers to do their dirty job.
>
> They are no longer dependent on just intelligence services. Instead, they utilize people from all walks of life—hackers, businesspeople, academics, researchers, diplomats, tourists—and anyone else who can get their hands on something of value.
>
> FBI Director Christopher Wray,
> Boston Conference on Cyber Security (BCCS) 2018

Cyber Security: An Overview

Cyber security is "the supply chain of the supply chains" due to its unique ability to monitor and control all supply chain operations in real time. It pertains to the defense of data, critical information and programs from different cyber threat types, encompassing cyber espionage, identity theft, cyber terrorism, cyber warfare, etc.

As the "big data universe" grows exponentially, governments and industries alike are presented with tremendous opportunities, but also risks in case hackers compromise their systems.

As depicted in Figure 11.1, cyber security in the energy industry safeguards the big data universe, government, banking, and corporate data, as well as the energy grid. The figure also suggests that certain energy segments (e.g., nuclear) involve innovative science and technology, and are used for defense, space exploration, and other critical purposes. As such, data leaks may be even more critical compared with those in conventional energy segments. Indeed, from 2010 to date hackers target nuclear facilities, yet, fortunately, no serious breach has been reported despite the hackers' persistent attempts.

Cyber-security threats and the fear of compromised big-data systems are shared among all energy companies. Cyber-security protocols safeguard energy networks, systems, and assets from both malicious and unintentional attacks, while ensuring that critical and confidential information is not compromised.

Although more than half of the global companies have to some extent been attacked by cyber criminals, it is a nation's critical infrastructure entities, in both the public and private sector, whose potential IT compromises will have significant impact at a national security level, including in the areas of energy, electric grid, mining, manufacturing, transport, banks and financial institutions. Cyber-security systems are designed to safeguard data, networks, and utilities from large-scale cyber attacks. Criminal groups often seek to attack high-value targets, including energy infrastructures, communication and network systems, hardware and software, and so on. The goal of blackmail hackers is to takeover cyber systems for exaction.

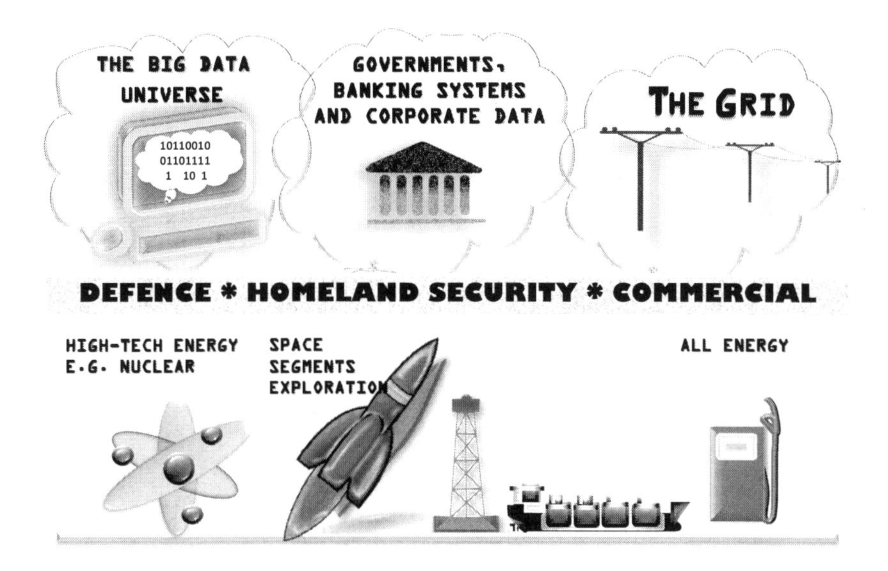

Figure 11.1 Cyber security in the energy supply chain.

Source: the Author.

Cyber Warfare

Cyber warfare is a new form of war in which the internet becomes the actual battlefield. It involves the activities by a country, global organization, or terrorist group to attack and make an attempt to destroy, harm, or sabotage another country's computer systems or critical information networks, by the means of computer viruses or denial-of-service (DoS) assaults (RAND, 2016).

It refers to government authorities utilizing information technology (IT) systems to penetrate another nation's cyber networks with the purpose of manipulation, incapacitation, and/or destruction.

Cyber warfare is considered as the fifth war domain, after sea, land, air, and space. It entails political, socioeconomic, religious, or other illegitimate entities attacking a nation or a region, and affecting its energy and other critical infrastructure entities.

Its tactics are orchestrated by experienced entities who possess the resources and know-how, but, most importantly, that work under the aegis of enemy states, who have the motive, the means, and the funds to attack another nation's cyber systems.

When lone wolf entities are employed by enemy governments or states, it is difficult for the nation/entity attacked to trace the original source of the attack. Hence this makes it difficult for attacked entities to trace the true source of the attack.

Cyber Espionage

Cyber espionage entails the usage of satellite systems and internet and computer platforms to acquire confidential information without having authorization from its proprietors, developers, or users. It pertains to gaining unauthorized and illegal access to computer networks in order to obtain confidential information typically retained by a governing administration, political party, or other institution. Cyber espionage is most commonly practiced to obtain political, military, strategic, financial, or technological confidential data. It is executed by making use of cyber-cracking techniques and malware.

A malicious ransomware code is a deceptive form of malware that encodes and obtains important electronic files and only releases them if a ransom is paid. The penetration of such systems has political, socioeconomic, commercial, and other repercussions. At a government level, cyber attacks that originate from particular nations are considered a more critical threat than warfare or economic competition.

Examples of Energy Industry Cyber Attacks

The energy industry's supply chains involve technology-intensive large-scale projects. Supercomputer algorithms are used in the global energy industry to monitor and control their upstream, midstream, and downstream flows, while tracking systems and electronic sensors can control safety, security, and operating procedures. To verify the magnitude, it is worth noting that the oil and gas industry's cyber security system includes more than a million active oil and gas wells, and several thousand miles of pipelines. On one hand, hackers may be driven by financial, commercial, political, or other motives, yet on the other hand the intricate global operations of energy companies make it hard to monitor all the regional risks and protect their entire geographical scope. Energy supply chains entail hundreds if not thousands of companies, making it easy for security offenders to infiltrate and attack their selected target from within.

The following list demonstrates some significant energy-related attacks within the energy industry.

The Dragonfly Cyber Espionage Group Attacks the Western Energy Sector

From 2011 to date, the Dragonfly cyber espionage missions are expressed through consecutive cyber-attack waves targeting western hemisphere energy companies. According to ongoing Symantec research, the hackers seek to access energy operational systems in order to gather knowledge on innovative technical and operational practices. The hackers' intensified activities in 2017 and 2018 gave away their *modus operandi*, while showing their ability to sabotage but also gain control over targeted energy companies (Symantec, 2017). Upon examining their cyber tactics and operational footprint, it is important for companies to understand that such sophisticated cyber groups do not attack a system by chance, but select their targets strategically, i.e., for very specific reasons.

Massive Electric Grid Blackout Disrupts Supply Chains and Impacts Public Safety

In 2015 and 2016 Russian hackers attacked electricity grids in Ukraine. In 2015 a Trojan horse malware, i.e., the "Black Energy," was used to conduct DDoS attacks specifically on energy companies' Supervisory Control & Data Acquisition (SCADA) and Industrial Control Systems (ICS). The blackout impacted about 225,000 people.

The 2016 cyber attack in Ukraine initially appeared as a single-hour power outage, but in fact was a test drive of a highly sophisticated malware, the so-called "Industroyer" or "Crash Override," with disturbing capacities to disrupt or damage the grid.

Over past years the malware has been reported to attack other critical infrastructure segments including the electric grid, petrochemical supply chains, transportation, water and waste treatment facilities, etc.

Hacking of Water Dams

In 2016, seven Iranian hackers carried out a series of organized cyber attacks against U.S. banks, dams, and companies. The hackers obtained access to a U.S. dam cyber network in New York, yet mistakenly attacked the small out-of-service Bowman Avenue Dam.

Most importantly, they also attempted to compromise the systems of banks and companies via DDoS attacks and SCADA. If the hackers had not been apprehended, some of America's 600 critical dams could have been at risk.

Malicious Virus Wipes Out Data from 30,000 Computers and Halts Energy Production

In 2012 the Shamoon virus targeted Saudi Aramco, a leading global energy corporation. When an IT employee opened a malicious email, the virus erased critical data from 30,000 computers (75% of the company computers), and temporarily locked down the company's production.

Yahoo: The World's Greatest Cyber Attack to Date

The impact of a large-scale cyber-security attack can be observed in the Yahoo case. In 2013, a cyber attack compromised the tech giant's three billion email customers. As the company verified, all three billion Yahoo user accounts were impacted, with their personal information stolen. This massive data breach has inspired industries and IT professionals to adopt a master recovery plan, and rethink the odds of a future cyber attack involving big data, where critical information is compromised.

Equifax: Two Cyberattacks have Impacted Millions of Customers

Equifax is a credit risk assessment company (i.e., credit report service provider) that offers fraud prevention services. It collects and handles the financial and credit data of over 800 million clients, with an annual revenue exceeding $3.1 billion.

In 2016 the company announced that the personal data of 430,000 clients was compromised. In 2017 another leak affected the personal data of 145.5 million people (driver's license and social security information), while the credit card information of over 209,000 clients was also stolen. In 2018 it became known that the past account breaches had impacted another 2.4 million people, including British and Canadian nationals.

The brief references to these case studies aim to provide valuable lessons regarding cyber security. First, these attacks may be targeting governments, companies, and individual users alike. Second, cyber security is a global phenomenon and, as such, it may involve perpetrators and victims from any global region. Third, no government or company is immune to the attacks. It appears that hackers typically attack high-profile targets. Here the high-risk and high-profit principal applies. Fourth, hackers may attack a target for many different reasons, i.e., sabotage, obtaining financial and commercial benefits, ideological or political conflicts, activism, or the acquisition of personal data (not necessarily financial) from millions of users, for identity theft, future marketing, or for other reasons. Fifth, and most alarming, is the fact that as energy companies seek for knowledge fusion platforms (i.e., big data, blockchain, etc.) to better monitor and control their supply chains, certain illegitimate groups may target these ground-breaking processes and may seek to sabotage them.

Profiling the Modern Hackers: Trends and Patterns

The hackers' capabilities grow each year, and the main factors that determine the cyber attacks are motive, opportunity, and the cyber means. To quote Mr. Robert Mueller, former FBI Director, "there are only two types of companies: Those that have been hacked, and those that will be" (Mueller, 2012).

Over the years hackers have differentiated from the lone-wolf stereotype; modern hackers have become better networked, more knowledgeable, and capable of targeting a company's "center mass" or most crucial information.

As for the methods hackers use, the most common types of attacks have been fraud, intrusion, spam, malware, and denial of service. The trends changed in 2018, with the most common types of cyber attacks pertaining to browser, brute force, denial of service, worm, and malware. Phishing and cross-site scripting (XSS) are also common. The past year has seen a dramatic shift from ransomware-based attacks to cryptominer-based attacks.

Evaluating the Impact of Cyber Attacks

Based on the "knowledge is power" aphorism,[1] big data is a critical asset for nations and industries. When energy companies fail to effectively safeguard, monitor, and control their big data universe, the smallest systemic vulnerability becomes a gate to attract cyber attackers. The impacts of hacking incidents may include bankruptcy, loss of reputation, reduced market share, reduced stock prices, but also loss of trust among partners and loss of morale among the workforce.

Data Laundering and Intelligence Outsourcing

An alarming dimension of cyber attacks pertains to one perpetrator sharing data with other interest groups. It is not unknown for cyber attackers to actually "sell" or "make available" the data retrieved to other illegitimate entities that are equally interested in attacking the same target, let's say for commercial or economic reasons. When they access data of national and/or industrial significance, hackers typically maximize benefits by re-selling the data to potential parties of interest. This complex hacking supply chain is what is called data laundering.

Here are some common scenarios that increase the risk of data outsourcing or data laundering:

- Data may be compromised when companies change legal ownership, e.g., due to bankruptcy or merger and acquisition.
- A common scenario encountered in the energy industry pertains to extending data-sharing services to third parties, e.g., outsourcing IT services to contractors.
- The likelihood increases when the company fails to monitor these global service providers. As the data is being distributed to global cyber channels, it is not uncommon for the same data to be insourced or re-utilized by political or commercial competitors (Burns, 2017a).

It is worth noting that cyber crimes are rarely isolated illegitimate activities. Hence, based on the principal that illegitimate supply chains are often interconnected, hackers may become involved in other crimes related to data laundering, such as patent infringement, money laundering, etc. (Burns, 2018).

The next section will lead us to another conclusion: As the big data universe grows exponentially, there is a compelling need for the energy industry to invest more in cybersecurity and to adopt a proactive stance.

The Global Market of Cyber Security: Facts and Figures

This section provides statistical information on global computer users, annual data on investments, actual cyberattacks, and damages incurred. Most importantly, this section will reveal that although cyber threats increase in magnitude and frequency, the energy industry is not adequately investing in cyber security training, technologies, and processes. The facts and figures of this section show that companies need to bravely adopt a more proactive stance as the aversion of an attack is less costly and more beneficial. On the contrary, a reactive culture entails high risk and high cost as it may expose company data, reputation, and finances in an irreversible manner.

It is estimated that, as of 2018, there are 23.14 billion computers connected to the internet globally. It is also known that a cyber attack occurs every 20 seconds, impacting one in three people. By 2025 the number will triple to 75.44 billion computers (Statista, 2018).

We must use empathy to better understand the above statistics: These incidents involve identity theft, commercial espionage, and loss of intellectual property. There are resources, ideas, and innovative projects that are stolen and implemented by other stakeholders "somewhere out there" (Burns, 2014).

When financial accounts are depleted, small–medium enterprises are more likely to collapse, as their resources and credit opportunities are limited. When an energy corporation's cyber system is severely compromised, its stock market value can be afflicted by about 33% within the next few days pursuant to the media coverage (Burns, 2015).

Based on the above facts, it is only logical to assume that, as the infrastructure expands, the cyber attacks will evolve and proliferate in frequency and impact. Hence, as the cyber universe grows (population, internet users, energy corporal activities), cyber attacks are likely to intensify in terms of likelihood and/or magnitude.

The following facts and figures seek to verify to what extent the energy industry is targeted by hackers. According to U.S. Government reports from 2014 to date, the energy sector tops the list of U.S. industries under cyber attack (US-CERT, 2018). As seen in **Figure 11.2**, energy is the second preferred cyber target, representing 14% of the global targets; the primary targets are aerospace and defense (17%), and the third main target is finance and banking institutions (11%) (Burns, 2014).

Figure 11.3 classifies cyber security cost into high, moderate, and low in order to better understand the financial impact of a cyber attack. Many of these costs are interrelated, e.g., loss of company reputation leads to loss of revenue, loss of stock prices, loss of market share, etc. However, additional cost occurs to rebuild reputation and re-establish the corporate image via media and public relations.

The figure also classifies cost into proactive, i.e., investment in cyber security, versus reactive, i.e., efforts for recovery from an attack. The figure shows that the cost for preparedness is rather moderate, whereas the vast majority of expenses pertain to losses incurred in the aftermath of the cyber attack.

What the figure reveals is that cyber security investment is the only expense type that pertains to preventive action, i.e., at a stage of preparedness, where the company aims to avert the risk of attack.

This moderate cost should make cyber security investment much more appealing to the industry, considering both the high cyber-attack risks (likelihood x impact) and the

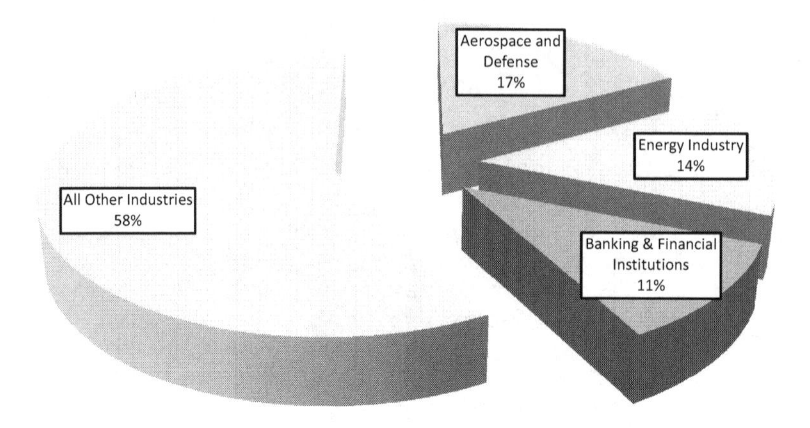

Figure 11.2 Global cyber attacks targeted to specific industries.

Source: Burns, 2017a. "Energy security: Supply chain and the paradox of change".

TRB National Academies, Special Committee on Transportation Security and Emergency Management (SCOTSEM), The American Association of State Highway and Transportation Officials (AASHTO). Conference proceedings. Available at: https://security.transportation.org/wp-content/uploads/sites/33/2017/04/Energy-and-the-Supply-Chain.pdf, last accessed: March 12, 2018.

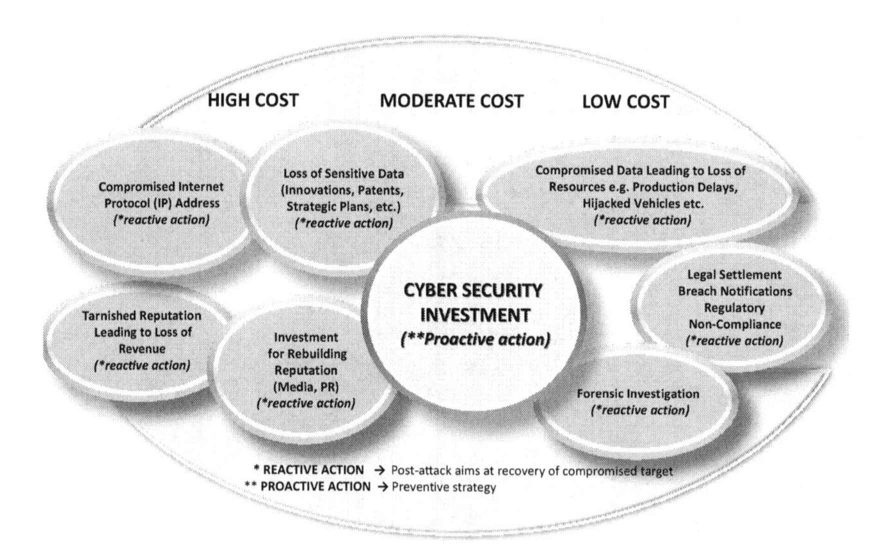

Figure 11.3 Measuring the cost of cyber attacks and the benefits of proactiveness.

Source: the Author, based on Maria Burns' presentation at National Academies (2017a) and Whitehouse, 2018. *Fighting Cybersecurity Threats to the Growing Economy Using Data From FBI (2017), OWESP (2014), OWESP (2006)*, February 21. Available at: www.whitehouse.gov/articles/fighting-cybersecurity-threats-to-the-growing-economy, last accessed: March 25, 2018.

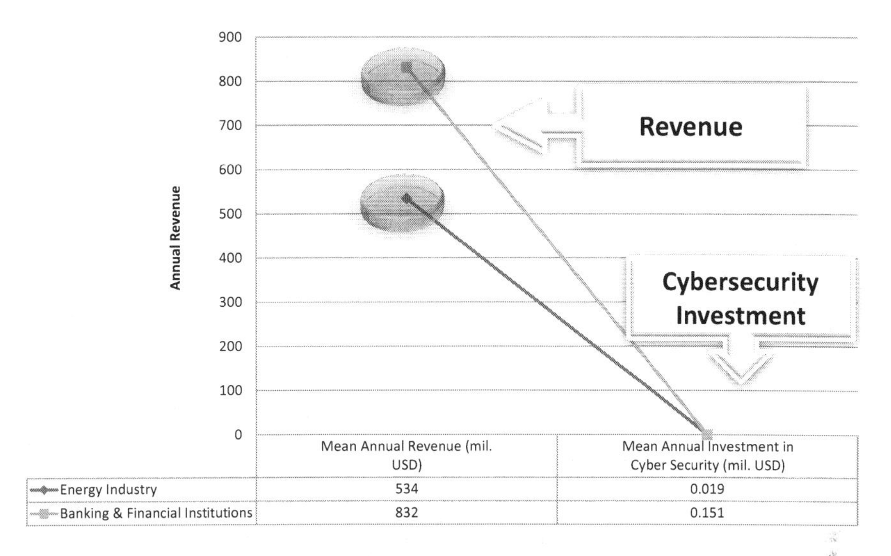

Figure 11.4 Annual revenue vs investment in cyber security.

Source: the Author, based on figures by Malik, Naureen S., 2018. "Energy companies aren't doing much to defend against soaring cyber attacks". *Bloomberg*, April 27. Available at: www.bloomberg.com/news/articles/2018-04-27/-cyber-blindspot-threatens-energy-companies-spending-too-little, last accessed: May 19, 2018.

exorbitant reactive/recovery costs. However, as seen in Figure 11.4, this is not the case, as the annual levels of investment are disproportionally low compared with both their revenue and the consequences of an attack.

Industrial Cyber Security and Investment Decisions

Figure 11.4 shows that the energy industry, at an aggregate level, does not seem to invest in energy security: Each year the average company spends under 0.2% of its revenue. This amount represents less than one third of the amount invested by banks and other financial establishments.

The root causes behind a "strategic inaction" cannot be interpreted as negligence: The energy industry is too serious and pragmatic to express excessive optimism or make an action-based slip. Therefore, one speculation pinpoints the human factor and its natural resistance to change. As thoroughly discussed in my *Supply Chain Security* book, risk managers find it challenging to request additional budget to mitigate potential risks that may never materialize. On the other hand, proposed investment in business development enjoys a positional leverage as it promises a tangible growth and return on investment (Burns, 2015).

Another paradoxical aspect of pivotal significance is the fact that global energy is controlled 80% by the industry and only 20% by governments. The industry's "strategic underinvestment" in cyber security shifts the burden of cyber defense to governments (Burns, 2017b; Burns, 2014).

Whatever the true root cause for underinvestment is, this decision leaves the energy industry unprepared for and vulnerable to potentially serious attacks with lasting consequences.

Evaluating Energy Cyber Security: External, Internal, and Hybrid Risks

Upon examining the big picture of cyber security and the energy industry, it is worth verifying if the findings of the previous section on cyber investment are valid. This section addresses the internal, external, and hybrid risks.

External Risks

Cyber Attacks

The global industry perceives that the trend of cyber attacks has increased, with seven out of ten organizations claiming that their security risk increased significantly in 2017. Companies have personal experience of these attacks, as 54% of companies encountered one or more cyber attacks that compromised data and/or IT infrastructures (Ponemon Institute, 2018).

It is estimated that by 2020 the cost of a data breach incident will surpass $150 million, due to the large-scale interconnectivity among supply chain infrastructures (Cybint, 2018).

Fileless Techniques: Being Overrun by the Invisible Enemy

Of the attacks, 77% utilized exploits or fileless techniques (Ponemon Institute, 2018). These "zero-footprint" methods allow hackers to overrun a system without having to place malware or install software. Their invisible footprint makes these attacks ten times more effective, and hence much more dangerous.

Internal Risks: Training and Recruitment Needs

Over 60% of IT professionals consider that security IT and cyber security groups are either underqualified, or undertrained, or understaffed (ISC², 2017). The number of positions not filled will grow by the year 2021. According to the ISC² report (2017), the number of unfilled positions will reach 1.8 million, whereas according to Comtech Networking (2018), the number will triple, i.e., will stretch to 3.5 million. Over 50% of IT professionals consider that their companies do not have or do not provide the resources for security training. Information technology professionals consider that, despite the 400,000 latent malware attacks on a daily basis, companies do not hire adequate workforce to close the security gaps (Comtech Networking, 2018).

Hybrid Risks: With a Little Help From . . . Within

About 60–80% of network misuse incidents originate from the inside network (Cybint, 2018). Hence the paraphrasing of the famous Beatles' song is meant to remind the industry that cyber security gaps allow hackers to partner with company employees, be it deliberate or undeliberate. Information technology abuse pertains to the inappropriate use of data, or the unauthorized misuse of the system, be it intentional or unintentional.

- If the act is intentional, a key root cause pertains to the HR recruitment process, and suggests an inadequate security background check.
- If the act is unintentional, the limited situational awareness suggests lack of IT familiarization and security training.
- In both cases, corporate leadership and HR initiatives should mitigate the challenge.

The Cost of Preparedness

According to the Information Systems Audit and Control Association (ISACA), 83% view cyber attacks as one of top three threats to business, while only 38% of global organizations claim they are prepared to handle a sophisticated cyber attack. The global cyber-security market reached $75–100 billion in 2015 and $122.45 billion in 2016. By 2020 it is anticipated to exceed $170 billion, while in 2021 it will reach $202.36 billion (ISACA, 2015; Cybint, 2018).

However, cyber preparedness seems to have become increasingly costly over the past few years. While industry preparedness is estimated by the billions, the actual cyber ransomware and recovery costs are estimated by the trillions: The annual cost of global cyber crime was approximately $400 billion in 2014, $500 billion in 2015, and rose to a whopping $2–3 trillion in 2016 (Siemens, 2017). It is estimated that by 2030 the losses will reach 90 trillion (NATO, 2017). The comparison between reactive vs proactive costs is also impressive: Proactive investment from 2017 to 2021 will reach $1 trillion annually, whereas reactive costs from 2018 to 2021 will exceed $6 trillion annually (Cybint, 2018).

Cyber Security Platforms and Technologies

> We're in the stone age of cyber security.
> Real learning will only come after the 1st major incident.
>
> Dr. Christopher Frei,Secretary General of World
> Energy Council, London, April 2016

Cyber Security and the Smart Power Grid

The electricity supply chain greatly depends on a country's electric grid. The grid is an intricate complex network comprising of the production communication supply monitoring and controlling systems. The grid's vulnerabilities are tangible and intangible. Safety risks (unintentional attacks, i.e., geophysical phenomena such as hurricanes, earthquakes, but also natural wear and tear requesting maintenance) and security risks (intentional attacks) may impact the physical network.

Cyber-security supply chains expand throughout a country's electric power grid. Hence the challenge for security professionals is to safeguard an intricate geographically expanding network with severe time and resource restrictions.

The contemporary electric grid sector has launched a plethora of measures to protect the grid from attacks. To ensure a wide-range all-encompassing protection, the industry has collaborated with federal, state, and tribal governments. "Smart Grid" (SG) and "Smart Homes" (SH) appliances are very promising for the monitoring and controlling of systems. However, the remote-access design allows unauthorized access from perpetrators.

Aging National Grid Infrastructure

Aging energy infrastructure causes systems to slow down or breakdown due to malfunctions and system failures. Since electricity networks and power lines were first used in developing nations, especially in America and Europe, these regions are using aging grid and transformers. Hence, the commercial life expectancy of an electricity grid is 40 years, while the average age of substation transformers (that are used as voltage modifiers among different transmission levels or among high transmission and lower distribution) is 42 years – i.e., has exceeded the life expectancy. And while these systems need to be maintained, repaired, or retrofitted or replaced, there is a need for these networks to be advanced enough, with their cyber-security issues being addressed.

Open Internet Accessibility

Open internet accessibility enables hackers to identify systemic performance and vulnerabilities. Hackers now have a competitive edge that eliminates, or puts at a disadvantage, government initiatives for cyber security, as well as control systems such as:

a Supervisory Control and Data Acquisition (SCADA) systems;
b Distributed Control Systems (DCS); and
c Industrial Control Systems (ICS).

Wi-Fi Risks

The widespread use of wi-fi networks in public locations exposes cyber data and imposes additional system vulnerabilities. When hackers penetrate such public networks they can obtain usernames and passwords, and can therefore obtain and manipulate corporation-related information.

Energy professionals should encrypt their information by the use of virtual private networks (VPNs), which are systems that encompass the private networks throughout public access systems, hence allowing internet users to transmit and accept data across shared or public networks.

Big Data

One of the top opportunities and challenges of the energy industry and cyberspace relates to big data and ways of gathering, analysing, and accurately interpreting pertinent information. Energy companies and governments are concerned about the exponential growth of cyberspace, and consequently the challenges in managing and protecting big data.

At the end of 2017, 4,156,932,140 internet users were registered (Internet world stats, 2017). This number exceeds 52% of the global population. It is worth noting that the number of internet users and the volume of big data grows each month, and so does world population.

As big data grows in volume, the task of finding, evaluating, and preserving accurate and pertinent information becomes more challenging.

Big Data and the Six Vs

From an organizational perspective, there are six ways of classifying and measuring big data, i.e., as per volume, value, velocity, variety, veracity, and variability. These are demonstrated in Figure 11.5 and duly explained below.

1 **Volume**: Big data grew at 40% annually, with 90% of the world's data being produced over the past two years. Therefore, while in 2012 global big data was 1.2 zettabytes, by 2020 it will exceed 100.2 zettabytes (Jawad, 2017).

   The global volume of big data surpasses 2.7 zettabytes of data in the digital universe. Table 11.1 demonstrates the volume, i.e., the memory units of measurement for big data, classified as size and in chronological order. It is worth observing the rapid growth of the big data volume.

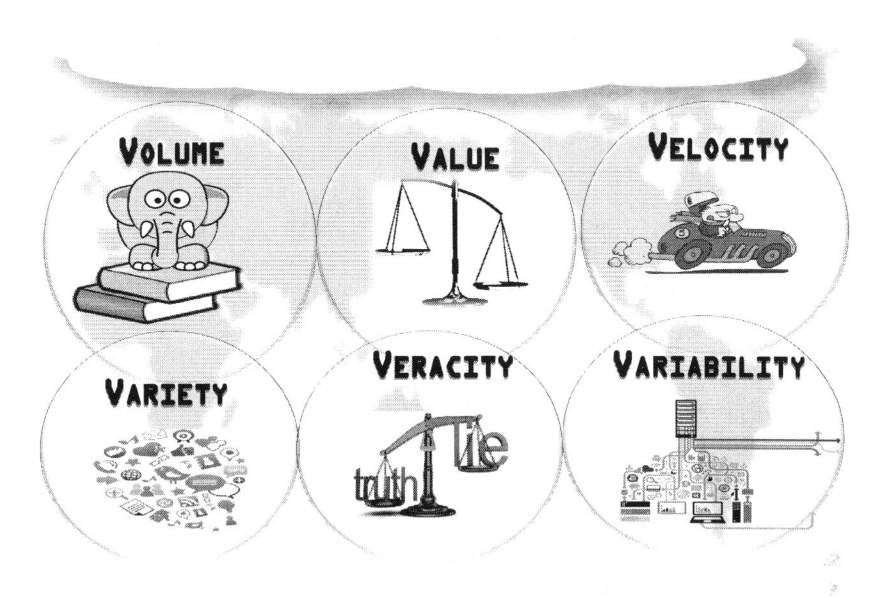

Figure 11.5 Big data and the six Vs.

Source: the Author.

Table 11.1 Big data volume and units of measurement.

Memory unit	*Abbreviation*	*Prevailing unit of measurement in year*	*Equals to*	*Decimal size*	*Binary size*
Binary digit	Bit	After 1960s	Basic unit 0 or 1		
Byte	b		8 bits		
Kilobyte	Kb	1970s – 1990s	1,024 bytes	10^3	2^{10}
Megabyte	Mb		1,024 kilobytes	10^6	2^{20}
Gigabyte	Gb	2005	1,024 megabytes	10^9	2^{30}
Terabyte	Tb	2010	1,024 gigabytes	10^{12}	2^{40}
Petabyte	Pb		1,024 terabytes	10^{15}	2^{50}
Exabyte	Eb		1,024 petabytes	10^{18}	2^{60}
Zettabyte	Zb	2020	1,024 exabytes	10^{21}	2^{70}
Yottabyte	Yb		1,024 zettabytes	10^{24}	2^{80}
Brontobyte	Bb		1,024 yottabytes	10^{27}	2^{90}
Geopbyte	Gpb or Gb	2025	1,024 brontobytes	10^{30}	2^{100}

Source: the Author.

I hereby suggest the following subcategories:

- Usage-based class can be broken down into: i) government, ii) industrial, and iii) domestic and personal use.
- Geographical-based class may be divided into: i) continents, ii) countries, and iii) states or municipalities.

- Time series data can be classified in chronological order, i.e., data that is annual, monthly, weekly, hourly, etc.

2. **Value**: By 2025 the market value of world data will exceed US$ 122 billion. The value may be classified in accordance to the user's benefits.
    - Information sources may be classified as: i) primary, ii) secondary, and iii) tertiary data.
    - Knowledge-based classification may include data that offers benefits of one or more of the following categories: e.g., financial, commercial, geopolitical, scientific, technological, etc.
3. **Velocity**, i.e., the speed of IT services may be classified into: i) streaming, ii) real time, iii) near real time, and iv) batch.
4. **Variety** of data is classified into three categories according to Human-Machine Interaction (HMI), namely, according to the identity of sender and receiver: i) people-to-people, ii) people-to-machines, and iii) machines-to-machines.
5. **Veracity** pertains to the use of valid, accurate, and true information from reliable sources. The quality, pertinence, and reliability of data input helps energy companies create a successful output, derived from strategic decisions such as investment, operations, risk management, etc.
6. **Variability** may pertain to:
    i the disparity of data based on the sources, the content, and the form, e.g., primary vs secondary vs tertiary data;
    ii emails vs contracts vs reports vs customer surveys;
    iii the pertinence or usefulness to industry-specific, function-specific;
    iv the volume, consistency, and geospatial coverage of data.

Note

1 Latin aphorism: *Scientia potential est.*

References

Burns, M. (2014). "Building resilient supply chains through policies, partnerships and technologies". DHS Supply Chain Security Workshop, Houston. April 23.

Burns, M. (2015). *Logistics and Transportation Security: A Strategic, Tactical, and Operational Guide to Resilience*. London: CRC Press, Taylor & Francis.

Burns, M. (2017a). "Energy security: Supply chain and the paradox of change". TRB National Academies Special Committee on Transportation Security and Emergency Management (SCOTSEM), The American Association of State Highway and Transportation Officials (AASHTO). Conference proceedings. Subcommittee on Highway Transport (SCOHT) and TRB Critical Infrastructure Committee (ABR 10) of the National Academies. Houston, Texas, July 30–August 3. Available at: https://security.transportation.org/wp-content/uploads/sites/33/2017/04/Energy-and-the-Supply-Chain.pdf, last accessed: April 27, 2018.

Burns, M. (2017b). "Energy security". Cyber Security Conference Proceedings. 11th Mare Forum USA 2017 in Houston, Texas. Available at: https://usa17.mareforum.com, last accessed: May 19, 2018.

Burns, M. (2018). "Participatory operational & security assessment on homeland security risks: An empirical research method for improving security beyond the borders through public/private partnerships". *Journal of Transportation Security* 11(3): 85–100.

Comtech Networking (2018). *5 Startling Statistics about Cybersecurity in 2018*. Comtech Networking. March 1. Available at: www.comtech-networking.com/blog/item/622-5-startling-statistics-about-cybersecurity-in-2018, last accessed: May 19, 2018.

Cybint (2018). *12 Alarming Cyber Security Facts and Stats*. Cybint. March 16. Available at: www.cybintsolutions.com/cyber-security-facts-stats, last accessed: May 19, 2018.

Federal Bureau of Investigation (FBI) (2017). *Intellectual Property Theft/Piracy*. Federal Bureau of Investigation. Available at: www.fbi.gov/investigate/white-collar-crime/piracy-ip-theft, last accessed: May 19, 2018.

Information Systems Audit and Control Association (ISACA) (2015). *Global Cybersecurity Status Report, 2015*. Information Systems Audit and Control Association. Available at: www.isaca.org/pages/cybersecurity-global-status-report.aspx, last accessed: May 19, 2018.

Internet world stats (2017). *Internet Users in the World*. December 31. Miniwatts Marketing Group. Available at: www.internetworldstats.com/stats.htm, last accessed: May 19, 2018.

ISC² (2017). *Cybersecurity Workforce Shortage Continues to Grow Worldwide, To 1.8 Million in Five Years. Survey of 3,000 IT Employees from the 2017 Global Information Security Workforce*. ISC².Available at: www.isc2.org/news-and-events/press-room/posts/2017/02/13/cybersecurity-workforce-shortage-continues-to-grow-worldwide, last accessed: May 19, 2018.

Jawad, I. (2017). "Mega trends driving our future: new business models, connectivity, and convergence". Presented by Iain Jawad for *Frost & Sullivan*. IBM, Cisco, *Frost & Sullivan* Analysis. Available at: https://tapahtumat.tekes.fi/uploads/68053789/GlobalMegaTrends-2339.pdf, last accessed: May 19, 2018.

Malik, N. S. (2018). "Energy companies aren't doing much to defend against soaring cyber attacks". *Bloomberg*, April 27. Available at: www.bloomberg.com/news/articles/2018-04-27/-cyber-blindspot-threatens-energy-companies-spending-too-little, last accessed: May 19, 2018.

Mueller, R. S. III (2012). "Combating threats in the cyber world: Outsmarting terrorists, hackers, and spies". RSA Cyber Security Conference, San Francisco, CA, March 1, Federal Bureau of Investigation (FBI). Available at: https://archives.fbi.gov/archives/news/speeches/combating-threats-in-the-cyber-world-outsmarting-terrorists-hackers-and-spies, last accessed: May 19, 2018.

NATO (2017). "Spending for success on cyber defence". *NATO Review Magazine*, June 4. Available at: www.nato.int/docu/review/2017/also-in-2017/nato-priority-spending-success-cyber-defence/en/index.htm, last accessed: May 19, 2018.

Open Web Application Security Project (OWASP) (2006). *OWASP 10 Most Common Backdoors*. Open Web Application Security Project. Available at: www.owasp.org/index.php/File:OWASP_10_Most_Common_Backdoors.pdf, last accessed: May 19, 2018.

Open Web Application Security Project (OWASP) (2014). *CISO Survey and Report 2013*. Open Web Application Security Project. Available at: www.owasp.org/images/2/28/Owasp-ciso-report-2013-1.0.pdf, last accessed: May 19, 2018.

Ponemon Institute (2018). *The 2017 State of Endpoint Security Risk Report. Evaluating the Largest Threats and True Cost of Attacks*. Barkly & Ponemon 2018 Endpoint Security Statistics Trends. Available at: www.barkly.com/ponemon-2018-endpoint-security-statistics-trends, last accessed: May 19, 2018.

RAND (2016). *Cyber Warfare*. RAND Corporation. Available at: www.rand.org/topics/cyber-warfare.html, last accessed: March 25, 2018.

Siemens (2017). *IT Security, Facts and Forecasts: Rapidly Growing Market for IT and Cyber Security*. Siemens. Available at: www.siemens.com/innovation/en/home/pictures-of-the-future/digitalization-and-software/it-security-facts-and-forecasts.html, last accessed: May 19, 2018.

Statista (2018). *Internet of Things and Number of Connected Devices Worldwide*. Statista. Available at: www.statista.com/statistics/471264/iot-number-of-connected-devices-worldwide, last accessed: May 19, 2018.

Symantec (2017). *Dragonfly: Western Energy Sector Targeted by Sophisticated Attack Group*. Threat Intelligence, Symantec Report of September 6. Symantec. Available at: www.symantec.com/blogs/threat-intelligence/dragonfly-energy-sector-cyber-attacks, last accessed: February 26, 2018.

United States Computer Emergency Readiness Team (US-CERT) (2018). *About Us*. United States Computer Emergency Readiness Team. Available at: www.us-cert.gov/about-us, last accessed: May 12, 2018.

Whitehouse (2018). *Fighting Cybersecurity Threats to the Growing Economy, Using FBI (2017) and Open Web (2006) Data*, February 21. Available at: www.whitehouse.gov/articles/fighting-cyber security-threats-to-the-growing-economy, last accessed: March 25, 2018.

12 Energy Security Forecasting

> I always avoid prophesying beforehand because it is much better to prophesy after the event has already taken place.
>
> Sir Winston Churchill

This chapter demonstrates a set of interrelated instruments used in the energy industry by governments and companies to attain goals of sustainability, independence, efficiency, affordability, and accessibility.

Strategic planning entails a company's continuous goal-setting, based on internal factors such as the corporate mission and vision, available resources, etc., and external factors such as the global economy, geopolitical influences, supply chains, etc. A market forecast aims to interpret the market trends and define the possible outcomes in the foreseeable future. Risk management is created based on the company's goal-setting process and market forecasting outcomes. The chapter's conclusive position entails the significance of training for an energy institution to reach its full potential of growth, motivation, and achievements.

Figure 12.1 underlines the interrelation between these tools and extended processes that are being used in every energy segment by governments, companies, and market investors alike to ensure energy security and process optimization. Figure 12.2 demonstrates how strategic goals, forecasting, risk management, and optimized training instruments can shape the future of energy, based on the pillars of energy security.

Elements of Strategic Planning for Energy Security

Strategic planning for energy security is defined as the method of envisioning and strategically shaping the future. Energy security strategies can be set by nations, the energy industry, and individual organizations. When constructing a strategic plan for energy, the three key questions that need to be answered are demonstrated in Figure 12.3 and pertain to 1) the current national or corporate position, 2) the future vision and pertinent strategic goals, and 3) the tactic goals, i.e., what steps we need to take, and what resources are needed to get there.

Each of the following questions must be answered by examining the pillars of energy growth and security as covered in Chapter 3, Figure 3.1 of this book: The structure of energy security is based on the pillars of geopolitics, partnerships, economics, trade, transport, technologies, and policies.

And while the above process entails the fundamental direction of strategic planning, the extended process is demonstrated in Figure 12.4.

Figure 12.1 Instruments for energy security and performance optimization.

Source: the Author.

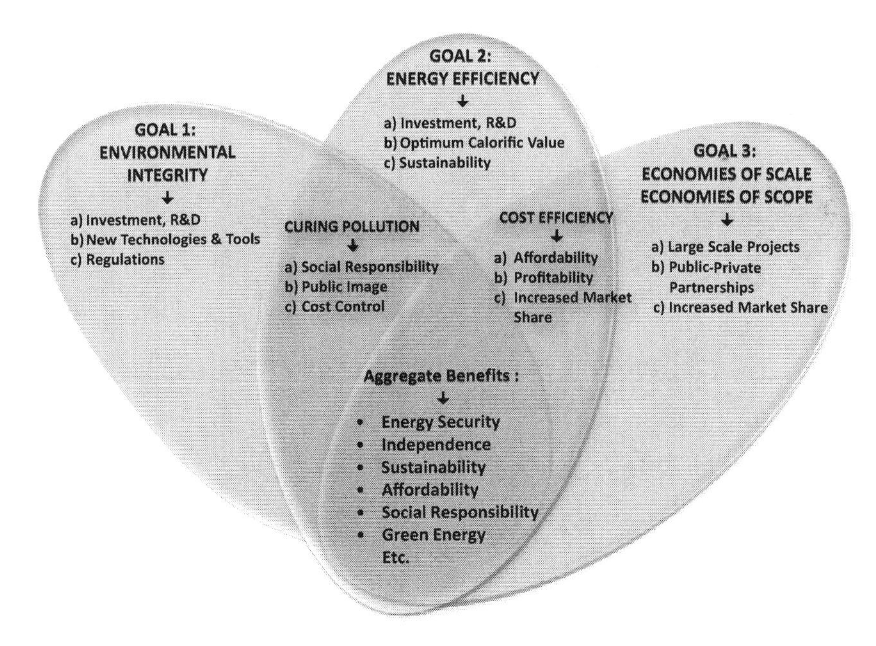

Figure 12.2 How strategic goals can shape the future of energy.

Source: the Author.

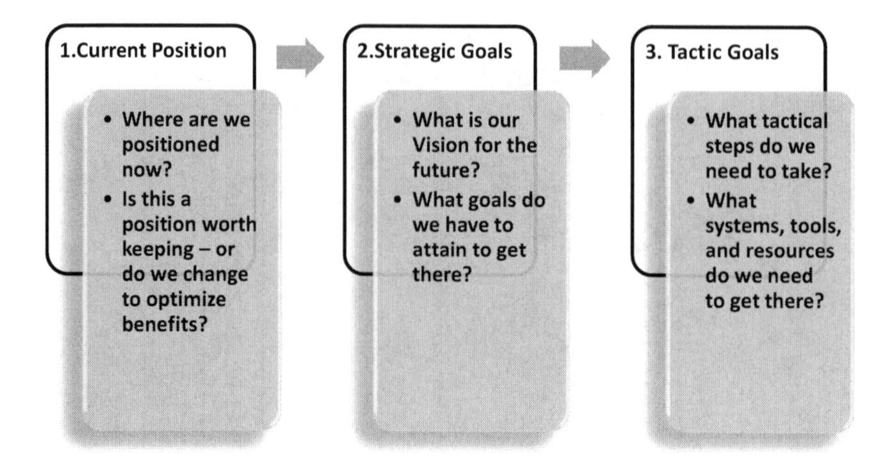

Figure 12.3 How company positioning questions lead towards effective strategic planning.

Source: the Author.

Figure 12.4 The stages of strategic planning.

Source: the Author.

Nations and companies must revisit their mission and vision to establish short-term goals (i.e., encompassing the next one–three years), medium-term goals (five-year plans) and long-term goals that exceed five-year plans (e.g., Asian nations and organizations have established 500-year plans). An optimum method that can be used to construct long-term strategic planning configuration is to configure several five-year plans, i.e., encompassing several five-year plans (Burns, 2013; Burns, 2014a).

Frequently a SWOT analysis is needed, i.e., to verify the Strengths, Weaknesses, Opportunities, and Threats. As a next step, it is important to break down the goals, budget, and resources at a division/department level. Finally, Key Performance Indicators (KPIs) must be set at a project level and individual employee level.

Long-term strategies might be difficult to attain as their success relies on future events that have not yet been materialized. The three critical areas to define the feasibility of future strategies entail 1) policies, 2) partnerships, and 3) technologies (Burns, 2014b; Burns, 2015).

These focal areas can be further distinguished into the seven pillars of energy security of Chapter 3.

A successful strategic plan is a feasible plan, hence forecasting potential future trends will help speculate how the current market trends are likely to evolve.

The next section of this chapter focuses on the significance of forecasting, and posits the optimum strategic goals that can be attained during the key energy/market cycle stages.

Forecasting Security Threats and Patterns, Based on the Energy Cycles

This section aims to forecast the future of the energy sector, and identify emerging security trends during different market cycles. Since the global energy market consists of multiple key players (both governments and industries), forecasting is an essential instrument for strategic planning: The more accurate the forecasts, the more effective and successful the energy strategy.

There are several forecasting techniques, essentially distinguished into three main classifications:

1 **Intuitive predictions** are based on professional expertise used to either interpret current trends leading to future events, or to compare and contrast past events and market cycles to interpret market outcomes.

The strengths of this type of prediction entail a possible interpretation of relatively new events even in conditions of i) limited data, ii) missing data, or iii) without an exact match or direct comparison to prior events, i.e., a currency war between conventional currencies and crypto-currencies, or a trade war between the U.S. and China.

The weaknesses of this type of prediction include: i) biased interpretation, e.g., due to affiliation with a key energy player, ii) judgment error – which is possible in every type of forecasting.

Intuitive predictions are distinguished into the following categories:

a experience-based, i.e., forecasting the future based on remembrance or observation of past market behaviors;
b the notorious "gut feeling" that many global moguls possess;
c the "market sentiment," i.e., a keen observation and behavioral interpretation of the movements, actions, and reactions of key market players.

2 **Statistical data analysis** is a classification distinguished in a plethora of methods and tools.

The strengths of this type of prediction entail: i) the increased ability to track data sources and reproduce findings based on methods that have been tested in a scientific manner; ii) the ability to fine-tune and improve the methodology based on retrospective evaluations.

The weaknesses of this type of prediction pertain to: i) the increased reliance on past data; ii) the dependence on accurate and consistent data input to develop accurate output; iii) biased interpretation; and iv) judgment error.

Statistical analysis tools are distinguished as follows:

a Data is categorized into primary, secondary, and tertiary to determine the originality and direct access or vicinity of the sources. Needless to say, the more accurate the data input is, the more accurate the forecasting output will be.

b The methods used to construct forecasting tools based on this data include:

 i) market-research-based data analysis; ii) cause and effect models, i.e., time series modeling (e.g., Box-Jenkins processes); smoothing techniques (e.g. exponential statistical analysis); and iii) artificial intelligence, i.e., software-generated predictions.

Forecasting Based on Different Stages of Business (and Energy Security) Cycles

This section posits that effective forecasting techniques must always take into consideration the market cycles and stages of energy security, as depicted in Figure 12.5.

The peak stage:

During the peak stage, companies are likely to possess more resources but limited time. They are struggling to reap maximum benefits before the next market contraction. This is the time to sell high, so proactive companies have already produced and are ready to deliver commodities or services that are worth buying. There are two pitfalls to watch during this stage. First, companies must cautiously operate on a day-to-day basis: Lessons learned from the 2018 and 2014 peaks are that the higher the market gets, the more abruptly it may collapse. Hence governments and companies must have a safety net in place.

Second, these cycles may be good for one market segment, but not for all. For example, if this is a sellers' market, obviously the buyers may not be equally thrilled to see the prices skyrocket. For this reason, the privileged market segment must remain humble and wise enough not to alienate their clients through win/lose negotiations.

The contraction stage:

The peak stage eventually comes to an end, when supply reaches or exceeds demand. At the earlier stage, when high demand causes high prices, several market players invest time and resources in order to meet that demand and reap the pertinent financial benefits. However, the essence of market stages is that they do not last forever. Hence, over-optimism leads to over-investment and oversupply, signifying the end of market boom and leading to the stage of contraction.

Contraction is "the status where the 'universe of economic and market opportunities' shrinks" due to oversupply.

Oversupply is "the status where too many market players have the same great idea leading to over-investment."

Over-investment is "instigated by the convincing culprit of **over-optimism** during the peak stage" (Burns, 2017).

Over-optimism leads to over-investment, oversupply, and eventually to market contraction.

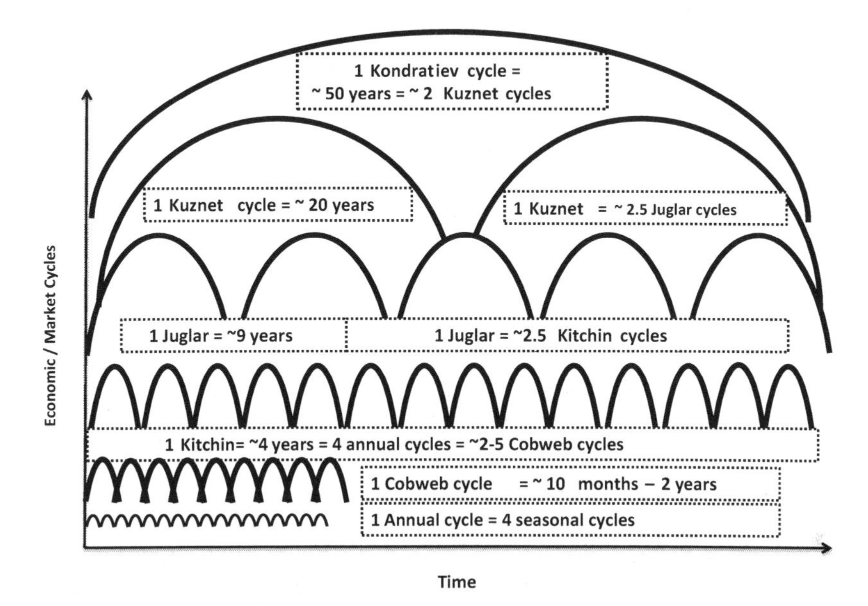

Figure 12.5 Forecasting the stages of the business (and energy security) cycles.

Source: the Author.

As we recall from the 2008 global economic meltdown and the 2015 oil price collapse, the faster and/or most impressive the oversupply, the more unexpected the contraction stage (Burns, 2016a).

The trough stage:

During the trough stage companies struggle to retain their market share while they are strategizing for the next recovery stage. The first corporate concern is to retain the market share, utilize resources wisely, and generally "do more with less." Cutting losses is a necessary evil during this stage, which may involve working on a tighter budget, and eliminating projects or segments of projects that do not break even. Sometimes companies decide to jettison modest projects in order to retain sufficient resources for their most innovative ventures. Often companies undertake business with limited return on investment, where they hardly break even, just in order to retain a strategic market share.

This is a challenging stage where companies sell their aging assets, fire employees, or form mergers and acquisitions to remain afloat. Although this is a painful process for individual companies, there are long-term benefits through disposal of processes and assets that are no longer useful.

The recovery stage:

When the market reaches its lowest point it provides an excellent opportunity for the market to reshuffle and key players to reposition towards new leveraging and niche markets. Some companies do not survive the trough stage challenges and declare bankruptcy, however other companies have the great opportunity to buy cheap and to initiate mergers and acquisitions. The new market balances caused by the market reshuffling provide new opportunities.

Hence, the recovery stage represents the light at the end of the tunnel: This is a cleansing renewal point where the older assets and dysfunctional schemes have given way to robust, durable entities, capable of contending, and ascending, throughout the next market peak.

The following sections will address the challenges and opportunities of risk management and forecasting the energy market and energy performance, both intuitively and by the use of econometric formulas.

Forecasting the Risks

> Foreknowledge of the future makes it possible to manipulate both enemies and supporters.
>
> Raymond Aron, *The Opium of the Intellectuals*

Energy Security and Risk Management

The energy industry entails a myriad of risks throughout its global supply chains and the key stages of upstream, midstream, and downstream. Literally, the essence of risk has two meanings: On the one hand, as the exposure to danger, and, on the other hand, as taking chances in a climate of uncertainty because the rewards are too high to ignore. Indeed, in our modern era of global uncertainties, opportunities and losses are greater than ever. This is a time where operating within our comfort zones actually compromises our success, and risking nothing is actually the greatest risk possible. Instead, what energy stakeholders need to do is to use a set of tools and processes pertaining to risk management, in order to be prepared for and deal with the risk in a constructive manner.

Risk management pertains to recognizing, assessing, and prioritizing risks, with the consequent tasks of managing resources and finances in a synchronized manner, in order to eliminate, monitor, and control the likelihood and impact of disastrous incidents, while optimizing positive and beneficial ventures. Risk management aims towards: i) risk avoidance, ii) risk mitigation, iii) risk acceptance, iv) risk outsourcing, and v) risk sharing.

Forecasting the Energy Industry: Challenges and Opportunities

Despite the plethora of intuitive and statistical tools, a number of risks are associated with forecasting, for a number of reasons. First, the science of forecasting bears great risks for the professional forecasters: Governments and industries act upon these findings by making significant decisions related to geopolitical strategies, concession agreements, investment, research and development, and so on. To alleviate the risk, decision makers should ideally combine a number of methods and tools, e.g., data analytics, software predictions, geopolitical trend analyses, etc.

Second, the energy industry is extremely complex: Not only is there a myriad of global key players, but, most important, crucial segments of their strategies, alliances, and activities remain confidential or difficult to track. Affluent organizations with a healthy budget may have unlimited resources and think tanks to consult, thus obtaining forecasts of adequate accuracy. On the other hand, medium and small-sized organizations, or entities with inadequate funding, may find forecasting an expensive and frequently elusive task.

When forecasts are based on missing, inconsistent, or inaccurate data, forecasters do not have the big picture in its multi-dimensional entirety. When dealing with statistical data analysis methods, a typical solution is to increase the sample size in order to optimize the outcome validity and reduce the margin of error. However, when the input consists of greatly inaccurate data, a multi-disciplinary analysis can shed light on the missing pieces of the puzzle. Ideally, a combination of intuitive and statistical analysis may be interpreted from a multidisciplinary perspective. An optimum approach is to employ a think-tank of interdisciplinary professionals capable of covering the commercial, financial, technological, geopolitical, and other aspects of energy security. For example, the Participatory Operational Assessment method is the ideal environment for project-specific energy-specific think-tanks (Burns, 2018a; Burns, 2018b).

Third, a number of forecasting methods use historical data in order to estimate future trends (e.g., seasonal demand). This is another example where the data input may not be accurate, as global energy is characterized by market cycles whose fluctuations are triggered each time by a unique combination of internal and external factors. In other words, despite the similarities, no market cycle is identical to another over a period of time. Again, an optimum forecast would provide a combination of intuitive and statistical analysis that should be interpreted in a multidisciplinary manner.

Fourth, while the majority of government departments and energy companies incorporate forecasting services as a consistent systematic part of their strategy, few actually evaluate the validity of these forecasts retrospectively. As a result, valuable information is underutilized.

Finally, problems may also arise when government security departments, policymakers, investors and economists act upon market forecasts that are inaccurate, biased or single-disciplinary. As a result, inaccurate, irrelevant, or useless information is overutilized.

The following section provides a concise forecast of the future energy market trends that combines energy market recommendations with intuitive interpretations and a set of econometric formulas.

Forecasting Energy Trends via Statistical Data Analysis Methods

This section demonstrates a set of data analysis tools that can be used to measure critical aspects of energy security.

Performance in this section is estimated in terms of energy efficiency, daily hot water energy load, Savings-to-Investment Ratio, etc.

Energy efficiency is estimated by the formula:

$$\textbf{Energy efficiency ratio} = \frac{\textbf{Joules of total useful output energy}}{\textbf{Joules of total input energy}}$$

Whereas:

- Useful output energy is energy produced (unit of measurement is in watts or joules per second).
- Total input energy is energy absorbed.
- The energy efficiency outcome is not measured in units, as it is a ratio ranging from 0 to 1.

To estimate the percentage of energy efficiency, the formula reads as follows:

$$\textbf{Energy efficiency}\,(\%) = \frac{\textbf{Joules of total useful output}}{\textbf{Joules of total input energy}} \times \textbf{100}$$

Whereas:

The outcome of energy efficiency is a percentage that is not measured in units.
Wind energy is measured as follows:

$$\textbf{WE} = \textbf{Ad X TD}^2 \times (\textbf{WV})^3 \times \textbf{c}$$

Whereas:

- WE is wind energy.
- Ad is air density.
- TD is the diameter of turbine blade.
- WV is the wind velocity.
- c stands for the dimensional physical constant.

The maximum value of solar energy efficiency is estimated as follows:

$$\eta\textbf{max} = \frac{\textbf{Pmax} \times \textbf{100}}{\textbf{E * Ac}}$$

Whereas:

- ηmax is the maximum value of solar energy efficiency.
- Pmax is the maximum power output in Watts.
- E is the radiant energy in W/m^2, whereas radiant flux is the radiant power that is produced, reflected, and conducted per unit time.
- Ac is the area of solar collector, measured in m^2.

The daily hot water energy load is measured as follows:

$$\textbf{L} = \frac{\textbf{MC}(\textbf{Thot} - \textbf{Tcold})}{\eta\textbf{boiler}}$$

Whereas:

- L is the daily hot water energy load (kWh/day).
- M is the daily mass of water (kg/day).
- C is the specific heat of water.
- Thot is the hot water delivery temperature (set in °F or °C).
- Tcold is the cold water temperature (set in °F or °C).
- ηboiler is the auxiliary heater efficiency.

The Savings-to-Investment Ratio is measured as follows:

$$\text{SIR} = \text{S} * \text{cvf} / \text{C}$$

Whereas:

- SIR reflects the savings-to-investment ratio.
- cvf is the current value factor of the energy technology investment.
- C is the cost.

In the equation's outcome, the energy project appears cost-effective when SIR>1.

The Linear Regression Analysis is an econometric formula that uses the independent variables (x) in order to predict the dependent variable (y).

$$y = \beta 0 + \beta_1 x_1 + \beta_2 x_2 + \beta_n x_n + \varepsilon$$

Whereas:

- y is the dependent variable.
- x is the independent (explanatory) variable.
- β0 is the population Y intercept.
- β_1, β_2 represent the regression coefficients (population slope coefficients).
- ε represents the random error component.

It is worth noting that each independent variable x is connected with a regression coefficient β to reflect the strength and the sign of that variable's relationship to the dependent variable.

Furthermore, the positive and negative signs in the variables will demonstrate the direction of the correlation, i.e., will show the linear association among these variables.

The linear regression formula can be used in several cases, e.g., to verify the annual energy consumption, as demonstrated in the following example:

$$\mathbf{EnCons} = \beta 0 + \beta_1 * \mathbf{NPop} + \beta_2 * \mathbf{GDP} + \beta_3 * \mathbf{EnProd} + \beta_4 * \mathbf{EnImp} + \varepsilon$$

Whereas:

- EnCons = energy consumption (dependent variable).
- NPop = national population (independent variable).
- GDP = gross domestic product (independent variable).
- EnProd = energy production (independent variable).
- EnImp = energy imported (independent variable).

Forecasting with the Moving Average Techniques

When it comes to forecasting, moving average techniques are among the prevailing methods used by market analysts and forecasters in order to verify the market cycle trends, while smoothing out any minor spikes and "the noise" of insignificant variables or factors in our data sample. These methods utilize a window of time period to estimate the average. A deciding factor for accurate outcomes entails selecting the length of time period (window or the exponential moving average) that best fits the forecasting event. Typically, moving average techniques are a variation among the simple moving average (SMA), which entails an average value over a specific time frame, or the exponential moving average (EMA), which focuses on current market fluctuations.

The following moving average techniques are most commonly used:

a The **Autoregressive Moving Average (ARMA)** technique is used to comprehend and forecast future trends in a time series. The first order regression ARMA (1,1) model is depicted as follows:

$$\mathbf{Y}_t = \alpha_0 + \alpha_1 \mathbf{Y}_{t-1} + \varepsilon_t - \omega_1 \varepsilon_{t-1}$$

A technique combined p autoregressive terms and q moving average terms appears as an ARMA (p,q) model and is depicted as follows:

$$Y_t = \alpha_0 + \alpha_1 Y_{t-1} + \alpha_2 Y_{t-2} + ... + \alpha_p Y_{t-p} + \varepsilon_t - \omega_1 \varepsilon_{t-1} - \omega_2 \varepsilon_{t-2} -...- \omega_q \varepsilon_{t-q}$$

b The **AutoRegressive Integrated Moving Average (ARIMA)** technique encompasses an array of standard time series when non-stationarity appears in the data. It relies on the level of moving averages, auto-regression, and integration. For example,

The first order regression ARIMA (1,0,0) = AR(1) model is depicted as follows:

$$Y_t = \alpha_0 + \alpha_1 Y_{t-1} + \varepsilon$$

The second order regression ARIMA (2,0,0) = AR(2) model is depicted as follows:

$$Y_t = \alpha_0 + \alpha_1 Y_{t-1} + \alpha_2 Y_{t-2} + \varepsilon$$

Whereas:

- $\mathbf{Y_t}$ represents the dependent variable at a specific timeframe (t).
- $\mathbf{Y_{t-1}}$ represent the independent variables of past timeframes.

c **The Box-Jenkins** estimation depends on autocorrelation configurations in the data, and combines either the ARMA or ARIMA techniques to verify if stationarity is present in the selected time series, and if seasonality must be modelled.

Holt's Linear Trend Method

Holt's technique is a trend-correcting double exponential smoothing method frequently used as a forecasting tool when the data sample is characterized by linear trends. The Holts and Holt-Winters techniques are extremely popular due to their accuracy and the allowance of fluctuations over time.

The following sets of formulas demonstrate a demand forecast based on the parameters of level, trend, and seasonality:

A demand forecast for a specified period of time is reflected as follows:

$$\text{Demand forecast } (DF): DF_{t+1} = \left[L + (t+1)T\right] S_{t+1}$$

A level forecast that incorporates demand for a period of time is:

$$\text{Level forecast } : L_{t+1} = \alpha D_{t+1} + (1-\alpha)(L_t + T_t)$$

An alternative set of formulas combines three significant parameters: Level, trend, and seasonality:

$$\text{Level forecast } : L_t = \alpha(x_t - S_{t-s}) + (1-\alpha)(L_t - 1 + T_t - 1)$$

$$\text{Trend forecast } : T_t = \beta(L_t - L_{t-1}) + (1-\beta)(T_t - 1)$$

$$\text{Seasonal forecast} : S_t = \gamma(x_t - L_t) + (1-\gamma) S_{t-s}$$

Whereas the variables for the above five formulas read as follows:

- L_t represents level of demand irrespective of seasonal demand. It estimates forecast averages over a time period (t).
- D_t stands for demand over a period of time.
- T stands for Trend of demand, i.e., increased or reduced.
- S stands for seasonal fluctuations, hence for S one may use s averages over the specific season.
- α represents a smoothing constant for level "L".
- β represents a smoothing constant for trend "T".
- γ represents a smoothing constant, deterministic for the seasonal cycle.

Forecasting Energy Security Trends by Examining the Triad of Technologies, Policies, and Processes

This section demonstrates the impactful interrelation between technologies, policies, and processes.

Technology Risks

Energy is by nature a capital-intensive technology-intensive industry segment. When obsolete technologies are used, there are risks of limited efficiency, which may impact – directly or indirectly – the areas of health, safety, security, environment, and/or quality. Typically novel technologies are greener, faster, safer, more effective, etc. Therefore, as soon as an innovative technology is launched, it impacts the input/output of the market, with financial, commercial, and likely regulatory effects.

Patent infringement is a key challenge for energy innovators that perfectly demonstrates the interrelation between technologies, regulations, and processes. Patent infringement pertains to the unauthorized use, manufacturing, sale, or copied patent, without permission by the patent owner. This illegitimate act is facilitated due to the inherent vulnerabilities pertaining to the regulatory, technological, and procedural vulnerabilities. Namely:

a The perfection of reverse engineering techniques enables copyright and patent infringement.

b Inadequate regulations fail to protect energy innovators and investors, when global compliance is either relaxed or inconsistently interpreted by global courts (Burns, 2016b).

The impact upon the energy industry is tremendous, as entities who invest in research and development do not seem to benefit financially or commercially from their innovations. To remedy this vulnerability there is a need for new policies.

Regulatory Risks

The history of the energy industry demonstrates that as new technologies bring about new capabilities, new laws and new regulatory frameworks are implemented to embrace the impacts of this industry change in terms of safety, security, environmental, occupational health, and other aspects. However, the caveat here is that the policy-making mechanism moves at a slower pace, as it takes months, if not years, to be conceived, submitted, evaluated, re-modified, and finally approved. Hence, the two critical stages of a policy are first during the introduction of a new policy, and, second, as an older policy becomes obsolete. Both these stages reflect the component of transition from the older policy to the new (Burns, 2016c). Transition times typically create vulnerabilities due to uncertainty as to compliance and the optimum technologies, processes, and tools needed.

Procedural Risks

Organizational processes are significant aspects of energy security as their optimization ensures regulatory compliance (health, safety, security, quality, environment, etc.), as well as operational efficiency, financial, and commercial success. The human factor and its effective interaction with technologies is pivotal in industrial procedures. Optimum procedures must be standardized, i.e., consistent and sustainable performance, based on a set of rules, protocols, and methods that are followed with limited deviation. Lack of standardization and process optimization imply lack of sustainable processes and lack of efficiency. Processes are a significant part of corporate performance, hence companies who manage to address vulnerabilities have the opportunity to also benefit from regulatory compliance and technological performance (i.e., output, efficiencies).

Train – Learn – Lead

Effective training for energy security must go beyond learning: It must be strategically designed to encompass team-building, brainstorming, role-playing tools, to fully capitalize on the human factor. After all, human resources have the potential to become a company's greatest strength and its greatest vulnerability.

Human force equals to a universe of talents, skills, and creative diversity that must be accepted and embraced. Each employee is a leader and an expert in their field. Motivated teams can attain the impossible and defy all laws of reason, whereas an unmotivated team can miss even the most modest goals.

Situational awareness (SA) is a term used frequently by military and security experts in order to denote the value of human alertness. It reflects the need for environmental sensitivity and discernment in real time. Most important, it reflects the prompt understanding of an imminent risk, especially when one of the conditions or factors changes for the worse, thus bringing volatility or unreliability.

Energy security training holds specific learning objectives, depending on the organization and department goals that holds the training. Every government and energy-related organization holds energy security training, hence the following examples are presented to demonstrate the structure of certain significant organizations related to energy security.

The NATO School Oberammergau (NSO)

The NATO School Oberammergau is an organization established by the North Atlantic Treaty Organization (NATO) in order to mitigate the training needs of NATO's military and civilian personnel at an operational level, while remaining aligned with NATO's strategy, operations, policies, and protocols.

Its energy security training program was designed in a manner that:

- promotes awareness of present energy advances and weaknesses as part of evolving security challenges;
- creates a mutual perception of the global energy security agenda;
- brings together NATO with its global and government partners in the area of energy security (NATO, 2018).

The International Energy Agency's Training Program

The International Energy Agency's training program includes workshops and short courses offered to government officials and significant stakeholders in various countries. The goal of the training is to help country members mitigate their energy goals and become harmonized with the international developments and other country missions (IEA 2018).

- From 2010 to 2016, over 2,875 government officials and statisticians have been trained globally.
- The training commenced in the late 1970s with Emergency Response Exercises (ERE); preparing for energy supply disruptions; technologies, policies, etc.

The National Training Center (NTC)

The National Training Center (NTC) is the U.S. Department of Energy's Center of Excellence for Security and Safety Training and Professional Development. Its role is to plan, produce, and carry out advanced security and safety training programs for government employees and domestic contractors of pivotal role, such as the National Nuclear Security Administration (NNSA) (DOE NTC, 2018). The training goals of the NTC include:

- Carrying out training that encompasses all security and safety components as needed to safeguard the Department of Energy and its critical components such as amenities, technologies, infrastructure, information, and other resources.
- Preserving the expertise, capabilities, and overall excellence of security and safety employees via standardized safety and security training, life-learning and professional development opportunities (DOE NTC, 2018).

The U.S. Office of Energy Efficiency and Renewable Energy in the Department of Energy

The U.S. Office of Energy Efficiency and Renewable Energy in the Department of Energy aims at the designing of training pertaining to energy efficiency and emergency management. It encompasses all aspects of energy security, such as, for example, technologies. Its goals are to produce and promote a deeper knowledge on significant aspects pertaining to short- and long-term energy interruptions, as well as maintenance and storage concerns.

The training has the following key learning objectives, among others:

- Designing location-specific, facility-specific energy security plans.
- Increasing energy security in zones of potential risks and vulnerabilities within Government facilities.

- Alleviating threats and taking security measures.
- Designing security plans to sustain mission-critical actions in case the energy infrastructure systems are disrupted (DOE OEERE, 2018).

A well-designed energy security training package must contain the following features:

- All encompassing, thus including the physical, cyber, and supply-chain aspects of security in a holistic manner.
- Situational awareness is especially significant in a new era of high-tech and artificial intelligence. Here, effective training serves as the fusion that brings together human force and advanced technologies.
- Effective and penetrating enough to educate in geographically remote areas.
- Flexible to meet working shifts.
- Concise and to-the-point, to respect employees' limited time. It must be a hybrid of company-specific, region-specific, industry-specific material.
- Pragmatic and timely, to address the industry's novel technologies, regulations, and methods.
- Effective training ensures knowledge retention and a hands-on environment with simulated risks presented.

Training is significant in the fast-paced global industries. Knowledge-based platforms have transformed training into a virtual environment of geospatial flexibility, e.g., online classrooms, self-paced exams, etc. A new era of unconventional training is especially needed in the energy industry, with a large segment of its professionals working in remote areas, frequent travelling, long working hours, and possible fatigue.

Under the principle that true leaders don't create followers but generations of new empowered leaders, modern training sessions should offer the platforms of knowledge and inspiration where a new generation of leaders is created in every energy segment.

The author, editor, and publishers of this book hope that the contents of this book will contribute towards the creation of new energy security training programs.

References

Burns, M. (2013). *Ensuring Optimum Resilience in Marine Transportation: Extended Applications of the Maritime Security Risk Analysis Model & the Dynamic Risk Management Model*. Washington: TRB National Academies.

Burns, M. (2014a). *Port Management and Operations*. Boca Raton: CRC Press, Taylor & Francis

Burns, M. (2014b). "Building resilient supply chains through policies, partnerships and technologies". DHS Supply Chain Security Workshop, Houston, April 23.

Burns, M. (2015). *Logistics and Transportation Security: A Strategic, Tactical, and Operational Guide to Resilience*. Boca Raton: CRC Press, Taylor & Francis.

Burns, M. (2016a). "Chair's welcome speech". TRB Committee on Critical Transportation Infrastructure Protection, Annual Meeting, Supply Chain Security Subcommittee. Available at: https://sites.google.com/site/trbcommitteeabe40/Welcome/subcommittees/supply-chain-security-subcommittee-abe40-8, last accessed: March 12, 2018.

Burns, M. (2016b). "Global projections for shipping. Pathways to sustainability". NAMEPA Conference, ExxonMobil Campus, Houston, Texas, February 5.

Burns, M. (2017). "Energy and the supply chain". The 2017 Special Committee on Transportation Security and Emergency Management (SCOTSEM) Annual Meeting was held in conjunction with the American Association of State Highway and Transportation Officials (AASHTO) Subcommittee on Highway Transport (SCOHT) and TRB Critical Infrastructure Committee (ABR 10) of the National Academies, Houston, Texas, July 30–August 3.

Burns, M. (2018a). "Participatory Operational Assessment (POA): Evaluating and predicting the operational effectiveness of Cargo Security Process at Ports of Entry (POEs)". University of Houston. U.S. Department of Homeland Security under Grant Award Number DHS-14-ST-061-COE-00. P.I. Maria Burns. The grant is awarded to the Borders, Trade, and Immigration (BTI) Institute: A DHS Center of Excellence led by the University of Houston. Available at: www.uh.edu/bti/news/stories/2017/October/Participatory%20Operational%20Assessment.php, last accessed: June 1, 2018.

Burns, M. (2018b). "Participatory operational & security assessment on homeland security risks: an empirical research method for improving security beyond the borders through public/private partnerships". *Journal of Transportation Security*, 11(3): 85–100.

Department of Energy: Office of Energy Efficiency and Renewable Energy (DOE OEERE) (2018). *Training. Achieving Energy Security of Federal Facilities*. U.S. Office of Energy Efficiency and Renewable Energy, Department of Energy, Federal Energy Management Program. Available at: www4.eere.energy.gov/femp/training/training/achieving-energy-security-federal-facilities, last accessed: March 12, 2018.

Department of Energy: National Training Center (DOE NTC) (2018). *The National Training Center. Center of Excellence for Security and Safety Training and Professional Development*. U.S. Department of Energy. Available at: www.energy.gov/ea/services/training, last accessed: June 1, 2018.

International Energy Agency (IEA) (2018). *Training*. International Energy Agency. France. Available at: www.iea.org/training, last accessed: June 1, 2018.

NATO School Oberammergau (NSO) (2018). NATO School Oberammergau, NATO International. Available at: www.natoschool.nato.int/Academics/Resident-Courses/Course-Catalogue/Course-description?ID=134&TabId=155&language=en-US, last accessed: June 1, 2018.

Index